Lecture Notes in Computer Science

Edited by G. Goos and J. Hartmanis

T0226243

214

CAAP '86

11th Colloquium on Trees in Algebra and Programming
Nice, France, March 24–26, 1986
Proceedings

Edited by P. Franchi-Zannettacci

Springer-Verlag
Berlin Heidelberg New York Tokyo

Lecture Notes in Computer Science

Vol. 117: Fundamentals of Computation Theory. Proceedings, 1981. Edited by F. Gécseg. XI, 471 pages. 1981.

Vol. 118: Mathematical Foundations of Computer Science 1981. Proceedings, 1981. Edited by J. Gruska and M. Chytil. XI, 589 pages. 1981.

Vol. 119: G. Hirst, Anaphora in Natural Language Understanding: A Survey. XIII, 128 pages. 1981.

Vol. 120: L. B. Rall, Automatic Differentiation: Techniques and Applications. VIII, 165 pages. 1981.

Vol. 121: Z. Zlatev, J. Wasniewski, and K. Schaumburg, Y12M Solution of Large and Sparse Systems of Linear Algebraic Equations. IX, 128 pages. 1981.

Vol. 122: Algorithms in Modern Mathematics and Computer Science. Proceedings, 1979. Edited by A. P. Ershov and D. E. Knuth. XI, 487 pages. 1981.

Vol. 123: Trends in Information Processing Systems. Proceedings, 1981. Edited by A. J. W. Duijvestijn and P. C. Lockemann. XI, 349 pages. 1981.

Vol. 124: W. Polak, Compiler Specification and Verification. XIII, 269 pages. 1981.

Vol. 125: Logic of Programs. Proceedings, 1979. Edited by E. Engeler. V, 245 pages. 1981.

Vol. 126: Microcomputer System Design. Proceedings, 1981. Edited by M. J. Flynn, N. R. Harris, and D. P. McCarthy. VII, 397 pages. 1982.

Voll. 127: Y.Wallach, Alternating Sequential/Parallel Processing. X, 329 pages. 1982.

Vol. 128: P. Branquart, G. Louis, P. Wodon, An Analytical Description of CHILL, the CCITT High Level Language. VI, 277 pages. 1982.

Vol. 129: B. T. Hailpern, Verifying Concurrent Processes Using Temporal Logic. VIII, 208 pages. 1982.

Vol. 130: R. Goldblatt, Axiomatising the Logic of Computer Programming. XI, 304 pages. 1982.

Vol. 131: Logics of Programs. Proceedings, 1981. Edited by D. Kozen. VI, 429 pages. 1982.

Vol. 132: Data Base Design Techniques I: Requirements and Logical Structures. Proceedings, 1978. Edited by S.B. Yao, S.B. Navathe, J.L. Weldon, and T.L. Kunii. V, 227 pages. 1982.

Vol. 133: Data Base Design Techniques II: Proceedings, 1979. Edited by S.B. Yao and T.L. Kunii. V, 229–399 pages. 1982.

Vol. 134: Program Specification. Proceedings, 1981. Edited by J. Staunstrup. IV, 426 pages. 1982.

Vol. 135: R.L. Constable, S.D. Johnson, and C.D. Eichenlaub, An Introduction to the PL/CV2 Programming Logic. X, 292 pages. 1982.

Vol. 136: Ch. M. Hoffmann, Group-Theoretic Algorithms and Graph Isomorphism. VIII, 311 pages. 1982.

Vol. 137: International Symposium on Programming. Proceedings, 1982. Edited by M. Dezani-Ciancaglini and M. Montanari. VI, 406 pages. 1982.

Vol. 138: 6th Conference on Automated Deduction. Proceedings, 1982. Edited by D.W. Loveland. VII, 389 pages. 1982.

Vol. 139: J. Uhl, S. Drossopoulou, G. Persch, G. Goos, M. Dausmann, G. Winterstein, W. Kirchgässner, An Attribute Grammar for the Semantic Analysis of Ada. IX, 511 pages. 1982.

Vol. 140: Automata, Languages and programming. Edited by M. Nielsen and E.M. Schmidt. VII, 614 pages. 1982.

Vol. 141: U. Kastens, B. Hutt, E. Zimmermann, GAG: A Practical Compiler Generator. IV, 156 pages. 1982.

Vol. 142: Problems and Methodologies in Mathematical Software Production. Proceedings, 1980. Edited by P.C. Messina and A. Murli. VII, 271 pages. 1982.

Vol. 143: Operating Systems Engineering. Proceedings, 1980. Edited by M. Maekawa and L.A. Belady. VII, 465 pages. 1982.

Vol. 144: Computer Algebra. Proceedings, 1982. Edited by J. Calmet. XIV, 301 pages. 1982.

Vol. 145: Theoretical Computer Science. Proceedings, 1983. Edited by A.B. Cremers and H.P. Kriegel. X, 367 pages. 1982.

Vol. 146: Research and Development in Information Retrieval. Proceedings, 1982. Edited by G. Salton and H.-J. Schneider. IX, 311 pages. 1983.

Vol. 147: RIMS Symposia on Software Science and Engineering. Proceedings, 1982. Edited by E. Goto, I. Nakata, K. Furukawa, R. Nakajima, and A. Yonezawa. V. 232 pages. 1983.

Vol. 148: Logics of Programs and Their Applications. Proceedings, 1980. Edited by A. Salwicki. VI, 324 pages. 1983.

Vol. 149: Cryptography. Proceedings, 1982. Edited by T. Beth. VIII, 402 pages. 1983.

Vol. 150: Enduser Systems and Their Human Factors. Proceedings, 1983. Edited by A. Blaser and M. Zoeppritz. III, 138 pages. 1983.

Vol. 151: R. Piloty, M. Barbacci, D. Borrione, D. Dietmeyer, F. Hill, and P. Skelly, CONLAN Report. XII, 174 pages. 1983.

Vol. 152: Specification and Design of Software Systems. Proceedings, 1982. Edited by E. Knuth and E. J. Neuhold. V, 152 pages. 1983.

Vol. 153: Graph-Grammars and Their Application to Computer Science. Proceedings, 1982. Edited by H. Ehrig, M. Nagl, and G. Rozenberg. VII, 452 pages. 1983.

Vol. 154: Automata, Languages and Programming. Proceedings, 1983. Edited by J. Díaz. VIII, 734 pages. 1983.

Vol. 155: The Programming Language Ada. Reference Manual. Approved 17 February 1983. American National Standards Institute, Inc. ANSI/MIL-STD-1815A-1983. IX, 331 pages. 1983.

Vol. 156: M.H. Overmars, The Design of Dynamic Data Structures. VII, 181 pages. 1983.

Vol. 157: O. Østerby, Z. Zlatev, Direct Methods for Sparse Matrices. VIII, 127 pages. 1983.

Vol. 158: Foundations of Computation Theory. Proceedings, 1983. Edited by M. Karpinski, XI, 517 pages. 1983.

Vol. 159: CAAP'83. Proceedings, 1983. Edited by G. Ausiello and M. Protasi. VI, 416 pages. 1983.

Vol. 160: The IOTA Programming System. Edited by R. Nakajima and T. Yuasa. VII, 217 pages. 1983.

Vol. 161: DIANA, An Intermediate Language for Ada. Edited by G. Goos, W.A. Wulf, A. Evans, Jr. and K.J. Butler. VII, 201 pages. 1983.

Vol. 162: Computer Algebra. Proceedings, 1983. Edited by J.A. van Hulzen. XIII, 305 pages. 1983.

Vol. 163: VLSI Engineering. Proceedings. Edited by T.L. Kunii. VIII, 308 pages. 1984.

Vol. 164: Logics of Programs. Proceedings, 1983. Edited by E. Clarke and D. Kozen. VI, 528 pages. 1984.

Vol. 165: T.F. Coleman, Large Sparse Numerical Optimization. V, 105 pages. 1984.

Vol. 166: STACS 84. Symposium of Theoretical Aspects of Computer Science. Proceedings, 1984. Edited by M. Fontet and K. Mehlhorn. VI, 338 pages. 1984.

Lecture Notes in Computer Science

Edited by G. Goos and J. Hartmanis

214

CAAP '86

11th Colloquium on Trees in Algebra and Programming
Nice, France, March 24–26, 1986
Proceedings

Edited by P. Franchi-Zannettacci

Springer-Verlag
Berlin Heidelberg New York Tokyo

Editor

Paul Franchi-Zannettacci
I.S.I. c/o I.N.R.I.A.
Route des Lucioles, Sophia Antipolis, 06560 Valbonne, France

CR Subject Classifications (1985): D.1, E.1, E.2, F.1, F.2, F.3, F.4, G.2, G.3

ISBN 3-540-16443-X Springer-Verlag Berlin Heidelberg New York Tokyo
ISBN 0-387-16443-X Springer-Verlag New York Heidelberg Berlin Tokyo

Printing and binding: Beltz Offsetdruck, Hemsbach/Bergstr.
2146/3140-543210

This volume contains the proceedings of the 11th Colloquium on Trees in Algebra and Programming (CAAP'86).

From 1976 to 1980, this Colloquium was held in Lille, France, under the name CLAAP for *Colloque de Lille sur les Arbres en Algèbre et en Programmation*. Since 1981, the Colloquium has begun a more european life with the name CAAP; and as M. Dauchet (one of the founders with A. Arnold) used to say in French : *Il a perdu son L pour mieux voler*.

After l'Aquila, Italy (1983), Bordeaux, France (1984), and Berlin, Germany (1985), the CAAP comes this year to NICE on MARCH, 24th-26th 1986.

The trees are well-known to be the basic structure for Computer Science, so usually the topics covered by the Conference are beyond the strict domain of trees and extend to many areas in Computer Science.

Following the CAAP tradition, the 20 accepted papers for CAAP'86, selected from a total of 44 papers, are representative of the main topics in theoretical Computer Science.

The selected papers and invited lectures are presented in 7 sections :

- Language theory
- Algebraic theory of semantics
- Graphs and grammars
- Program schemes and programming
- Tree-automata and transducers
- Probability on trees
- Logic for computing

The Program Committee for CAAP'86 consists of :

A. Arnold (Univ., Bordeaux)
E. Astesiano (Univ., Genova)
G. Boudol (INRIA, Sophia Antipolis)
G. Cousineau (INRIA, Paris)
M. Dauchet (Univ., Lille)
M. Dezani-Ciancaglini (Univ., Torino)
J. Diaz (Univ., Barcelona)
H. Ehrig (Univ., Berlin)
J. Engelfriet (Univ., Leiden)
Ph. Flajolet (INRIA, Paris)
P. Franchi-Zannettacci (Univ., Nice, Chair)
H. Ganzinger (Univ., Dortmund)
G. Kahn (INRIA, Sophia Antipolis)
F. Oles (IBM, Yorktown Heights)
P. Mosses (Univ., Aarhus)
M. Nivat (Univ., Paris)

We would like to thank all Program Committee members and referees of CAAP'86 and all the contributors listed below for their help in reviewing the submitted papers:

J. Beauquier
J. C. Bermond
G. Berry
D. Bert
C. Boehm
F. Boussinot
F. J. Brandenburg
P. Casteran
J. Chazarin
G. Costa
B. Courcelle
P. Deransart
J. Despeyroux
G. File
M. Fontet
J. Françon
D. Gouyou-Beauchamps
H. J. Hoogeboom
J. P. Jouannaud
M. Jourdan
H. C. M. Kleÿn
M. Koenig
H. J. Kreowski
P. Lescanne
J.J. Levy
E. Madelaine
Y. Metivier
J. Morgenstern
F. Orejas
P. Padawitz
E. Pin
J. Pouget
J.C. Raoult
J. L. Remy
F. Rodriguez
E. M. Schmidt
S. Schwer
J. M. Steyaert
H. Vogler

We highly appreciate the financial support provided by the following organisms :

I.S.I. (Informatique et Sciences de l'Ingénieur)
University of Nice
I.N.R.I.A.
GRECO de Programmation
Conseil Général des Alpes Maritimes
Conseil Régional Provence Alpes Côte d'Azur
Délégation Régionale à la Recherche et à la Technologie
I.B.M. FRANCE

Finally, we thank L. Galligo and C. Juncker for their friendly cooperation, and we wish to express our gratitude to Isabelle Attali from I.S.I. who assumed the complete organization of the Conference.

Paul Franchi-Zannettacci
Université de Nice
Sophia Antipolis, Mars 1986

CONTENTS

A Categorical Treatment of Pre- and Post Conditions

(Extended Abstract)

Eric G. Wagner
Exploratory Computer Science
IBM T.J. Watson Research Center,
Yorktown Heights, N.Y. 10598 / U.S.A.

1. INTRODUCTION

The use of pre- and post-conditions as the basis for program specification methodologies is well-known, see [1], [2], [5], [6], [7], and others. Generally, in these treatments the pre- and post-conditions are viewed as assertions concerning programs, the developments have a distinctly logical flavor and there is a concern with proofs of program correctness. The work we want to talk about today comes originally from looking at pre- and post-conditions, not so much as a means for making assertions about programs, but rather as a means for writing loose semantic specifications for programs. That is, we view a pair of pre- and post-conditions as specifying a set of functions, and a program is said to satisfy the specification if the program realizes one of the functions in the set. One consequence of this small shift to a view emphasizing the semantical is that the new view lends itself naturally to abstraction in category theoretic terms. As we will show, such an abstraction leads to a number of new results concerning pre- and post-conditions and to a generalization of Dijkstra's predicate transformers, [1].

2. PRELIMINARIES

Given a category **C** we write $|\mathbf{C}|$ to denote its class of objects and **C** to denote its class of morphisms. Composition in a category is written in diagramatic order: given $f:A \to B$ and $g:B \to C$, their composite is written as $f \bullet g:A \to C$. Given a product $A \times B$ of two objects in **C** with projections $\pi_A:A \times B \to A$ and $\pi_B:A \times B \to B$ and given morphisms $f:X \to A$ and $g:X \to B$, we write $(f, g):X \to A \times B$ for the corresponding mediating morphism, i.e., the unique morphism $u:X \to A \times B$ such that $u \bullet \pi_A = f$ and $u \bullet \pi_B = g$. Finally, let **Set** denote the category of sets and total functions, and let **Pfn** denote the category of sets and partial functions. For definitions of other categorical concepts see [8].

3. PRE- AND POST-CONDITIONS

We start by replacing the notion of a predicate by a relativized notion of an equalizer.

DEFINITION 3.1. Let C be a category and let $T \subseteq C$ be a subcategory with $|T| = |C|$. Let $f,g:A \to B$ in C, then we say that a morphism $e:E \to A$ in T is <u>an equalizer for f and g w.r.t. T</u> iff $e \bullet f = e \bullet g$ and for each $\alpha \epsilon T$ such that $\alpha \bullet f = \alpha \bullet g \epsilon T$, there exists a unique $h \epsilon T$ such that $h \bullet e = \alpha$. We say that $e:E \to A$ is an <u>equalizer on A w.r.t.</u> T if there exist $f,g:A \to B$ such that e is an equalizer for f and g w.r.t. T. □

An abstraction of pre- and post-conditions that captures the usual applications is

DEFINITION 3.2. Let C be a category, let $T \subseteq C$ with $|T| = |C|$, let $A_1, A_2 \epsilon |C|$, and let $q:E_1 \to A_1$ be an equalizer on A_1 w.r.t. T, and let $r:E_2 \to A_1 \times A_2$ be an equalizer on $A_1 \times A_2$ w.r.t. T. Then for $k:A_1 \to A_2$ in C we say that e_1 is <u>a pre-condition for k with respect to post-condition e_2 over</u> T, (and write $\{e_1\}k\{e_2\}$) if there exists $h:E_1 \to E_2$ in T such that $h \bullet e_2 = e_1 \bullet (1_{A_1}, k)$.

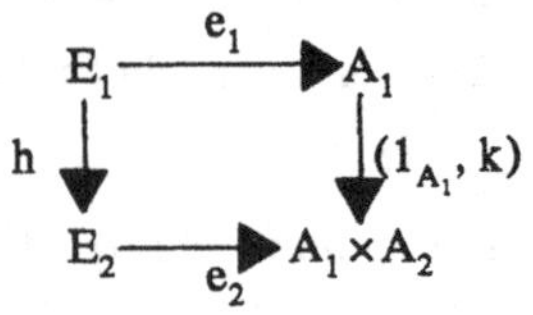

Further, let

$$\mathbf{Mng}(e_1, e_2) = \{ k:A_1 \to A_2 \mid \{e_1\}k\{e_2\} \}.$$

□

EXAMPLE 3.3. Let $C=\mathbf{Pfn}$, $T=\mathbf{Set}$, let $A_1=A_2=\mathbf{R}$, the set of real numbers, let E_1 be the set of non-negative reals and let e_1 be the inclusion of E_1 into $\mathbf{R}$, let E_2 be the set

$$E_2 = \{ \langle x,y \rangle \epsilon \mathbf{R} \times \mathbf{R} \mid |(x-y^2)/y| \leq 2\epsilon \}$$

and let e_2 to be inclusion of E_2 into $\mathbf{R} \times \mathbf{R}$. Then $\mathbf{Mng}(e_1, e_2)$ is the set of all $k:\mathbf{R} \to \mathbf{R}$ such that for all $r \geq 0$, $k(r)$ is the "square root of r up to ϵ". Note that the choice $T=\mathbf{Set}$ forces the elements of $\mathbf{Mng}(e_1, e_2)$ to be defined on all $r \geq 0$, and is thus, in effect, a condition guaranteeing that the "computation terminates" for all arguments satisfying the precondition. Taking $T=\mathbf{Pfn}$ would remove this "termination condition". □

In the above definition we need the $(1_{A_1}, k):A_1 \to A_1 \times A_2$ to capture the effect that the post-condition, as a predicate, involves both the argument and the result of the specified function(s). We will now investigate a more general, and weaker, notion.

DEFINITION 3.4. Let C be a category, Let $T \subseteq C$ with $|T| = |C|$, let $A_1, A_2 \varepsilon |C|$, and let $e_i: E_i \to A_i$ be an equalizer on A_i w.r.t. T, i=1,2. Then for $k: A \to B$ in C we say that e_1 is a simple pre-condition for k with respect to post-condition e_2 over T, (and write $[e_1]k[e_2]$) if there exists $h: E_1 \to E_2$ in T such that $h \bullet e_2 = e_1 \bullet k$.

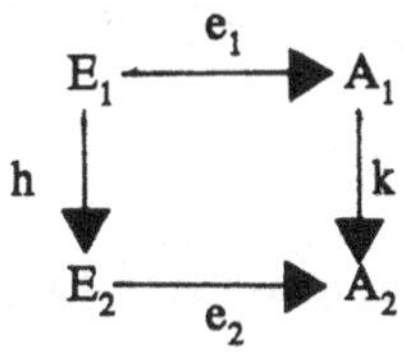

□

From here on let us assume that the category C is well-powered (for any object A the class of subobjects of A is a set) and that C is locally small (for all A and B, C(A,B) is a set). Given any object A in C, let P(A) denote the set of all (isomorphism classes of) equalizers on A, and given any objects A and B in C let $P[A \to B]$ denote the set of all sets of morphisms from A to B in C.

DEFINITION 3.5. Let C be a category and let $A, B \varepsilon |C|$. By a (generalized) predicate transformer we mean a mapping $\tau: P(B) \to P(A)$. Given two predicate transformers τ_1, $\tau_2: P(B) \to P(A)$ we say $\tau_1 \sqsubseteq \tau_2$ iff for each $e \varepsilon P(B)$, there exists h_e such that $\tau_1(e) = h_e \bullet \tau_2(e)$.

Given a category C and subcategory $T \subseteq C$ with $|T| = |C|$, and a predicate transformer $\tau: P(B) \to P(A)$, we say that w.r.t. T τ specifies the set of morphisms

$$\mathbf{Fns}(\tau) = \{ k: A \to B \mid \forall e \varepsilon P(B)\ (\exists h \varepsilon T\ (h \bullet e = \tau(e) \bullet k)) \}.$$
$$= \{ k: A \to B \mid \forall e \varepsilon P(B)\ ([\tau(e)]k[e]) \}.$$

□

Calling such a $\tau: P(B) \to P(A)$ a predicate transformer is a direct generalization of Dijkstra's concept of a predicate transformer as given in [1]. Indeed, we can capture Dijkstra's notion of a "weakest precondition" in a particularly nice way as a predicate transformer.

DEFINITION 3.6. Given C and $T \subseteq C$ as above, and given $k: A \to B$ in C, and given an equalizer $e: E \to B$ on B w.r.t. T, then by a weakest precondition for e and k, denoted wp(k, e), we mean an equalizer $w: W \to A$ on A w.r.t. T such that there exists $h_w: W \to E$ in T such that $h_w \bullet e = w \bullet k$ (i.e., $[w]k[e]$), and for any pair $\langle h: U \to E, g: U \to A \rangle$ in T such that $h \bullet e = g \bullet k$, there exists a unique $p \varepsilon T$ such that $p \bullet w = g$ and $p \bullet h_w = h$.

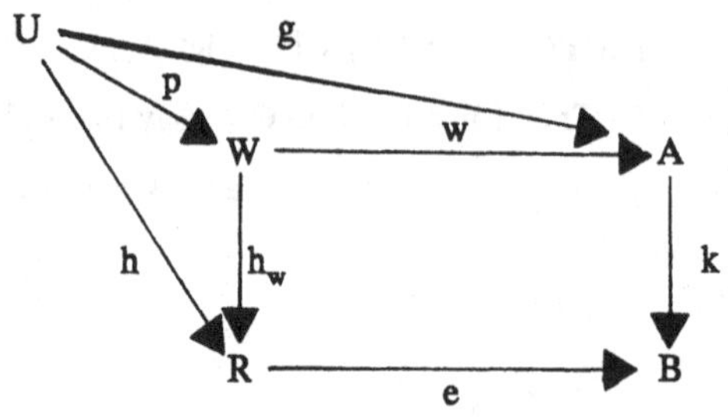

□

Note that in the case where T = C the above says that <w, h_w> is a pullback for <e, k>.

We now have two constructions, **Mng** and **Fns**, for specifying subsets of P[A→B]. The following proposition shows a strong connection between these two constructions in the cases of greatest interest.

PROPOSITION 3.7. Let C = **Pfn** and T = **Set** or **Pfn**, let A and B be sets and let K be a set of partial functions from A to B. Then there exists equalizers e_1 on A and e_2 on A×B such that **Mng**(e_1, e_2) = K, iff there exists a predicate transformer τ:P(B)→P(A) such that **Fns**(τ) = K.

□

However this strong connection appears to break down in general categories. While the concept of pre- and post-conditions is more accessible to informal arguments than that of predicate transformers, the predicate transformers appear to be the better mathematical concept. One illustration of this is the following result.

THEOREM 3.8. Let C be a category and let T⊆C such that |T| = |C|, both T and C are small complete, well-powered, and locally small, T contains all the equalizers of C, f•g∈T implies f∈T, and m is product mediator in C all of whose components are in T then m∈T, and if f∈C then there exists g such that g•f∈T such that, if k•f∈bT then k factors through h, then the above defined construction

Fns:[P(B)→P(A)]→P[A→B]

is part of a Galois connection, i.e., there exists a construction

Spc:P[A→B]→[P(B)→P(A)]

such that both **Fns** and **Spc** are monotonic (with respect to ⊆ and ⊑) and for all K∈P[A→B], and $\tau\in$[P(B)→P(A)],

K⊆**Fns**(τ) iff **Spc**(K)⊑τ.

□

This says that the **Fns** construction (as a construction on posets) is part of an adjoint situation. From the well-known properties of Galois connections (and adjoints) it follows that the constructions **Fns** and **Spc** have nice closure properties, and that **Fns** preserves limits and **Spc** preserves colimits. It appears that the **Mng** construction does not have such good properties in the general setting.

A paper containing the proofs of the above results, and additional results (for both the general case and the special cases involving **Set** and **Pfn**), and comparing this approach to some related approaches [3] and [4], is now in preparation.

4. BIBLIOGRAPHY

[1] Dijkstra, E. W., A Discipline of Programming. Prentice Hall, Englewood Cliffs, 1978.

[2] Gries, D. The Science of Programming. Springer-Verlag, New York, Heidelberg, Berlin, 1981.

[3] Plotkin, G.D., "Dijkstra's Predicate Transformers and Smyth's Powerdomains," Abstract Software Specifications, Proceedings of the 1979 Copenhagen Winter School, LNCS 86 (Edited by D. Bjorner) 527-553.

[4] Smyth, M.B., "Power Domains and Predicate Transformers: a topological view," Proceedings of ICALP'83 (Barcelona), LNCS 154, pp 662-675.

[5] Reynolds, J.C., The Craft of Programming, Prentice-Hall International (1981).

[6] Hoare, C.A.R., "An Axiomatic Basis for Computer Programming," Comm. ACM 12 (October 1969) 576-580 and 583.

[7] Jones, C.B., Software Development: A Rigorous Approach," Prentice-Hall International (1980).

[8] MacLane, S., Categories for the Working Mathematician, Springer-Verlag, New York, 1971.

COLOURINGS OF PLANAR MAPS

AND THE EQUALITY OF TWO LANGUAGES

R. Cori and S. Dulucq
Université de Bordeaux I
U.E.R. de Mathématiques et d'Informatique
33405 Talence-Cedex

The investigation of the four coulour problem inspired the developement of a lot of important combinatorial theories ; in the paper of T. Saaty [6] thirteen statements equivalent to the existence of a four colouring for planar maps are quoted. These statements are not only concerned with graph theory and some of thems seem far from graphs like those looking as number theory problems. In what follows, we add a statement in the field of formal language theory. We show that the 4-colouring of planar maps is equivalent to the equality of two subsets in the cartesian product of two free monoïds. Unfortunately this doesn't give immediately new results in the field of map colouring as the question of the equality of two subsets in the cartesian product of two free monoïds is undecidable.

Also, the fact that any planar map has a five coulouring, which is not so difficult to prove, has a formal language theory version which seems untractable.

In order to transform four colourings of maps into words of a free monoïd we use the intermediate notion of tree. The first construction we give was proposed by F. Jaeger, it allows the decomposition of a cubic planar map into two trees, that we discribe in part I. The next step consists in coding a tree by a word of the Lukasiewicz language giving the way to transform problems on trees into problems on words.

This communication is concerned with maps, trees and words ; it is difficult to give all the definitions concerning these notions, we will only restate a few of thems. The books of O. Ore [5], D.Knuth [2], M. Lothaire [3] and J. Berstel [1] contain more detailed presentation of maps, trees, words and formal languages respectively.

I - Planar maps and flows

Let us state a few definitions concerning maps. A planar map determines a partition of the plane in a set S of vertices, a set A of edges and a set F of faces . Each vertex is a point of the plane, each edge is an open curve having two vertices as end points and each face is an open simply connected domain bounded by edges and vertices. Each edge is in the boundary of two faces ; in the sequel we suppose that these faces are distinct thus the map has no isthmus.

A colouring of the faces of a map consists in a mapping of F in a finite set of colours in such a way that any two faces having a common edge in their bourdaries are mapped into different colours.

The degree of a vertex of a map is equal to the number of edges having it as end point. A map is said cubic if all its vertices have degree three.

An orientation of a map is determined by choosing for each edge a an initial end point $i(a)$ and a terminal end point $t(a)$.

A k-flow on an oriented map is a mapping φ of the set of edges into the ring $\mathbb{Z}/k\mathbb{Z}$ of integers modulo k such that, for any vertex s, we have :

$$\sum_{i(a)=s} \varphi(a) \equiv \sum_{t(a)=s} \varphi(a) \qquad (\mathrm{mod}\ k)$$

A non-zero k-flow is a flow φ for which $\varphi(a)$ is different from 0 for any edge a. In fact, it is better to say a nowhere zero k-flow, but non zero is shorter.

Remark that the existence of a non-zero k-flow for an oriented map depends only on the map itself and not on the orientation choosen. As if M' is obtained from M by changing the orientation of the edge a_0 then, if φ is a non-zero k-flow on M, the mapping φ' defined below is a non-zero k-flow on M' :

$$\varphi'(a) = \varphi(a) \quad \text{for each} \quad a \neq a_0 \quad \text{and} \quad \varphi'(a_0) = -\varphi(a_0) .$$

If an oriented map has a colouring consisting in a mapping of the set of faces in $\{1,2,...,k\}$ then one can construct a non-zero k-flow by defining $\varphi(\underline{a})$ as the difference (molulo k) between the coulours of the faces containing $\underline{a}$ in their boundary (the orientation of $\underline{a}$ allows to distinguish between the face at the left of $\underline{a}$ and that at the right of $\underline{a}$). In fact, this construction can be reversed giving the following theorem which seems to be due to W.T. Tutte [7].

Theorem 1 : The following statements are equivalent :

(1) Every planar map with no isthmus is face colourable in four colours.

(2) Every planar map with no isthmus has a non-zero 4-flow.

(3) Every cubic planar map with no isthmus has a non-zero 4-flow.

II- Flows in binary trees :

In what follows we consider that a binary tree is determined by a set of nodes, one of them being the root r, such that each node, except the root, has two sons or no son at all ; between the two sons of a node one is distinguished as the left son, the other being the right one. A node which has no son is called a leaf. Generally, the root of a binary tree has two sons, in order to simplify further developments we suppose that it has only one son (or no son if the tree reduces to r). The size of a binary tree is equal to the total number of nodes ; with our conventions concerning the root this size equals twice the number of leaves of the tree.

Figure 1 gives an example of a binary tree of size 12.

The left (resp. right) subtree below node n is recursively defined in the following way :

If n is a leaf then this left (resp. right) subtree is empty, if it is not a leaf then it consists of the left son of n (resp. the right son of n) and of the left and right subtrees below this left (resp. right) son.

Below the root there is a subtree consisting of all the other nodes.

The preorder sequence of the vertices of a tree is obtained by concatenating :

the root r, its son s, the preorder sequence of the left subtree below s, the preorder sequence of the right subtree below s. This sequence determines a total order on the set of nodes, in the sequel when we will speak of ith leaf this will mean the ith leaf in the preorder sequence. In figure 1 the nodes are numbered as they appear in the preorder sequence : the first leaf is numbered 3, the second one is numbered 6 and the 6th one 12.

A k-flow in a binary tree is given by a mapping Ψ of the set of nodes in the set Z/kZ of integers moludo k, such that the flow in a node which is not a leaf is equal to the sum (modulo k) of the flows of its sons. Note that a flow is completely determined by its value on the set of leaves . A flow Ψ is non zero if $\Psi(n)$ is different from zero for any node n. Figure 2 gives a non zero 4-flow for the tree in figure 1.

Given two binary trees $\mathcal{A}_1$ and $\mathcal{A}_2$ of the same size, two flows Ψ_1 on $\mathcal{A}_1$ and Ψ_2 on $\mathcal{A}_2$ are said to be compatible if the flows on the roots are the same ($\Psi(r_1) = \Psi(r_2)$) and if the flow on the ith leaf of $\mathcal{A}_1$ is equal to the flow on the ith leaf of $\mathcal{A}_2$ for all leaves. Figure 3 gives a binary tree of same size than that of figure 2 and a flow compatible with that previously given.

Theorem 2 ([4], see also [6]) - The two following statements are equivalent :

(1) Every planar map with no isthmus is face colourable in 4 colours.

(4) Two binary trees of the same size have two non-zero 4-flows which are compatible.

Proof. We use the condition (3) of theorem 1. We consider here planar cubic maps with no isthmus such that any union of two or three faces (with their boundaries) which are two by two adjacent (i.e. with a common boundary edge) is a simply connected domain. Then, it' s easy to prove that if any planar cubic map with no isthmus verifying this condition is face colourable in four colours (or admits a non-zero 4-flow), it's the same for any planar cubic map with no isthmus. Let C be a planar cubic map with no ithmus verifying the previous condition and let C^* be the dual of C (its vertices correspond to the faces of C and two vertices are joigned by an edge if the corresponding faces have a boundary edge in common). In C^* all the faces are bounded by three edges as C was cubic and there is no cycle which contains one, two or three edges excepted the cycles composed by the boundaries of the faces which are triangles. Thus by a theorem of Whitney [8], C^* has a hamiltonian cycle.

This hamiltonian cycle of C^* determines in C a closed curve Γ which go through any face cutting some edges : thus C is decomposed in a binary tree in the interior of Γ and a binary tree in its exterior. The leaves of these binary trees are the points in which Γ intersects the edges of C, and a root is choosen among them.

Then, it is clear that the existence of a non-zero 4-flow for C is equivalent to that of two compatible non zero 4-flows for the trees obtained in the decomposition :

The flow of a leaf of the trees is equal to the flow in the corresponding edge intersected by Γ, and the flow (in the tree) of the other nodes is determined by the flow (in the map) in the edge joining it to its "father". This construction can clearly be reversed. □

III - Coding binary trees

Let X and Y be the following alphabets :

$$X = \{x, \bar{x}\} \qquad Y = \{x, x_1, x_2, x_3\}$$

To any binary tree is associated a word of X^* considering its preorder sequence of nodes and writing an $\bar{x}$ for any leaf and an x for a node which is not a leaf, nothing being written for the root. Clearly the word corresponding to a tree of size $2k$ has length $2k-1$, k occurrences of $\bar{x}$ and $k-1$ of x. Moreover we have :

Proposition 1 - The set of words associated with binary trees is a context free language L generated by the grammar :

$$\xi \to x\xi\xi \qquad \xi \to \bar{x} \ .$$

Proof. This proposition is deduced from the fact that the coding process considers first the son s of the root; then, if s is not a leaf, the left subtree below this son, then the right subtree below it (explaining the first rule) ; when s is a leaf then the code is $\bar{x}$ (explaining the second rule).

In order to code a binary tree with a flow it is only necessary to remind the value of this flow on the leaves, as remarked in paragraph II. Then we use the following process : from the preorder sequence of the vertices of the tree, write x for any node which is not the root and x_i for a leaf having a flow equal to i.

This gives :

Proposition 2 - the set of words coding binary trees having a non zero 4-flow with value equal to 1 on the root is a context free language L_1 given by the grammar (ξ_1 is the axiom) :

$$\xi_1 \to x\ \xi_2\ \xi_3 \quad \xi_1 \to x\ \xi_3\ \xi_2 \quad \xi_1 \to x_1$$

$$\xi_2 \to x\ \xi_1\ \xi_1 \quad \xi_2 \to x\ \xi_3\ \xi_3 \quad \xi_2 \to x_2$$

$$\xi_3 \to x\ \xi_1\ \xi_2 \quad \xi_3 \to x\ \xi_2\ \xi_1 \quad \xi_3 \to x_3 \ .$$

Proof - Similar considerations to that given for the proof of proposition 1 give the result. It suffices to note that the words generated using axiom ξ_i code the binary trees having a flow with value i on the root ; and that $\xi_i \to x\ \xi_j\ \xi_k$ is a rule then $j+k \equiv i$ (modulo 4) as the flow in a node is equal to the sum of the flows of its sons.

IV – A language in $X^* \times X^*$

In this part we use the definitions and notations of [1]. Let φ be the morphism of Y^* in itself defined by $\varphi(x) = 1$, $\varphi(x_i) = x_i$: φ forgets the x's and conserves the x_i . This morphism can be extended in a morphism of $Y^* \times Y^*$ in itself by :

$$\varphi((f,g)) = (\varphi(f),\varphi(g)).$$

Let Ψ be the morphism of Y^* in X^* defined by $\Psi(x) = x$ and $\Psi(x_i) = \bar{x}$ for $i=1,2,3$: Ψ replaces all the letters x_i by $\bar{x}$, thus if f is an element of L coding a binary tree with a 4-flow then $\Psi(f)$ code the same binary tree forgetting the flow.

Let Δ be the subset of $Y^* \times Y^*$ consisting of all the pairs (u,u) ; clearly Δ is a rational subset of $Y^* \times Y^*$ as :

$$\Delta = \{(x_i,x_i) \mid i=1,2,3\}^* .$$

Let D be the subset of $X^* \times X^*$ consisting of all the pairs (f,g) such that $|f| = |g|$, D is also a rational subset of $X^* \times X^*$ as

$$D = \{(x,\bar{x}) , (\bar{x},x),(x,x), (\bar{x},\bar{x})\}^*$$

Theorem 3 – The two following statements are equivalent :

(1) Every planar map with no isthmus has a face colouring in four colours.

(5) The two following subsets of $X^* \times X^*$ are equal :

$$P = \Psi\ (\varphi^{-1}(\Delta)\ \cap\ (L_1 \times L_1))$$
$$Q = (L \times L)\ \cap\ D.$$

Proof. Of course we will show that (5) is equivalent to the statement (4) of theorem 2.

Two binary trees $\mathcal{A}_1$ and $\mathcal{A}_2$ of the same size are coded by two words f and g such that $|f| = |g|$ thus (f,g) is an element of Q.

A non zero 4-flow on a binary tree coded by f gives a code u of Y^* such that $\Psi(u) = f$; the same is true for a binary tree coded by g giving v such that $\Psi(v) = g$. The order in which are met the x_i in u and in v is the same as that of the leaves in the preorder sequences of the vertices of the two trees ; the two 4-flows are thus compatible if $\varphi(u) = \varphi(v)$ or equivalentely if $\varphi((u,v)) \in \Delta$.

Thus two binary trees coded by f and g admit compatible non-zero 4-flows if and only if it exists two words u and v of L such that

$$\Psi(u,v) = (f,g) \quad ; \quad \varphi(u,v) \in \Delta$$

which ends the proof.

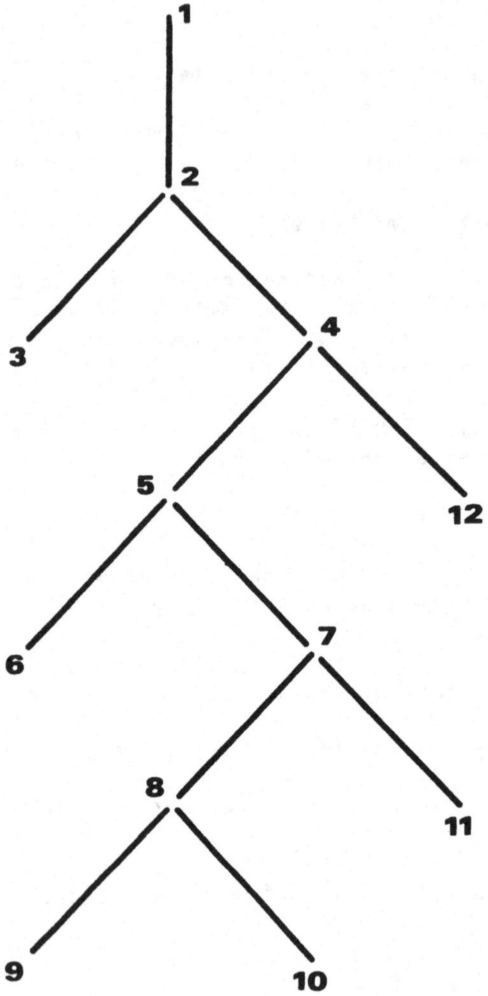

Figure 1 : A binary tree of size 12 in which the vertices are numbered as they appear in the preorder sequence.

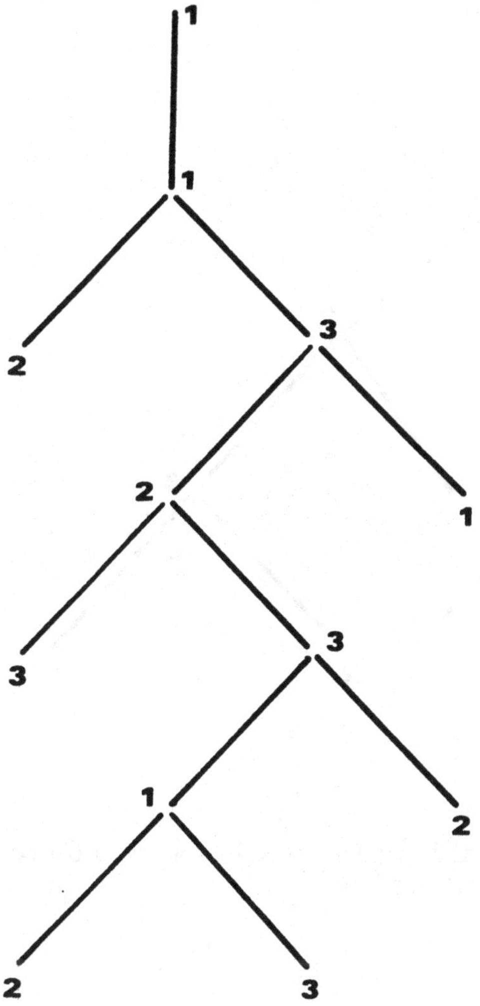

Figure 2 : A non-zero 4-flow for the tree given in figure 1.

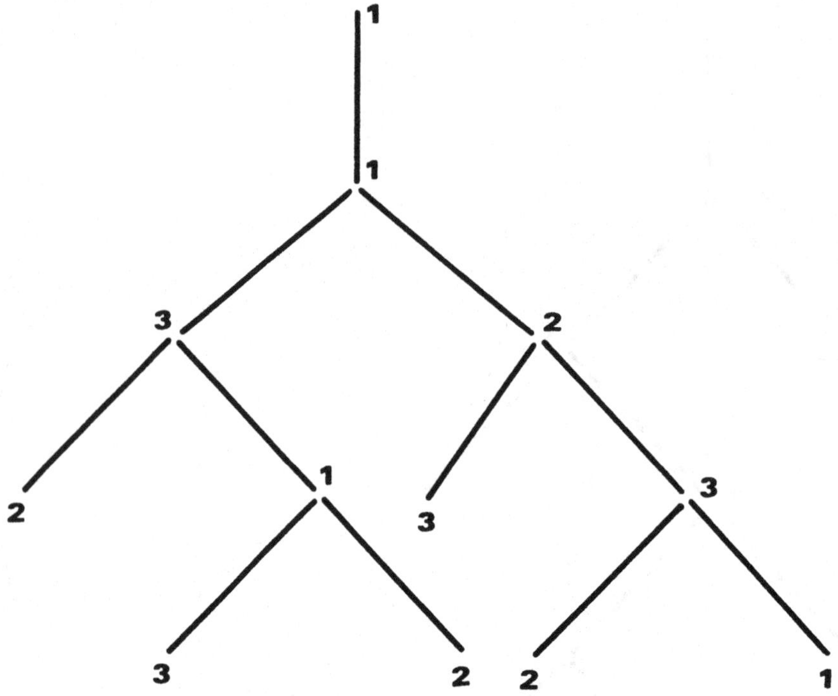

Figure 3 : A 4-flow compatible with that given in figure 2 for a tree of the same size.

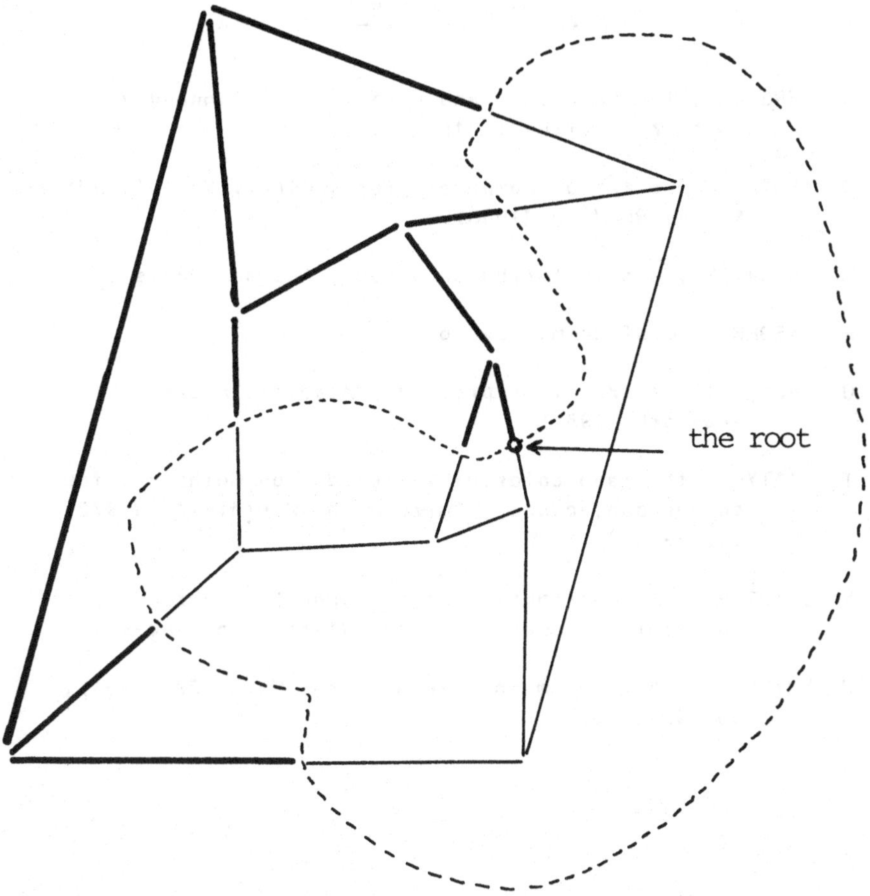

Figure 4 : A cubic map C and the trees obtained by its decomposition using a hamiltonian cycle of C^*. Note that the two trees obtained are those of figures 2 and 3.

REFERENCES

[1] J. BERSTEL , Transductions and context-free languages ; Teubner, Stuttgart (1979).

[2] D. KNUTH , The art of computer programming ; Vol. 1, Addison Wesley Reading (1968).

[3] M. LOTHAIRE , Combinatorics on words ; Addison Wesley.

[4] F. JAEGER , Oral communication.

[5] O. ORE , the four colour problem ; Academic Press, New-York (1967).

[6] T. SAATY , thirteen colorful variations on Guthrie's four colour conjecture ; Amer. Math. Morthly 79 (1972), pp. 2-43.

[7] W.T. TUTTE , A contribution to the theory of chromatic polynomials ; Car. J. Math. (1954) ; pp. 80-91.

[8] H. WHITNEY , A theorem on graphs ; Ann. Math. 32 (1931), pp. 378-390.

ON THE EQUIVALENCE OF SYNCHRONIZATION SETS

J. Beauquier and B. Bérard
Université de Paris Sud - Centre d'Orsay -
LRI Bât.490 - 91405 ORSAY Cedex - France.

ABSTRACT

In the framework of the description of processes behaviours by words and languages, conditions of synchronization can be expressed by the means of regular sets. Two problems about such synchronization sets are studied here: the decidability of the equivalence of two of them and the obtention of minimal ones.

INTRODUCTION

In [1] , the behaviour of a process is considered as a word, representing the sequence of actions performed by this process. The extension to several processes executing concurrently is then quite easy. Let $f_1, f_2, \dots, f_n$ be the words associated to processes $p_1, p_2, \dots, p_n$, when each of them is executed alone on a single processor. The parallel execution of $p_1, p_2, \dots, p_n$ on n different processors is then the cartesian product $f_1 \times f_2 \times \dots \times f_n$. But real things are not so simple. There could be less processors than processes or some actions could require mutual exclusion. To the previous model, we will first add, everywhere in a word f_i, a special letter # , corresponding to the action "wait" for a process. We will also introduce n-tuples of letters representing forbidden simultaneous actions. For instance, the n-tuple (#, #, ..., #) will mean that processes are never together idle. We call synchronization set a set of such n-tuples. In [2] and [8] , it has been shown that this model allows to treat a number of problems in parallel computation.

Generally, it is possible to know in advance that the future behaviour of a given process will be necessarily of a certain type. In the model, this will mean that the word representing the behaviour of a pro-

-cess is known, a priori, to belong to some language. Now, we can state the problem we are interested in : knowing the language related to each process, is it possible to decide whether two sets of forbidden simulta-neous actions produce the same result or not ? A consequence of a po-sitive answer is then that a minimum synchronization set can be built.

PART 0 - PRELIMINARIES

We assume classical notions of formal language theory to be known ([3] or [4]). In particular, we will handle with regular sets and finite automata, context-free languages and grammars and deterministic context-free languages. We just give here some specific points.
Let X be an alphabet, let $n \geq 2$ be an integer and let $f_1, \dots, f_n$ be n words in X^* of the same length k. ($|f|$ will denote the length of the word f .)
If $k > 0$, $f_j = f_j^{(1)} f_j^{(2)} \dots f_j^{(k)}$, where $f_j^{(i)}$ is in X , for each i,j, $1 \leq i \leq k$, $1 \leq j \leq n$.
The Interleaving of the words $f_1, \dots, f_n$ is defined by :
$I_n(f_1, \dots, f_n) = f_1^{(1)} \dots f_n^{(1)} f_1^{(2)} \dots f_n^{(2)} \dots f_1^{(k)} \dots f_n^{(k)}$.
If $k = 0$, ε will denote the empty word in X^* and we set :
$I_n(\varepsilon, \dots, \varepsilon) = \varepsilon$.

Let $n \geq 1$ be an integer and let $Y_n = \{ a_1, \dots, a_n \}$ be an alphabet.
Define $\bar{Y}_n = \{ \bar{a}_i , 1 \leq i \leq n \}$ and $Z_n = Y_n \cup \bar{Y}_n$.
The restricted Dyck set D'^*_n over Z_n is the context-free language gene-rated by :

$S \to a_i S \bar{a}_i$, $1 \leq i \leq n$
$S \to S S$
$S \to \varepsilon$

The Dyck set D^*_n over Z_n is the context-free language generated by :

$S \to a_i S \bar{a}_i$, $1 \leq i \leq n$
$S \to \bar{a}_i S a_i$, $1 \leq i \leq n$
$S \to S S$
$S \to \varepsilon$

The symmetric language S_n is the context-free language generated by :

$S \to a_i S \bar{a}_i$, $1 \leq i \leq n$
$S \to \varepsilon$

PART 1 - SYNCHRONIZATION OF LANGUAGES BY REGULAR SETS

Definition 1.1 Let X be an alphabet and let $L_1,\dots,L_n$ be languages in X^*, $n \geq 2$. We note $//(L_1,\dots,L_n)$ the language defined by :

$//(L_1,\dots,L_n) = \{ I_n(f_1,\dots,f_n) \ / \ f_i \in L_i,\ 1 \leq i \leq n,\ |f_1| = \dots = |f_n| \}$.

Definition 1.2 Let X be an alphabet and let $n \geq 2$ be an integer. A n-synchronizing set (respectively regular set) on X is a subset of X^n (resp. a regular set of $(X^n)^*$).

This terminology comes from the following idea. A word in L_i can be viewed as the history of a process p_i, a word in $//(L_1,\dots,L_n)$ as the history of n parallel processes $p_1,\dots,p_n$. A word in X^n can then be viewed as a set of simultaneous forbidden actions, that is a mutual ex--clusion condition.

For example, let $X = \{a,\bar{a}\}$ and suppose that the letter "a" means "read in file A" and the letter "$\bar{a}$" means "write in file A". Let p_1,p_2 and p_3 be three processes that successively read and write in file A. Then, $L_1 = L_2 = L_3 = X^*$. Consider the word $f = a\,a\,\bar{a}\,\bar{a}\,a\,a\,\bar{a}\,a\,a$ in $//(L_1,L_2,L_3)$. This word can be viewed as the following sequence : during the first cycle of the clock, p_1 and p_2 read in file A while p_3 writes on A; du-ring the second cycle, p_1 writes on A while p_2 and p_3 read in A, etc.. Now, if we wish accesses to file A in read and write mode to be in mu--tual exclusion, we introduce the subset S of X^3 defined by :

$S = \{ x\,y\,z \ / \ x,y,z \text{ in } X,\ x \neq y \text{ or } y \neq z \text{ or } x \neq z \}$.

Thus, the set of behaviours which satisfy this mutual exclusion requi-rement is exactly : $//(L_1,L_2,L_3) \cap (X^3 - S)^*$.

Remark: It should be noticed that the model concerns synchronous pro--cesses with lock-step progress. Meanwhile, the introduction of a spe--cial letter in X, figurating the empty action, allows to modelize pro--cesses with asynchronous progress.

The replacement of $(X^n - S)^*$ by an arbitrary regular set is a natural extension that allows to express relations between synchronization re--quirements.

Definition 1.3 Let X be an alphabet and let $L_1,\dots,L_n$ be languages in X^* ($n \geq 2$). Let S (resp. K) be a n-synchronizing set (resp. regular set) on X. The subset of $//(L_1,\dots,L_n)$ synchronized by S (resp. K) is:

$$\Lambda(S, //(L_1,\dots,L_n)) = //(L_1,\dots,L_n) \cap (X^n - S)^*$$

(resp. $\Lambda(K, //(L_1,\dots,L_n)) = //(L_1,\dots,L_n) \cap K$)

In this part, we will be mainly interested in the following problem :

can it be decided whether two synchronizing sets (resp. regular sets) S_1 and S_2 (resp. K_1 and K_2) are or not equivalent, that is wether or not:

$$\Lambda (S_1, //(L_1,\dots,L_n)) = \Lambda (S_2, //(L_1,\dots,L_n))$$

(resp. $\Lambda (K_1, //(L_1,\dots,L_n)) = \Lambda (K_2, //(L_1,\dots,L_n))$.

We call this problem the synchronization equivalence problem and we note it: $\Sigma (S_1, S_2, L_1,\dots,L_n)$ (resp. $\Sigma (K_1, K_2, L_1,\dots,L_n)$).

The first result is easy :

Theorem 1.1 Let $L_1,\dots,L_n$ be regular sets in X^* ($n \geq 2$) and let K_1 and K_2 be two n-synchronizing regular sets (all given by finite automata). Then, the synchronization equivalence problem $\Sigma (K_1, K_2, L_1,\dots,L_n)$ is recursively solvable.

In this particular case, the synchronization equivalence problem is reduced to the problem of the equality of two regular sets, that can be solved by an exponential algorithm, according to the number of states of finite automata recognizing L_1, L_2, ... , L_n, K_1 and K_2 .

The second result is negative : theorem 1.1 cannot be extended to the general context-free case.

Theorem 1.2 Let $L_1,\dots,L_n$ be context-free languages in X^* ($n \geq 2$), gi--ven by context-free grammars, and let S_1, S_2 be two n-synchronizing sets on X. Then, the synchronization equivalence problem $\Sigma (S_1, S_2, L_1,\dots,L_n)$ is recursively undecidable, even if n=2.

Proof: the problem is encoded into the problem of the emptiness of the intersection of two context-free languages, which is known to be recursively undecidable.

Let L and L' be two context-free languages. We define $L_1 = L$, $L_2 = L'$ and, if $n > 2$, $L_3 = \dots = L_n = X^*$. We also define :

$S_1 = X^n$, $S_2 = \{ x_1 \dots x_n \ / \ x_i \text{ in } X , \exists\, i,j,\ 1 \leq i \leq n,\ 1 \leq j \leq n,\ x_i \neq x_j \}$, so that $X^n - S_2 = \{ x^n \ / \ x \in X \}$ and $\Lambda (S_2, //(L_1,\dots,L_n)) = \{ I_n(f,\dots,f) / f \in L_1 \cap L_2 \}$ while $\Lambda (S_1, //(L_1,\dots,L_n)) = \emptyset$.

Clearly, S_1 and S_2 are equivalent if and only if $L \cap L' = \emptyset$.

Facing to this fact, we can try to weaken the problem. There are at least two ways to do it. The first one leads to the following proposition, the second one to Part 2.

Theorem 1.3 Let $L_1,\dots,L_n$ be languages in X^* ($n \geq 2$), that are regular, excepted at most one which is context-free non regular, and let K_1 and K_2 be two n-synchronizing regular sets on X (all given by finite automata or context-free grammar). Then, the synchronization equivalence problem $\Sigma(K_1, K_2, L_1, \ldots, L_n)$ is recursively solvable.

Sketch of the proof: the proof consists in building two context-free grammars generating $\Lambda(K_1 - K_2, //(L_1, \ldots, L_n))$ and $\Lambda(K_2 - K_1, //(L_1, \ldots, L_n))$ and to check the emptiness of these two languages. Here again, the algorithm is exponential.

PART 2 - SELF-SYNCHRONISATION OF A LANGUAGE BY REGULAR SETS

In this part, we restrict our attention to a special case. First, we will assume that languages $L_1, \ldots, L_n$ of Part 1 are all equal to a same language L. Secondly, we only study synchronizing (regular) sets that force the words $f_1, \ldots, f_n$ in, now, L to be equal each to another one. We will call such (regular) sets, (n-)self-synchronizing (regular) sets for L.

A motivation for such a study does exist. In some distributed operating systems, reliability is guaranteed by means of several processors computing redundantly (for instance in the US Space shuttle system [6]). Then, a control must be provided to ensure that the outputs of processors are identical. The minimization of this control is a problem. Same things appear, always in distributed systems, in order to ensure information consistency. If copies of a file are stored at different sites, a control has to check that a modification of a copy at one site has been registered at other sites. Here, maximal cost concerns transmission by messages and the more local to the sites is the control, the more economical is the system. Assuming that possible configurations of a file are described by words of a language L, the idea is to ensure consistency in periodically checking:

(i) that some particular points are identical in each copy of the file (by messages);

(ii) that each copy corresponds to a word in L (by local control).

Notation $//(L,n)$ will denote $//(\underbrace{L, \ldots, L}_{n \text{ times}})$.

Definition 2.1 A n-synchronizing set S on X is n-self-synchronizing for a language L in X^* if and only if

$$\Lambda(S, //(L,n)) = \{ I_n(\underbrace{f, \ldots, f}_{n \text{ times}}) \ / \ f \text{ in } L \} .$$

Remark For any language L in X^*, there exists a trivial n-self-synchronizing set S : $S = \{ x_1 x_2 \ldots x_n \ / \ x_i \text{ in } X \text{ and } \exists\, i,j,\ x_i \neq x_j \}$.

The cardinality of S is $(\text{card}(X))^n - \text{card}(X)$. We will first study n-self-synchronizing sets of minimal cardinality.

Definition 2.2 Let L be a language in X^* and let $n \geq 2$ be an integer. We call n-synchronization-optimum of L and we note $C_n(L)$ the integer:
$C_n(L) = \text{Min}\{\text{card}(S)$, S is a n-self-synchronizing set for L$\}$.
We call synchronization-optimum of L and we note C(L) the integer:
$C(L) = \text{Min}\{C_n(L) , n \geq 2\}$.

The following proposition will simplify our study.

Proposition 2.1 $C(L) = C_2(L)$.

Then, in the sequel, we will simply say synchronizing instead of 2-syn--chronizing. The synchronization-optimum as defined above has a draw--back. For a sequence of languages with the same structure but on al--phabets of growing cardinality, it takes growing values. Therefore, we introduce an other notion :

Definition 2.3 Let X be an alphabet and $m = \text{card}(X)$. Let L be a lan--guage in X^* (X of minimal cardinality). We call synchronization-indi--cator and we note I(L) the following integer:

$$I(L) = \frac{2C(L)}{m(m-1)}$$

We now present the computation of synchronization parameters for some classical languages. The proof that a given set is self-synchronizing for L is easy. The proof that this set is minimum is more involved.

Proposition 2.2 Let $n \geq 1$ be an integer and let $D_n'^*$, D_n^* and S_n be res--pectively the restricted Dyck set, the Dyck set and the symmetric lan--guage over Z_n. Let # be a new symbol not in Z_n. We have :

1) $C(D_n'^*) = n(2n-1)$ $\quad I(D_n'^*) = 1$

   $C(D_n'^* \#^*) = n(2n+1)$ $\quad I(D_n'^* \#^*) = 1$

2) $C(D_n^*) = n(2n-1)$ $\quad I(D_n^*) = 1$

   $C(D_n^* \#^*) = n(2n+1)$ $\quad I(D_n^* \#^*) = 1$

3) $C(S_n) = n(n-1)$ $\quad I(S_n) = \frac{n-1}{2n-1}$

   $C(S_n \#^*) = n(n+1)$ $\quad I(S_n \#^*) = \frac{n+1}{2n+1}$

Remark We also consider $L\#^*$ together with L in order to obtain va--lues for the synchronization parameters, independant from the distri--bution of lengths of the words in L.
For instance, $L = \{a^n b^n , n \geq 1\}$ is self-synchronized by the empty set, since there is at most one word of length p ($p \geq 1$) in L, while $L\#^*$ is

not synchronized by $\emptyset$.

We will now study the structure of self-synchronizing sets. Synchronization parameters C(L) and I(L) lead to deal with the minimality of the cardinal of self-synchronizing subsets. We also consider the minimality for the inclusion relation of subsets.

Definition 2.4 A self-synchronizing set S for a language L is minimal if and only if any proper subset of S is not self-synchronizing for L. S is minimum if and only if it is of minimum cardinality.

The two results about this point are :

Proposition 2.3 There exist a language L and minimal self-synchronizing sets for L that are not minimum.

Proof The language L is defined by $L = \{ c_1adbbc_2 , c_2aadbc_1 \}$.
The subset $S_1 = \{ ad , da \}$ is minimal but not minimum, since $S_2 = \{c_1c_2\}$ is self-synchronizing for L.

Proposition 2.4 There exist a language L and two minimum self-synchronizing sets S_1 and S_2 for L, that are not isomorphic.

Proof L is the context-free language generated by the grammar with productions :

$$S \to aSe + dTd$$
$$T \to bTc + d$$

where S is the start symbol.

$S_1 = \{ ad , da \}$ and $S_2 = \{ ad , bd \}$ are both self-synchronizing sets for L, but are not isomorphic.

We now state the decidability results of this section.
It should be first noticed that Theorem 1.1 has the :

Corollary 1 Let L be a regular set in X^* and let S be a subset of X^n ($n \geq 2$). It can be decided whether S is self-synchronizing for L or not.

From this corollary, one easily obtains :

Corollary 2 It can be recursively decided whether S is a minimum n-self-synchronizing set for L or not.

Corollary 3 For any regular set L, one can effectively obtain the minimum n-self-synchronizing set(s) for L.

Theorem 2.1 Let L in X^* be a deterministic, under prefix closed, context-free language and let S be a subset of X^n ($n \geq 2$).
It can be recursively decided whether S is n-self-synchronizing for L or not.

Proof see Theorem 2.2.

Theorem 2.1 involves :

Corollary 4 For any deterministic, under prefix closed, context-free language L, one can effectively obtain the minimum n-self-synchronizing for L.

We will now consider the same study, except the fact that an arbi--trary regular set K replaces $(X^n - S)^*$. In the proofs of the decidabi--lity results, the fact mainly used is that $(X^n - S)^*$ is regular rather than S is finite. Therefore, it is not surprising that these results can be extended. Concerning the amount of control needed for self-syn--chronization, we will choose as measure the minimum number of states of a finite automaton recognizing a regular set.

<u>Definition 2.5</u> Let L be a language in X^* and let K be a regular set in X^*. K is n-self-synchronizing for L if and only if :

$$\Lambda(K, /\!/(L,n)) = \{ I_n(\underbrace{f,\dots,f}_{n \text{ times}}) \ / \ f \text{ in } L \} .$$

As in the previous part, it should be noted that for any language L, the set $K = \{ x^n , x \text{ in } X \}^*$ is a trivial regular set n-self-synchronizing for L.

<u>Definition 2.6</u> Let A be a non deterministic finite automaton. A is n-self-synchronizing for a language L iff the language recognized by A is n-self-synchronizing for L.

For the sake of simplicity, we will actually restrict our study to 2-self-synchronizing automata and we will say synchronizing instead of 2-self-synchronizing. We note M(L) the minimum number of states of a fi--nite automaton synchronizing the language L.

We first show that, for particular languages, the size of the au--tomaton does not depend on the cardinality of the alphabet.

<u>Proposition 2.5</u> Let X be an alphabet and let L be a language such that $L \subseteq a_1^+ a_2^+ \dots a_k^+ \#^*$, where a_i are letters in X and # is a new symbol. Define $Y = \{ a_{2p} , 1 \le p \le [k/2] \}$ and $Z = \{ a_{2p+1} , 1 \le p \le [k/2] \}$, where $[k/2]$ denotes the integral part of k/2.
If $Y \cap Z = \emptyset$, there exists an automaton with three states, synchro--nizing L.

<u>Proposition 2.6</u> Let w_1 be a word in X^* and let $L \subseteq w_1^+ \#^*$ be a lan--guage, where # is not in X. There exists an automaton with two states synchronizing L.

<u>Proposition 2.7</u> Let w_1 and w_2 be words in X^*, satisfying the following condition : there are no integers p , q such that $w_1 = u^p$, $w_2 = u^q$, for some word u in X^*.
If L is a language, $L \subseteq w_1^+ w_2^+ \#^*$, then there exists an automaton with three states synchronizing L.

We now present a non trivial result about the restricted Dyck set over Z_n.

Proposition 2.8 There exists a finite automaton synchronizing $D_n'^*$, with a number of states at most of order $\sqrt{n}$.

Sketch of the proof

Define $p = \lceil \sqrt{n} \rceil$ and $q = \lceil n/p \rceil$, where $\lceil x \rceil$ denotes the smallest integer $\geq x$. Since $n \leq pq$, we may assume that Z_{pq} is an alphabet of $D_n'^*$. We can write the letters in Z_{pq} as :

$Z_{pq} = \{ x_{i,j} \ / \ 1 \leq i \leq p \ , \ 1 \leq j \leq q \} \cup \{ \bar{x}_{i,j} \ / \ 1 \leq i \leq p \ , \ 1 \leq j \leq q \}$.

For each i, $1 \leq i \leq p$ and for each j, $1 \leq j \leq q$, define :

$A_{i,o} = \{ x_{i,j} \ , \ 1 \leq j \leq q \}$ and $\bar{A}_{o,j} = \{ \bar{x}_{i,j} \ , \ 1 \leq i \leq p \}$.

We construct now a finite automaton A_{pq}, according to the figure below:

A_{pq} :

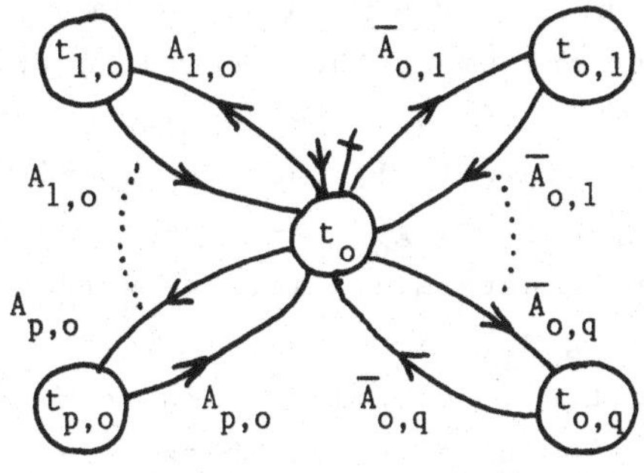

This automaton is synchronizing for $D_n'^*$ and its number of states is : $p + q + 1 \leq 2\sqrt{n} + 1$.

In some particular cases, the results are more precise.

Proposition 2.9 $M(D_1'^*) = M(D_1'^* \#^*) = 3$ and $M(S_1 \#^*) = 2$.

The following decidability result extends theorem 2.1 :

Theorem 2.2 Let L be a language in X^*, deterministic context-free, un--der prefix closed and let K be a regular set in $(X^n)^*$ $(n \geq 2)$. Then, it can be recursively decided whether K is n-self-synchronizing for L or not.

Proof

Since L is a context-free language, it satisfies the pumping Ogden's lemma ([9]) for some integer $p_o \geq 1$. Let q be the number of states of the minimal automaton accepting K and let M be the language defined by:

$M = \Lambda(K, //(L,n)) - \{ I_n(f,\ldots,f) \ / \ f \text{ in } L \}$.

Define $k = q(q+1)$ and $m = n(p_o^k + 1)$. We will prove that there exists a word of X^* in M if and only if there exists a word of X^* in M of length less than or equal to m .

So, let us assume that all words in M are of length greater than m and choose such a word f of minimal length. Since L is closed under prefix, we can write :

$f = f_o x_1 \ldots x_n$, where f_o is in X^* , x_i in X for each i, $1 \leq i \leq n$,

with : (i) f_o , of length $|f| - n$ is in $\{ I_n(f,\ldots,f) \ / \ f \text{ in } L \}$ (since f is minimal);

(ii) there exist $i \neq j$, such that $x_i \neq x_j$.

Therefore, there exists f'_o in L such that $f_o = I_n(f'_o,\ldots,f'_o)$.
Since $|f| > m$, $|f'_o| > p_o^k$, so that it can be obtained k non empty distinct nested iterative pairs in f'_o for L. These pairs obviously in--duce k non empty distinct nested iterative pairs in f_o for $//(L,n)$. Since $k = q(q+1)$, at least one among these pairs is also an iterati--ve pair for K , thus for $\Lambda(K, //(L,n))$. Let us erase (using the expo--nent 0) this pair in f_o (respectively the corresponding pair in f'_o), so obtaining a word g_o (respectively g'_o). We have : $g_o = I_n(g'_o,\ldots,g'_o)$.
Now, since L is supposed to be deterministic, the word

$$g = g_o x_1 \ldots x_n$$

is, likewise g_o, in $\Lambda(K, //(L,n))$. Clearly, it is in M. But the length of g is strictly less than the length of f, which is a contradiction with the minimality of f. So the theorem is proved.

PART 3 - PARALLEL OPERATION AND CONTEXT-FREENESS

In Part 1 and 2, when looking for the reasons for which decidabi--lity results concerning regular sets cannot be extended to general context-free languages, it appears that the main fact is that $//(L_1,L_2)$ is generally not a context-free language as soon as L_1 and L_2 are no more regular. Intuitively speaking, it comes from the possibility for $//(L_1,L_2)$ to have "crossed" iterative pairs (like in $a^n b^p c^n d^p$). The question we treat here is : does there exist context-free non regular languages L_1 and L_2 , such that $//(L_1,L_2)$ is context-free ? This ques--tion must be related to the negative answer by Latteux ([7]), when the shuffle operation replaces the operation // .
The main results of this part are :

Proposition 3.1 Let $L = \{ a^n b^n \ / \ n \geq 1 \}$ over the alphabet $X = \{ a,b \}$ and let # be a new symbol. Then, $//(L\#^*, L\#^*)$ is not context-free.

Proof

Assume $//(L\#^*, L\#^*)$ is context-free and let n_o be the integer related to Ogden's lemma. We consider the word:

$f = I_2(a^{2n_o}b^{2n_o}, a^{n_o}b^{n_o}\#^{2n_o}) = a^{2n_o}(ab)^{n_o}(b\#)^{2n_o}$, where the first $2n_o$ a's are marked positions.

Clearly, all possible factorizations lead to a contradiction. Thus, the language is not context-free.

Let G be the language (known as the Goldstine's language [5]) over $X = \{a,b\}$ defined by:

$$G = \{ a^{n_1}ba^{n_2}b\ldots a^{n_k}b \ / \ k \geq 1, n_j \geq 0 \text{ and } i \neq n_i \}.$$

G is known to be a linear context-free language.

Theorem 3.1 $//(G,G)$ is a context-free language.

Proof We will use two different versions of non-deterministic push-down automata recognizing G (by final states).

First version. The underlying idea of how this automaton works is the following: let w be a word in X^*. Non deterministically, a block of a's is chosen. The b's preceding this block are pushed into the stack. Then, each a in the chosen block makes a b to be popped from the stack. The word w is accepted if the number of a's in the chosen block does not match the number of b's in the stack.

Second version. It allows to keep in the stack, after checking, the rank of the chosen block of a's. It is based upon the fact that G is also defined by :

$$G = \{ a^{n_1}b\ldots a^{n_p}b \ / \ n_1 \neq 1 \text{ or there exists } k, 1 \leq k \leq p-1, \text{ such that } n_{k+1} \neq n_k + 1 \}.$$

The automaton first checks whether $n_1 = 1$ or not, or chooses a block of a's (let k+2 be its rank). In the second case, the first k b's are pushed into the stack, then the a's of the following block (their num--ber is n_{k+1}) are also pushed in stack. The b is skipped and it is then checked if the number of a's in the following block is different of $n_{k+1} + 1$ (by using the n_{k+1} a's on the top of the stack). If this test is positive, the word is accepted and the rank of the current block can be retrieved from the stack (number of b's plus 2).

Now, we can describe a non-deterministic pushdown automaton recognizing $//(G,G)$. Let w be a word in X^*. As long as couples of letters aa or bb are encountered, the automaton works as in the second version, pu--shing only one of the two letters in the stack. As soon as a couple of letters ab or ba is encountered (involving that one of the two subwords w' or w" constituted by the sequence of even or odd letters of w is in G), the automaton uses the a's at top of the stack for de--termining which word is in G (for instance w'). Using the b's in the stack and switching for first version, the automaton checks then that the other word (here w") does belong to G.

We end this paper by stating a conjecture.
Let # be a new symbol. By pumping arguments ([9]), it can be proved that $//(\#^*G\#^*, \#^*G\#^*)$ is not context-free.

<u>Conjecture</u>. Let L be a non regular language over X and let # be a let--ter not in X. Then, $//(\#^*L\#^*, \#^*L\#^*)$ is not context-free.

REFERENCES :

[1] ARNOLD A. et NIVAT M. , Comportements de processus, LITP Report n° 82-12, Univ. Paris VII, 1982.

[2] BEAUQUIER J. et NIVAT M. , Application of formal language theory to problems of security and synchronization, in Formal Language Theory, Perspectives and Open Problems, R.V. BOOK (ed.), Academic Press, 407-453.

[3] BERSTEL J. , Transductions and context-free languages, Teubner Verlag, 1979.

[4] GINSBURG S. , Algebraic and Automata-Theoretic Properties of Formal Languages, North-Holland, 1975.

[5] GOLDSTINE J. , Substitution and bounded languages, J. of Comput. and Syst. Sci. 6 , 9-29 , 1972.

[6] GIFFORD D. and SPECTOR A. , The Space Shuttle Primary Computer System, Commun. ACM , 27 , 9 , 872-900 , 1984.

[7] LATTEUX M. , Cônes rationnels commutatifs, J. of Comput. and Syst. Sci. 18 , 307-333 , 1979.

[8] NIVAT M. , Behaviours of synchronized systems of processes, LITP Report n°81-64 , Univ. Paris VII , 1981.

[9] OGDEN W. , A Helpful Result for Proving Inherent Ambiguity, Math. System Theory 2 , 191-194 , 1967.

INNER AND MUTUAL COMPATIBILITY OF BASIC OPERATIONS ON MODULE SPECIFICATIONS*

Francesco Parisi-Presicce
Department of Mathematics
University of Southern California
Los Angeles, California 90089-1113

ABSTRACT The category Mod of module specifications and module morphisms is introduced and used to show the inner compatibilities of union, composition and actualization of module specifications. Using algebraic techniques, we can provide a simple proof of the compatibility of actualization and union (i.e. that the union of actualized modules is an appropriate actualization of a union of the modules) and establish the compatibility of composition with parameterized actualization (i.e., that an actualization of the composite M1•M2 can be obtained as a composition of the actualized modules M1 and M2).
It is also shown that the operations of union, composition and actualization are monotone in all their arguments with respect to the partial order M1 ≤ M2 defined as "M1 is a submodule specification of M2".

1. INTRODUCTION

Algebraic and categorical methods have been used successfully in describing the denotational semantics of programming languages (/Mo84/, /Ol83/), and in specifying concurrent systems (/AMRW85/). A considerable amount of research (/EK83/, /EM85/, /G83/, their references, and many more) has dealt with algebraic specification techniques for data types at least as early as /LZ74/. This approach has evolved by isolating an export interface (/GM82/) and carried one step further by introducing module specifications. The modular development of large software systems (/Pa72/, /WE85/) is based on the concept of an abstract module, which combines the main ideas of parametrized specification, "information hiding" and implementation of abstract data types. An abstract module, as introduced in /EW85/, /WE85/ and /BEPP85/, is an abstract data type with two interfaces (import and export), a body and a parameter part. The body of the module (corresponding to the body specification of an Ada package) contains both interfaces and provides an implementation of the sorts and operations of the export interface by those of the import interface. The export interface (the visible part of an Ada package) represents the sorts and operations available to the user of the module, with the operations in the body not in the

---

*Supported in part of the National Science Foundation under Grant DCR-8406920.

export to be considered "hidden". The import interface (allowed in Ada using calls to procedures present in export interfaces of other modules, such as STANDARD) represents the operations to be provided to the module. The two interfaces are allowed to share a common parameter part (not present in Ada packages but provided for in Ada generics). All four parts of an abstract module are given by algebraic specifications, with the interfaces and parameter having loose semantics, while the semantics of the body is the free construction over import algebras.

Four basic constructors on module specifications have been introduced in /EW85/, /BPP85/ and /BEPP85/. The union of two modules provides a new module whose specifications are a "union" of the corresponding ones in the two modules, allowing for identification or duplication of common parts. The composition of two modules "matches" the import of one with the export of the other, with the two remaining ones giving the interfaces of the new module, whose parameter and body are the "intersection" and "union", respectively, of the corresponding ones in the two modules. The operation of actualization "replaces" the parameter part with an actual specification in both interfaces and in the body. The extension allows for sorts and operations to be added to all four parts in such a way that the original module is a submodule of the new one.

The compatibility of these constructors (as was the case for implementation and parameter passing in /EK83/, /G83/, /SW82/) is an important property for the correctness of a stepwise refinement strategy for module specifications. If we think of the final module as providing a stepwise implementation of the final export interface in terms of the import one, then the compatibility of the constructors "expresses a basic requirement of software engineering, that the order in which parts of a machine are implemented should not effect the correctness of the implementation" /GM82/.

Our objective is two-fold: on one side, to establish fundamental compatibilities of the operations, while on the other to show how these results can be obtained by manipulating equations representing properties of pushouts in the category Mod of module specifications. This is the first step toward an algebra of modules. In section 2, we define the category Mod, show that it is closed under pushout constructions and redefine the union of module specifications as the pushout of injective Mod-morphisms providing us with several "inner compatibilities". In section 3, we characterize actualization as a pushout construction allowing us to explicitly define its semantics. This characterization is also used to prove the associativity of iterated parameterized actualization and its compatibility with union. In section 4, we use a simple property of composition to extend its compatibility with standard actualization in /EW85/ to the parameterized case. We also show, in the different sections, that composition, actualization and union are monotone operations in all their arguments with respect to the partial order $M1 \leq M2$ defined as "M1 is a submodule specification of M2". Due to space limitations, all proofs are omitted and can be found in /PP85/.

2. MODULE SPECIFICATIONS AND THEIR UNION

In this section, we first review the notions of module specification and of submodule and then define the category Mod of module specifications and module morphisms. The category Mod is then used to show the inner and mutual compatibility of the notion of submodule and the operation of union of modules.

2.1 Preliminaries

An algebraic specification SPEC is a triple (S,OP,E) where S is a set of sorts, OP a family of sets of operator symbols indexed by S* x S, and E is a set of equations (or universal Horn clauses). A specification morphism f: SPEC1 → SPEC2 is a pair (f_S, f_{OP}) of functions f_S: S1 → S2, f_{OP}: OP1 → OP2 such that for each N: s1...sn → s in OP1, $f_{OP}(N)$: $f_S(s1)...f_S(sn) \to f_S(s)$ is in OP2 and $f\#(E1) \subset E2$, where f# is the obvious extension of f to (S,OP)-terms. The category of SPEC-algebras and SPEC-homomorphisms is denoted by **SPEC.** Any specification morphism f: SPEC1 → SPEC2 defines a forgetful functor V_f: **SPEC2** → **SPEC1**. The left adjoint of V_f is called the free functor associated with f. It is well known (/EM85/) that the category CATSPEC of specifications and specification morphisms is closed under the pushout construction. We will use SPEC1 $+_{SPEC0}$SPEC2 to denote the pushout object of fj: SPEC0 → SPECj, j = 1,2, where it is obvious from the context which specification morphisms fj are being used. For any pushout SPEC3 = SPEC1 $+_{SPEC0}$SPEC2, any SPEC3-algebra A3 (resp. SPEC3 - homomorphism h3) is the amalgamated sum A1 $+_{A0}$A2 (resp., h1 $+_{h0}$h2) of SPECi-algebras Ai (resp., SPECi-homomorphisms hi). We then write **SPEC3** = **SPEC1** $+_{\mathbf{SPEC0}}$**SPEC2** (see/BPP85/). If SPEC3 and SPEC3' are pushout specifications and Fi: **SPECi** → **SPECi'**, i = 0,1,2, are functors, we use F1 $+_{F0}$F2 to denote the functor F3: **SPEC3** → **SPEC3'** defined by F3(A1 $+_{A0}$A2) = F1(A1) $+_{F0(A0)}$F2(A2). For more details, see /EM85/.

2.2 Definition (Module Specification and Semantics)

A module specification M is a four-tuple (PAR, EXP, IMP, BOD) of algebraic specifications along with four specification morphisms i, v, e and s satisfying

1) Syntactical conditions: s and e are injective and the diagram

$$\begin{array}{ccc} PAR & \xrightarrow{e} & EXP \\ {\scriptstyle i}\downarrow & & \downarrow{\scriptstyle v} \\ IMP & \xrightarrow{s} & BOD \end{array}$$

commutes.

2) Semantical conditions: The free functor F: **IMP** → **BOD** associated with s is strongly persistent (resp. conservative), i.e. $V_s(F(A)) = A$ for all IMP-algebras A (resp., and F preserves injective morphisms).

The semantics of M is SEM = $V_v \bullet F$: **IMP** → **EXP** . The restricted semantics of M is RSEM = R • SEM: **IMP** → **EXP** where R_e: **EXP** → **EXP** is the restriction functor defined, for every EXP-algebra E by $R_e(E) = \bigcap \{E' \in \textbf{EXP}: E' \subset E, V_e(E') = V_e(E)\}$.

Interpretation The specifications IMP and EXP represent the import and export interfaces, respectively, PAR is the shared parameter part and BOD is the body of the module. The semantics SEM is a transformation from IMP-interface algebras to EXP-interface algebras and the strong persistency guarantees that the PAR part of the IMP-algebra is not modified. The restriction functor R reduces the carrier of the EXP-algebra SEM(A) to those data reachable from its parameter part.

2.3 Definition (Category of Module Specifications)

The category of module specifications Mod consists of the following:

1) The objects of Mod are the module specifications as in 2.2;

2) A Mod-morphism from M1 = (PAR1, EXP1, IMP1, BOD1) to M2 = (PAR2, EXP2, IMP2, BOD2) is a four-tuple $m = (m_P, m_E, m_I, m_B)$ of specifcation morphisms satisfying:

1) **Syntactical conditions** every side of the following diagram commutes

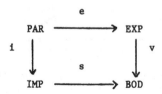

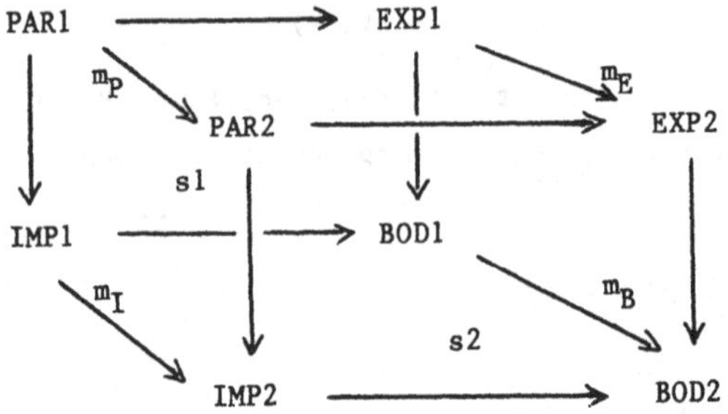

ii) **Semantical conditions** if Fj is the free functor associated with sj, j = 1,2, then $V_{m_B} \circ F2 = F1 \circ V_{m_I}$.
A Mod-morphism m is called an R-morphism if, in addition, $V_{m_E} \circ R_{e2} = R_{e1} \circ V_{m_E}$.

Remark The composition of two Mod-morphisms is given by componentwise composition of the specification morphisms. It is easy to show that the composite four-tuple satisfies the syntactical and semantical conditions if the components do and that the composition of R-morphisms in an R-morphism. A Mod-morphism m is called injective if each component is an injective specification morphism. It can be shown that if SEMj and RSEMj are the semantics of Mj, j = 1,2, and m: $M1 \to M2$ is a Mod-morphism, then $V_{m_E} \circ SEM2 = SEM1 \circ V_{m_I}$.
If m is an R-morphism, then $V_{m_E} \circ RSEM2 = RSEM1 \circ V_{m_I}$.

2.4 Definition (Submodule Specification)

A module specification M0 in a submodule specification of M1 if there exists an injective R-morphism m: $M0 \to M1$. In this case, we write $M0 <_m M1$, or just $M0 < M1$.

Remark Since the composition of injective morphisms is injective, we immediately have the inner compatibility of the notion of submodule, i.e.
if $M0 <_{m1} M1$ and $M1 <_{m2} M2$, then $M0 <_{m2 \circ m1} M2$.

2.5 Theorem

The category Mod is closed under pushout constructions.

2.6 Theorem

Let $M3 = M1 +_{M0} M2$ be the pushout object of $mj : M0 \to Mj$, $j = 1,2$, in the category Mod and let SEMj and RSEMj be the semantics of Mj, $j = 0, 1,2,3$.
Then $SEM3 = SEM1 +_{SEM0} SEM2$. If mj, $j = 1,2$, are R-morphisms, then $RSEM3 = RSEM1 +_{RSEM0} RSEM2$.

The operation of union of module specifications M1 and M2 with sharing submodule M0 was introduced first in /BPP85/and it suggested the pushout construction of 2.5, of which the union is a special case. For this reason, we have kept the same notation $M1 +_{M0} M2$ for the pushout object and for the union operation.

2.7 Definition (Union of Module Specification)

If $M0 <_{m1} M1$ and $M0 <_{m2} M2$, then the union $M1 +_{M0} M2$ of M1 and M2 with respect to M0, is the pushout object of $mj : M0 \to Mj$ in the category Mod.

2.8 Corollary

If $M0 <_{mj} Mj$, $j = 1,2$, then $Mj <_{nj} M1 +_{M0} M2$.

2.9 Corollary (Associativity of Union)

If $M0 <_{mj} Mj$, $j = 1,2$, and $M3 <_{n1} M2$, $M3 <_{n2} M4$, then
$(M1 +_{M0} M2) +_{M3} M4 = (M1 +_{M0} M2) +_{M2} (M2 +_{M3} M4) = M1 +_{M0} (M2 +_{M3} M4)$.

2.10 Theorem (Compatibility of Union and Submodule)

Let $M0 <_{mj} Mj$ and $N0 <_{nj} Nj$, $j = 1,2$ and $Mj <_{fj} Nj$, $j = 0,1,2$. If the Mod-morphisms are consistent, i.e. for $j = 1,2$, $nj \cdot f0 = fj \cdot mj$, then $M1 +_{M0} M2 < N1 +_{N0} N2$.

Several other properties of the union can be derived using the fact that it is a pushout in the category Mod. We single out two of them which are used in the next section in the more general context where the Mod-morphisms are not injective. The first one can be considered as some sort of distributivity of union over submodules.

2.11 Lemma

Let $M0 <_{mj} Mj$, $N0 <_{nj} Nj$, $Q0 <_{qj} Qj$, $j = 1,2$, and let $Qj <_{fj} Mj$, $Qj <_{gj} Nj$. If the Mod-morphisms are consistent i.e. $fj \cdot qj = mj \cdot f0$ and $gj \cdot qj = nj \cdot g0$ for $j = 1,2$, then $(M1 +_{M0} M2) +_{(Q1 +_{Q0} Q2)} (N1 +_{N0} N2) = (M1 +_{Q1} N1) +_{(M0 +_{Q0} N0)} (M2 +_{Q2} N2)$.

The second property can be thought of as a cancellation law of common submodule specifications in a union.

2.12 Lemma

Let $M0 <_{mj} Mj$, $N0 <_{nj} Nj$, $j = 1,2$, $M0 <_{m0} N0$ and $Mj <_{fj} Nj$ for $j = 0,1,2$. If $fj \circ mj = nj \circ m0$, $j = 1,2$, then

$$(N1 +_{M0} M2) +_{(M1 +_{M0} N0 +_{M0} M2)} (M1 +_{M0} N2) = N1 +_{N0} N2.$$

The last two Lemmas give an idea of the type of inner compatibility results for the operation of union which can be derived from properties of pushouts.

3. ACTUALIZATION OF MODULE SPECIFICATIONS

In this section we consider the operation of (standard and parametrized) actualization which, given a data type specification ACT1 and a module specification M, provides a new module whose parameter part in the interfaces and the body is replaced by ACT1. If ACT1 is parameterized, its parameter part Par1 becomes the parameter specification of the new module (parameterized actualization). Otherwise (standard actualization) the parameter specification is empty. We give an explicit characterization of the semantics of the (standard) actualized module and prove the compatibility of actualization with union and submodule. These results are based on a characterization of actualization as a pushout in Mod and the proofs are simpler then the direct ones contained in /PP 84/.

A parameterized specification PS1 is a pair (Par1, ACT1) of algebraic specifications with an injective specification morphism f : Par1 → ACT1 (/EKTWW 81/). The module specification (Par1, ACT1, ACT1, ACT1) with the obvious specification morphisms f and id, is denoted by MA1.

3.1 Definition (Actualization of Module Specifications)

Let M = (PAR, EXP, IMP, BOD) be a module specification, PS1 = (Par1, ACT1) a parameterized specification and h : PAR → ACT1 a "parameter passing" specification morphism. The actualization of M by PS1 via h, denoted by act_h(PS1, M), is the module specification M1 = (Par1, EXP1, IMP1, BOD1) where the last three components are given in the diagram

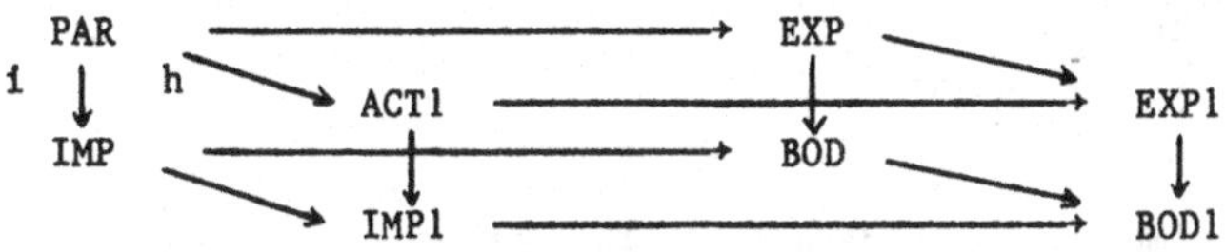

where the left, top and bottom (and, hence, right) square is a pushout in CATSPEC. The specification morphisms s1 : IMP1 → BOD1 and v1 : EXP1 → BOD1 are those induced by the diagram, while i1 and e1 are the compositions i'•f : Par1 → ACT1 → IMP1 and e'•f : Par1 → ACT1 → EXP1, respectively.

Remark Standard actualization, introduced in /EW 85/, corresponds to the case where Par1 is empty. Parametrized actualization has been shown to be well defined, i.e. to satisfy the conditions of 2.2, in /BEPP 85/. We now give an alternate characterization of act_h(PS1, M) in the context of Mod.

Notation For any module specification M = (PAR, EXP, IMP, BOD), we use $M\phi$ to denote the module (ϕ, EXP, IMP, BOD) and $MP\phi$ to denote (ϕ, PAR, PAR, PAR).

3.2 Lemma

Let M, PS1 and h be as in Definition 3.1. Then

$act_h(PS1, M) = MA1 +_{MP\phi} M\phi$.

The following result is a direct consequence of 3.2 and 2.6. An implicit characterization of the semantics of act_h(PS1, M) in terms of two functorial equations was given in /EW 85/, for Par1 = ϕ.

3.3 Theorem (Semantics of Actualization)

Let SEM be the semantics of M, and id_P: **PAR** → **PAR** and id_A : **ACT1** → **ACT1** identity fuctors. Then the semantics SEM1 of act_h(PS1, M) is given by $SEM1 = id_A +_{id_P} SEM$.

A similar characterization of the restricted semantics of act_h(PS1, M) as in 2.6 is not possible directly since m1 in the proof of 3.2 is in general not an R-morphism. We are able to establish the following general properties

3.4 Proposition

Let RSEM1 and RSEM be the restricted semantics of act_h(PS1, M) and M, respectively, and let $A +_P I$ be an $(ACT1 +_{PAR} IMP)$-algebra. Then

i) $R_f(A) \subset V_{e'}(RSEM1(A +_p I))$
ii) $V_{h'}(RSEM1(A +_p I)) \subset RSEM(I)$

The situation improves considerably if the parameter passing morphism $h : PAR \to ACT1$ is "parameter consistent", i.e. there is a morphism $p : PAR \to Par1$ with $h = f \cdot p$.

3.5 Theorem (Restricted Semantics of Actualization)
Let M, PS1 and h be as in 3.1, with h parameter consistent, and let $id_P : \mathbf{PAR} \to \mathbf{PAR}$ be the identity functor. Then $RSEM1 = R_f +_{id_P} RSEM$.

Remark The semantics of PS1 has not been taken into account in 3.4 and 3.5. In the special case where the semantics of PS1 is the free functor $F_f : \mathbf{Par1} \to \mathbf{ACT1}$, $RSEM1 = id_A +_{id_P} RSEM$, but the new import interface IMP1 must be equipped with additional constraints to restrict its (no longer) loose semantics. (See /EW85/ and /BEPP85/ for a discussion on module specifications with constraints.)

We now turn to the compatibility of actualization. The next result shows that the operation of actualization is monotone with respect to $\leq$.

3.6 Theorem (Compatibility of Actualization and Submodule).
Let $M1 \leq_m M2$, PSj, j = 1,2, be parameterized specifications and $q_P : Par1 \to Par2$ and $q_A : ACT1 \to ACT2$ injective morphisms such that $f2 \cdot q_P = q_A \cdot f1$. If $h1 : PAR1 \to ACT1$ and $h2 : PAR2 \to ACT2$ are parameter passing morphisms, such that $q_A \cdot h1 = h2 \cdot m_P$, then $act_{h1}(PS1, M1) \leq act_{h2}(PS2, M2)$.

3.7 Corollary
Let $M1 \leq_m M2$ and let $h : PAR2 \to ACT1$ be a parameter passing morphism. Then $act_{h \cdot m_P}(PS1, M1) \leq act_h(PS1, M2)$.

The result in the next Theorem is given also in /PP84/, where a direct proof is provided, starting with the simple case of two module specifications sharing a common parameter part, to the case of a shared subparameter, to the most general situation of a shared submodule. Here we can exploit the representation of union and actualization as pushouts in Mod and Lemma 2.11.

3.8 Theorem (Compatibility of Actualization and Union)

Let $M0 <_{mj} Mj$, $j = 1,2$ and let $(qj_P, qj_A) : PS0 \to PSj$, $j = 1,2$, be such that $fj \cdot qj_P = qj_A \cdot f0$. If $hj : PARj \to ACTj$, $j = 0,1,2$, are compatible parameter passing morphisms, then

$$act_{h1 +_{h0} h2}(PS1 +_{PS0} PS2,\ M1 +_{M0} M2) = act_{h1}(PS1,\ M1) +_{act_{h0}(PS0,M0)} act_{h2}(PS2,\ M2).$$

3.9 Corollary

Let $M0 <_{mj} Mj$, $j = 1,2$, $M3 = M1 +_{M0} M2$, $PS1 = (Par1, ACT1)$ a parametrized specification and $h : PAR3 \to ACT1$ a parameter passing morphism. Then there exist morphisms $hj : PARj \to ACT1$, $j = 0,1,2$ such that

$$act_h(PS1,\ M1 +_{M0} M2) = act_{h1}(PS1,\ M1) +_{act_{h0}(PS1,\ M0)} act_{h2}(PS1,\ M2).$$

We close this section by showing the inner compatiblity of actualization. More precisely, we show that actualizing M with PS1 and then the result with PS2 yields the same module specification obtained by actualizing M directly with PS1*PS2, where PS1*PS2 indicates the result of the parameterized parameter passing of PS1 and PS2. We briefly review this notion, referring the reader to /EM 85/ for details. The result $PS1*_{h1}PS2$ of the parametrized parameter passing of parameterized specifications PS1 and PS2 via $h1 : Par1 \to ACT2$ is the parameterized specification PS3 = (Par 3, ACT3) given by the pushout diagram

$$\begin{array}{ccccc}
 & & Par1 & \xrightarrow{f1} & ACT1 \\
 & & {\scriptstyle h1}\downarrow & & \downarrow \\
Par3 = Par2 & \xrightarrow{f2} & ACT2 & \xrightarrow{f1'} & ACT3.
\end{array}$$

The proof of the following result uses 3.2 and 2.9.

3.10 Theorem (Inner Compatibility of Actualization)

Let M = (PAR,EXP,IMP,BOD) be a module specification and PSj = (Parj, ACTj), j = 1,2, parametrized specifications. For any specification morphisms $h : PAR \to ACT1$ and $h1 : Par1 \to ACT2$ we have

$$act_{h1}(PS2,\ act_h(PS1,M)) = act_{h1'\cdot h}(PS1*_{h1}PS2,M).$$

Exploiting the associativity of parameterized parameter passing (/EKTWW 81/), we have the following "associative" property of actualization.

3.11 Corollary

Let M be a module specification, PSj = (Parj, ACTj), j = 1,2,3, parameterized specifications, and h : PAR → ACT1, h1 : Par 1 → ACT2 and h2 : Par2 → ACT3 specification morphisms. Then

$act_{h2}(PS3, act_{h1'\cdot h}(PS1*_{h1}PS2,M)) = act_{h2'\cdot h1}(PS2*_{h2}PS3, act_{h}(PS1,M))$.

4. COMPOSITION OF MODULE SPECIFICATIONS

The objective of this section is to extend a result in /EW85/ and prove that composition is compatible with parametrized actualization using the characterization of 3.2. In the process, we show that the treatment of composition can be restricted to the case where the interface morphism is the identity and this simplifies the proof of the associativity ("inner" compatibility) of composition. The following is the original definition of composition in /EW 85/ and /BEPP85/.

4.1 Definition (Composition of Module Specifications)

Given Mj = (PARj, EXPj, IMPj, BODj), j = 1,2, and an "interface" morphism h : IMP1 → EXP2, the composition $M1\cdot_h M2$ is the module specification M3 = (PAR3,EXP3,IMP3,BOD3) given by the diagram

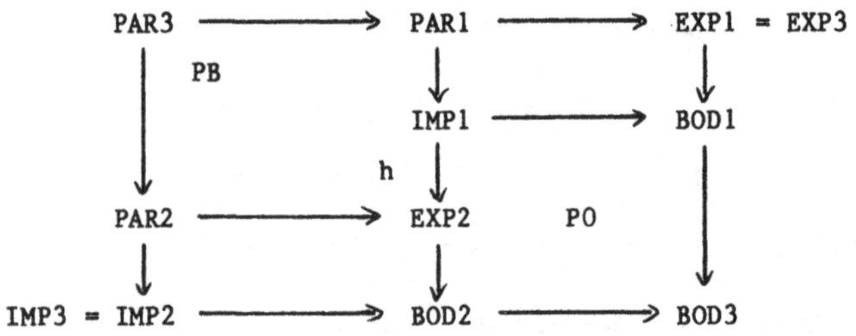

If h is "parameter consistent", i.e. there is a morphism p : PAR1 → PAR2 such that h∘i1 = e2∘p, then PAR3 = PAR1. The semantics has been shown (/EW85/, /BEPP85/) to be $SEM3 = SEM1\cdot V_h\cdot SEM2$, while $RSEM3 = RSEM1\cdot V_h\cdot RSEM2$ if h is parameter consistent. If h is the identity, the composition is denoted just by M1∘M2 without the subscript.

4.2 Lemma

a) if Mj, j = 1,2,3, are module specifications with EXP3 = IMP2 and EXP2 = IMP1, then $M1 \cdot (M2 \cdot M3) = (M1 \cdot M2) \cdot M3$

b) For $h : IMP1 \to EXP2$ an interface morphism, define $Mh = (IMP1, IMP1, EXP2, EXP2)$ with the obvious morphisms $i = v = h$ and $e = s = id$.
Then $M1 \cdot_h M2 = (M1 \cdot Mh) \cdot M2 = M1 \cdot (Mh \cdot M2)$.

A combination of the two parts of the previous lemma gives the associativity of composition for arbitrary interface morphisms h1 and h2.

4.3 Theorem (Associtivity of Composition)
Let $h1: IMP1 \to EXP2$ and $h2: IMP2 \to EXP3$ be interface morphisms. Then
$M1 \cdot_{h1} (M2 \cdot_{h2} M3) = (M1 \cdot_{h1} M2) \cdot_{h2} M3$.

The proof that composition is compatible with the notion of submodule is used in the proof of one of the main results of /EPP85/.

4.4 Theorem (Compatibility of Composition and Submodule)
Let $M1 \leq_{m1} M3$ and $M2 \leq_{m2} M4$ and let $h1 : IMP1 \to EXP2$ and $h3 : IMP3 \to EXP4$ be interface morphisms such that $h3 \cdot m1_I = m2_E \cdot h1$. Then
$M1 \cdot_{h1} M2 \leq M3 \cdot_{h3} M4$.

Before proving the compatibility of parametrization and composition, we need a lemma which describes a particular interaction of composition with the pushout of module specifications, in a more restrictive case than the one treated in /EPP85/. It can be shown that the restricted composition $M1 \cdot M2$ is a functor from the subcategory of Mod × Mod, consisting of (M1,M2) with IMP1 = EXP2, into Mod, which preserves pushouts. One of the possible proofs uses 4.5 and 4.6.

4.5 Lemma
Let $M3 \leq M4$, $M3 \leq M5$ and let Mj' denote the module specification with import and body equal to Mj and empty parameter and export. Let $fj : EXPj \to EXP4 +_{EXP3} EXP5$ $j = 3,4,5$ be the obvious canonical injections.
1) if $h1 : IMP1 \to EXP4$ is an interface morphism, then
$M1 \cdot_{f4 \cdot h1} (M4 +_{M3} M5) = (M1 \cdot_{h1} M4) +_{M3'} M5'$
2) if $h0 : IMP0 \to EXP3$ is an interface morphism, then
$M0 \cdot_{f3 \cdot h0} (M4 +_{M3} M5) = M4' +_{M3'} (M0 \cdot_{h0} M3) +_{M3'} M5'$

4.6 Theorem (Left Distributivity)

Let $M0 \leq Mj$, $j = 1,2$, $h : IMP1 +_{IMP0} IMP2 \to EXP3$ a parameter consistent interface morphism and $hj : IMPj \to IMP1 +_{IMP0} IMP2 \to EXP3$. Then

$$(M1 +_{M0} M2) \bullet_h M3 = (M1 \bullet_{h1} M3) +_{(M0 \bullet_{h0} M3)} (M2 \bullet_{h2} M3).$$

We can now state the main result of this section.

4.7 Theorem (Compatibility of Composition and Actualization)

Let $PS1 = (Par1, ACT1)$ be a parameterized specification, $h1 : PAR1 \to ACT1$ a parameter passing morphism, and $h : IMP1 \to EXP2$ an interface morphism parameter consistent with injective $p : PAR1 \to PAR2$. Then there exist $PS2 = (Par2, ACT2)$ and $h2: PAR2 \to ACT2$ such that

$$act_{h1}(PS1, M1 \bullet_h M2) = act_{h1}(PS1, M1) \bullet_k act_{h2}(PS2, M2).$$

5. CONCLUSION AND FURTHER DEVELOPMENTS

In this paper, we have introduced the category Mod of module specifications and module morphisms, proved that it is closed under pushouts (2.5), and used it to show the inner compatibility of submodule, union, actualization and composition. More precisely, we have shown that the union satisfies associative (2.9), submodule-distributive (2.11) and submodule-absorptive (2.12) laws and that iterated actualization is associative (3.10). The compatibility of union with submodule (2.10), and of composition with (parameterized) actualization (4.7) are also new. This framework provides a more algebraic proof than the one in /PP84/ of the compatibility of union and actualization (3.8) based on properties of the union and on the characterization of actualization as a pushout in Mod (3.2). The laws describing the interaction of the operations on modules, given here and in /EPP85/, provide a first step toward the development of an "algebra of modules" which could lead to a semi-automatic checking of equivalence of modular system specifications. One basic constructor which we have not studied here is the extension (/BEPP85/). The extension of M by E, $ext_E(M)$, is a module specification obtained by adding sorts, operators or equations (represented by E) to M in such a way that $M \leq ext_E(M)$. The results on submodules presented here state, for example, that the union of extensions is an extension of the union or that the composition of extensions is an extension of the composition: what needs to be done is to determine how to express the extending part E' in terms of E1, E2 and E0 for union or in terms of E1 and E2 for composition. This problem is currently under investigation.

We are also studying iterative and recursive interconnections of module specifications (the second type allowed, for example, in Ada (/B184/)) and we have some preliminary results on their constructions and induced semantics. An extension of the module concept to allow logical constraints for the interfaces (see discussion in /EW85/, /BEPP85/) is also under study, along with the effect of the four basic constructions of union, composition, actualization and extension on the constraints.

Acknowledgements This research is part of a joint project with E.K. Blum, of University of Southern California, and with H. Ehrig, of Technische Univ. Berlin: I am grateful to both of them for many helpful discussions. Thanks also to Leon Lemons for fast and accurate typing.

REFERENCES

/AMRW85/ Astesiano, E., Mascari, G.F., Reggio, G., Wirsing, M., On the Parameterized Algebraic Specification of Concurrent Systems, Proc. CAAP 85, LNCS 185(1985) 342-358.

/B184/ Blum, E. K., An Abstract System Model of Ada Semantics, TRW Technical Report, August 1984.

/BEPP85/ Blum, E.K., Ehrig, H., Parisi-Presicce, F., Algebraic Specification of Modules and their Basic Interconnections, to appear in JCSS Special Issue on TAPSOFT85.

/BPP85/ Blum, E. K., Parisi-Presicce, F., The Semantics of Shared Submodule Specifications, Proc. CAAP85, LNCS 185(1985) 359-373.

/BG77/ Burstall, R.M., Goguen, J.A., Putting Theories together to make Specifications, Proc. 5th Int. Joint Conf. on Artif. Intel., Cambridge, MA, (1977) 1045-1058.

/EK83/ Ehrig, H., Kreowski, H.-J. Compatibility of Parameter Passing and Implementation of Parameterized Data Types, Theor. Comp. Sci. 27(1983) 255-286.

/EKTWW81/ Ehrig, H., Kreowski, H.-J. Thatcher, J.W. Wagner, E.G., Wright, J.B., Parameter Passing in Algebraic Specification Languages, Proc. Aarhus Workshop on Prog. Spec., 1981, LNCS 134(1982) 322-369.

/EM85/ Ehrig, H., Mahr, B., Foundations of Algebraic Specifications 1, EATCS Monographs on Theor. Comp. Sci. Vol. 6, Springer-Verlag 1985.

/EPP85/ Ehrig, H., Parisi-Presicce, F., Union and Composition of Module Specifications for Software Systems are Compatible, Extended Abstract, Sept 1985.

/EW85/ Ehrig, H., Weber, H., Algebraic Specification of Modules, Univ. Dortmund Techn. Report 190, 1985, to appear in Proc. IFIP Working Conf. on Formal Models in Programming, Vienna 1985.

/G83/ Ganzinger, H., Parameterized Specifications: Parameter Passing and Implementation, ACM TOPLAS 5, 3(1983).

/GM82/ Goguen, J.A., Meseguer, J., Universal Realization, Persistent Interconnection and Implementation of Abstract Modules, Proc. ICALP 82, LNCS 140(1982) 265-281.

/HS73/ Herrlich, H., Strecker, G.E., Category Theory, Allyn and Bacon Inc., Boston, 1973.

/LZ74/ Liskov, B., Zilles, S.N., Programming with abstract data types, SIGPLAN Notices 9,4(1974) 50-59.

/Mo84/ Mosses, P., A Basic Abstract Semantic Algebra, Proc. Semantics of Data Types, LNCS 173(1984) 87-108.

/O183/ Oles, F.J., Type Algebras, Functor Categories, and Block Structure, in "Algebraic Methods in Semantics", eds. M. Nivat and J.C. Reynolds, Cambridge Univ. Press, 1985.

/PP84/ Parisi-Presicce, F., The Operations of Union and Actualization of Module Specifications are Compatible, Tech. Report, Univ. of Southern California, Sept. 1984.

/PP85/ Parisi-Presicce, F., Inner and Mutual Compatibility of Operations on Module Specifications, Tech. Report, Univ. of Southern California, Oct 1985.

/Pa72/ Parnas, D.L., A Technique for Software Module Specification with Examples, CACM 15, 5(1972) 330-336.

/SW82/ Sannella, D., Wirsing, M., Implementation of Parameterized Specifications, Proc. ICALP82, LNCS 140(1982) 473-488.

/WE85/ Weber, H., Ehrig, H., Specification of Modular Systems, Univ. Dortmund Tehcn. Report 198, 1985.

Exact Computation Sequences

Alex Pelin* and Jean H. Gallier**
*** Florida International University**
Department of Mathematical Sciences
Tamiami Campus, Miami, Florida 33199

**** University of Pennsylvania**
Department of Computer and Information Science
Philadelphia, Pa 19104

Abstract: Exact computation sequences are sequences of the form

$$<L_0,A_0> \xrightarrow{S_1} <L_1,A_1> \xrightarrow{S_2} \dots \xrightarrow{S_n} <L_n,\emptyset>,$$

where L_0 is a free algebra, A_0 is a set of conditional equations over L_0, S_i is a "step function", $L_i=S_i(L_{i-1})$, and $A_i = S_i(A_{i-1})$. Each step function is the *top-down reduction extension* of a set of confluent and noetherian rewrite rules. These sequences are used in solving the word problem for free algebras, since for any pair of terms t_1,t_2 in L,

$$t_1 =_{A_0} t_2 \text{ iff } S_n o \dots o S_1(t_1) = S_n o \dots o S_1(t_2).$$

We analyze properties of exact computation sequences such as: determining the relation between the sets $<L_{i-1},A_{i-1}>$ and S_i, and the output pair $<L_i,A_i>$, and we present criteria for choosing the equations E_i in $<L_{i-1},A_{i-1}>$ which are used to generate the reductions S_i. We also give examples showing how to construct exact computation sequences for several axiom systems by applying the properties and the criteria presented in the article.

1. Introduction

In Pelin and Gallier [14] and [15], we presented a method for computing normal forms with respect to a set of (conditional) equations. We assumed that we were given a *signature* Σ, and a countable set of *variables* V. The initial algebra over Σ is denoted by T_Σ, and the free algebra generated by V is denoted by $T_\Sigma(V)$. We are also given a set of (conditional) equations E over $T_\Sigma(V)$, and we want *to solve the word problem* for T_Σ, that is, the problem of determining for any two terms t_1 and t_2 in T_Σ, whether t_1 and t_2 are congruent modulo the congruence $=_E$ induced by the set of equations E. The object of our method is *to compute a representation function for* $<L,E>$. We say that a function $f:L \rightarrow L$ is a representation function for $<L,E>$, if for all terms t_1,t_2 in L, $t_1 =_E t_2$ if and only if $f(t_1)=f(t_2)$. In other words, t_1 and t_2 are equivalent modulo $=_E$ if and only if their representatives are identical. In our method, the representation function *Rep* is computed as *a composition* $S_n \text{o} \ldots \text{o} S_1$ *of step functions*. Each step function S_i takes as input a set of ground terms L_{i-1} and a set of equations A_{i-1} over L_{i-1}. The equations (conditional equations) in A_{i-1} are called *axioms*. Let $\text{Th}(A_{i-1})$ be the set of theorems derivable from A_{i-1}. The step function S_i selects a subset E_i of $\text{Th}(A_{i-1})$ which it attempts to "eliminate." The elimination is accomplished by transforming the equations into *reduction rules*. This is done by using a *complexity function* over L_{i-1} which *suits* E_i. A complexity function f over L_{i-1} assigns to each term t in L_{i-1} an element $f(t)$ in $(N^k,>)$, where N is the set of natural numbers, k a positive integer, and > a well ordering on the set N^k. A function $f{:}L_{i-1} \rightarrow (N^k,>)$ is a *complexity function* if it is recursive, "monotonic", and has the subterm property, that is , if t_1 is a subterm of t_2, then $f(t_2)>f(t_1)$. A complexity function f *suits an equation* $l=r$ *strongly*, if for all ground substitutions s, $f(s(l)) > f(s(r))$, or for all ground substitutions s, $f(s(r)) > f(s(l))$. A complexity function f *suits an equation* $l=r$ *weakly*, if for all ground substitutions s, either $f(s(l)) \geq f(s(r))$ or $f(s(l)) \leq f(s(r))$, and $f(s(l))=f(s(r))$ implies that $s(l)=s(r)$, that is, $s(l)$ and $s(r)$ are identical. Both strong and weak suitability are used to generate *meta-reductions*. A meta-reduction has the form $(C) \Rightarrow l \rightarrow r$, where C is a recursive predicate in a meta language, and l and r are terms in L_{i-1}.

The meta-language must contain names for well orders, complexity functions, *top down reduction extensions*, the equality predicate, and the boolean connectives $\neg, \wedge, \vee$. It must also contain predicates which determine if two terms t_1 and t_2 are unifiable. For all meta-reductions $(C) \Rightarrow l \rightarrow r$ and all ground substitutions s over L_{i-1}, $f(s(l))>f(s(r))$ whenever $s(C)$ evaluates to true. If a complexity function f suits an equation $l=r$ strongly over L_{i-1}, then we transform it into the reduction (true) $\Rightarrow l \rightarrow r$, if for some substitution s, $f(s(l))>f(s(r))$, and we transform the equation into the meta-reduction (true) $\Rightarrow r \rightarrow l$ if for some ground substitution s over L_{i-1}, $f(s(r))>f(s(l))$. If the complexity function f suits an equation $l=r$ weakly, then we obtain two metareductions: $(f(l)>f(r)) \Rightarrow l \rightarrow r$, and $(f(r)>f(l)) \Rightarrow r \rightarrow l$. We will write $l \rightarrow r$ instead of (true) $\Rightarrow l \rightarrow r$ and we call it a *pure* reduction. We say that a complexity function f suits a set of equations E_i if it suits all of the equations in E_i. If f suits E_i then E_i is transformed into a set of meta-reductions R_i.

We define the step function S_i to be the top-down reduction extension of the set of reductions R_i. The *top-down reduction extension* α of a set of meta-rules R_i is defined as follows: For every term t in L_{i-1}, we have the following cases:

1. If for a meta-rule $(C) \Rightarrow l \rightarrow r$ and a ground substitution s, $s(l)=t$ and $s(C)$ evaluates to true, then $\alpha(t)=\alpha(s(r))$.
2. If case 1 does not apply and t has the form $g(t_1,\ldots,t_n)$, then we compute recursively $\alpha(t_1)$, $\alpha(t_2),\ldots,\alpha(t_n)$. If for any i, $1 \leq i \leq n$, $\alpha(t_i) \neq t_i$, then $\alpha(t)=\alpha(g(\alpha(t_1),\ldots,\alpha(t_n)))$.
3. If neither of the above cases applies then $\alpha(t)=t$. We impose a ranking on the set of rules in R_i and we put the restriction that if more than one rule can be applied in (1), we choose the rule with the highest rank.

   We use the top-down reduction extension α as our step function S_i. S_i takes as input the pair $\langle L_{i-1}, A_{i-1} \rangle$. Its output is the set $\langle L_i, A_i \rangle$, where L_i is the range of α, and A_i is the set of equations $\{\alpha(l_i)=\alpha(r_i) \mid l_i=r_i \in A_{i-1}\}$.

We say that a top-down reduction-extension S_i:$<L_{i-1},A_{i-1}> \rightarrow <L_i,A_i>$ has the *α-property*, if for every operator f of rank m, and for every m terms $t_1,t_2,...,t_m$ in L_{i-1}, for every j, $1 \leq j \leq m$, if $f(t_1,...,t_m) \in L_{i-1}$, then $S_i(f(t_1,...,t_j,...,t_m))=S_i(f(t_1,...,S_i(t_j),...,t_m))$.
We assume that L_{i-1} has the *subterm property*, i.e., if $f(t_1,...t_m) \in L_{i-1}$ then $t_1,...t_m \in L_{i-1}$. If L_{i-1} does not have the subterm property, then S_i cannot have the α-property, since $S_i(t_j)$ may be undefined for some subterm t_j of a term $f(t_1,...,t_j,...,t_m) \in L_{i-1}$. The representation function Rep is then the composition of the functions given by the sequence:

4. $<L_0,A_0> \xrightarrow{S_1} <L_1,A_1> \xrightarrow{S_2} <L_2,A_2> \ldots \xrightarrow{S_n} <L_n,\emptyset>$.

We can show that if all sets of reductions R_i are pure, i.e. all reductions in R_i are pure, then all sets $L_0,L_1,...,L_n$ have the subterm property. It can also be shown that if the language L_{i-1} has the subterm property, then S_i has the α-property if and only if the set of reductions R_i is locally confluent.

We call a sequence such as (4) an *exact computation sequence* if each set of reductions R_i is confluent and Noetherian. If each of the sets of reductions R_i is finite, we say that (4) is a *finite exact computation sequence*.

We can show that if (4) is an exact computation sequence, then the composition $S_n o S_{n-1} o ... o S_1$ is a representation function for $<L_0,A_0>$. We can also find a set of conditional meta-reductions R which has the Church Rosser property for $<L_0,A_0>$ and is terminating. This is done by transforming each reduction $(C)=>l \rightarrow r$ in R_i into the reduction $(C \wedge l \in L_i)=>l \rightarrow r$ in R. Since the sets L_i are recursive and the conditions C are recursive, $(C \wedge l \in L_i)=>l \rightarrow r$ conforms to our definition of meta-reductions. If (4) is a finite exact computation sequence then R is a finite set of reductions.

In this approach, we have more flexibility than Knuth and Bendix [11] since we do not require that the sets of reductions R_i be confluent on L_0 but only on its subset L_i. In fact the extension of R_i to L_0 can be non-terminating or not having the local confluence and this does not affect the properties of the exact computation sequence: We call such a computation sequence exact because for each pair of consecutive step function S_i and S_{i+1}, the range of S_i is exactly the domain of S_{i+1}.

In [14] and [15] we have concentrated on finding complexity functions which suit various types of equations. In this paper we concentrate on exact computation sequences. We assume that we have step functions which handle certain types of equations, and we are interested in the way in which they are joined together to yield exact computation sequences. There are *four problems* which one has to consider in building exact computation sequences.

The first problem is *to obtain characterizations of* L_i from the properties of the set of reductions R_i and the input set of terms L_{i-1}. We can easily show that if R_i is a finite set of reductions and L_{i-1} is a recursive set of terms, then L_i is also a recursive set. There are some other properties which one may want to be preserved by the step funtions such as if L_i is a context-free language what conditions must be imposed on R_i such that L_i is a context-free language.

The second problem is to *obtain characterizations of* A_i from the descriptions of R_i, L_{i-1} and A_{i-1}. An equation in A_{i-1} may vanish under R_i, may be transformed into one equation in A_i, may be transformed into a finite set of equations in A_i, or may generate an infinite set of equations in A_i. For example let L_0 be the language generated by the context free grammar

$G=\langle\{S\},\{+,-,a,b,0\},P,S\rangle$ whose productions are shown in (5), let A_0 be set of equations shown in (6), and $R_1 = \{--x \rightarrow x, -+xy, \rightarrow +-y-x\}$

(5) $S \rightarrow 0|a|b|+SS|-S$

(6)

1. $++xyz=+x+yz$
2. $+x-x=0$
3. $+-xx=0$
4. $+x0=x$
5. $+0x=x$

Under R_1, axiom 1 becomes $++xyz=+x+yz$ (unchanged), axiom 2 becomes the set of axioms $+xR_1(-x)=0$, axiom 3 becomes the set of axioms $+R_1(-x)x=0$, axiom 4 is $+x0=x$, and axiom 5 is $+0x=x$. The variables in the set of axioms A_1 take values in the set $L_1 = R_1(L_0)$.

The third problem is to *obtain sufficient conditions under which a set of reductions* R_i *is complete for a set of equations E over* L_{i-1}. We say that R_i is complete for E if all equations in E vanish. Here we are looking for both methods of proof and sufficient conditions. A problem in this category of problems is to determine sufficient conditions under which R_i transforms an infinite set of equations E_i in A_{i-1} into a finite set of equations in A_i.

The fourth problem is to *obtain criteria for choosing the equations* E_i which are to be transformed into the set of reductions R_i. We try to choose E_i in such a way that L_i and A_i have "nice" characterizations.

Our paper is organized as follows: in section 2 we present definitions and technical results; in section 3 we discuss how these results relate to the four problems described above, pointing out difficulties, solutions and methods; in section 4, we present examples in some details; in section 5 we draw conclusions and point out directions for future research.

2. Definitions and Technical Results

We will follow the notations and definitions found in Huet and Oppen [6]. Given a finite signature (S,Σ,τ), the initial algebra is defined in the usual way ([4]). The terms in T_Σ will be represented in prefix form. Then, the set T^s_Σ of terms of sort s is a deterministic context-free language. Given an S-sorted set of variables V, the free algebra over V is denoted as $T_\Sigma(V)$ and consists of terms with variables. A Σ-equation of sort s is a pair $\langle M,N\rangle$ of terms in $T_\Sigma(V)$ (where Σ is a finite signature, and V an S-indexed set of variables). A *conditional equation* is an expression of the form $e_1,e_2,\ldots,e_n \Rightarrow e$, where $e_1,e_2,\ldots,e_n,e$

are equations. A ***presentation*** is a triple $P=<\Sigma,V,E>$, where Σ is a finite signature and E a finite set of (conditional) equations. Substitutions and E-unification are defined as in Huet and Oppen.

The notions of complexity function, suitability, meta-reduction and top down reduction extension were presented in the introduction and will not be repeated here. We say that a set of meta-reductions S is locally confluent if for any term t, meta-reductions $C_1 => l_1 \rightarrow r_1$, and $C_2 => l_2 \rightarrow r_2$ in S, substitutions s_1, and s_2, and tree-address u in t, if $C_1(t)$ and $C_2(t/u)$ are true and $t=s_1(l_1)$ and $t/u=s_2(l_2)$, then there is a term t' such that $t[u<-s_2(r_2)] \rightarrow^* t'$, and $s_1(r_1) \rightarrow^* t'$.

This notion is an extension of the notion of local confluence found in Huet [4] to meta-reductions. A system of meta-reductions S is *terminating* for a set of terms L if there is no infinite sequence $t_1 \rightarrow t_2 \rightarrow \ldots t_n \rightarrow t_{n+1} \rightarrow \ldots$ where all terms are in L and the reductions are in S.

Let $P=<\Sigma,V,E>$ be a presentation. An *exact computation sequence* is a sequence

$S: <L_0,A_0> \xrightarrow{S_1} <L_1,A_1> \ldots \xrightarrow{S_n} <L_n,\emptyset>$, where:

1. For each i, $1 \leq i \leq n$, S_i is a set of meta-reductions which is *both locally confluent and terminating.*
2. For each i, $1 \leq i \leq n$, $L_i = S_i(L_{i-1})$ and $A_i=S_i(A_{i-1})$.
3. L_0 is T_Σ, the set of terms in the initial algebra, and $A_0=E$.

We say that S is an exact computation sequence is *standard* if the sets S_i are *pure and finite.* We say that a set of meta-reductions S_i is *pure* if it contains only reductions of the type (true) => $l \rightarrow r$. The notion of representation function for $<L_0,A_0>$ and that of subterm property for a sublanguage $L \subseteq L_0$ were introduced in the introduction.

Lemma 1

If S is an exact computation sequence for $<L_0,A_0>$, the sequence $S_n o S_{n-1} o \ldots o S_1$ is a representation function for $<L_0,A_0>$.

Lemma 2

If S is a finite exact computation sequence, then all sets $L_1,L_2,\ldots,L_n$ are recursive sets.

We say that $<L_0,A_0>$ is *equational* if A_0 does not contain conditional axioms.

Lemma 3

If $<L_0,A_0>$ is equational and S is a standard exact computation sequence, then $L_1,L_2,\ldots,L_n$ have the subterm property.

Lemma 4

If S is an exact computation sequence for $<L_0,A_0>$, then the set of meta-reductions $\cup\{(l\in L_i\wedge C)=>l\rightarrow r|(C)=>l\rightarrow r\in S_i\}$ is confluent, terminating and has the Church Rosser property for $<L_0,A_0>$.

Lemma 5

Assume that L_{i-1} has the subterm property and f is an operator of type $\tau(f)=s_1\times s_2\times...\times s_n\rightarrow s$. If $f(v_1,...v_n)$ and l are not unifiable for any variables $v_1,v_2,...,v_n$ (with each v_i of sort s_i) and meta-rule $(C)=>l\rightarrow r$ in S_i, then $S_i(f(t_1,...,t_n)) = f(S_i(t_1),...,S_i(t_n))$.

The notion of α-property was defined in the introduction and will not be repeated here.

Lemma 6

Assume that L_{i-1} has the subterm property. Then S_i has the α property if and only if S_i is locally confluent.

Let S_i be a set of pure meta-reductions. We say that a rule $l\rightarrow r$ is *left-linear* if each variable occurs in 1 at most once. We say that S_i is left-linear if all rules occurring in the meta-reductions are left-linear.

Lemma 7

Let S_i be finite and left-linear and let L_{i-1} be a language accepted by a (deterministic) bottom up, finite tree automaton. Then L_i is accepted by a deterministic bottom-up automaton.

A consequence of the above lemma is that if L_{i-1} is (deterministic) context-free then L_i is also (deterministic) context-free.

The following Lemma is very useful in proving properties about terms in L_{i-1}.

Lemma 8

Assume that L_{i-1} has the subterm property and S_i has the α-property. Given equation $l=r$ with terms in L_{i-1}, $S_i(s(l))=S_i(s(r))$ for all substitutions s over L_{i-1}, if and only if $S_i(s(l)) = S_i(s(r))$ for all substitutions s over L_i.

Lemma 9 (The Splitting Lemma)

Assume that $S_i = S_i'\cup S_i''$, where each of the subsets of meta-reductions S_i', S_i'' are locally confluent, and assume that $S_i''(S_i'(L_{i-1}))\subseteq S_i'(L_{i-1})$. Then $S_i=S_i''\circ S_i'$.

We say that a language $L\subseteq L_0$ is *standard* if it satisfies the following properties:

1. For every sort $s\in S$ there is a subset $A^s\subseteq L$ of terms of sort s called *atoms* and a set of operators f of type $\tau(f)=s_1\times s_2\times...\times s_n\rightarrow s$, such that, any term t of sort s in L is either an atom, or there are terms $t_1,...,t_n$ in L such that $t=f(t_1,...,t_n)$.

2. The sets of atoms A^s are generated by deterministic context-free languages, and they are disjoint from the terms obtained from the application of the operators.

We can also view L as an initial algebra whose constants are the atoms and its operators are the ones described in (1) above.

L is not necessarily a subalgebra of L_0, since it does not have to contain all the operators in the signature (S,Σ,τ). Let A be a set of equations, and let g be an operator of type $\tau(g)=s_1\times\ldots\times s_n\to s$. We say that g is *globally quasidistributive* if for all sorts s_i, $1\leq i\leq n$, for every operator h of type $\tau(h)=s_1'\times s_2'\times\ldots\times s_m'\to s_i$, the equation $g(x_1,\ldots,x_{i-1},h(y_1,\ldots,y_m),x_{i+1},\ldots,x_n)=E_{i,h}$ is provable from A, where $E_{i,h}$ is an expression in which the operator g can have only arguments from the set $\{x_1,\ldots,x_{i-1},x_{i+1},\ldots,x_n,y_1,\ldots y_m\}$

Lemma 10

Let L be a standard language and g be an operator in L which is globally quasidistributive. Then the set of reductions

$R=\{g(x_1,\ldots,x_{i-1},h(y_1,\ldots,y_m),x_{i+1},\ldots,x_n)\to E_{i,h}|g(x_1,\ldots,x_{i-1},h(y_1,\ldots,y_m),x_{i+1},\ldots,x_n)=E_{i,h}\}$ is terminating and locally confluent. Moreover $R(L)$ is a standard language.

The proofs for Lemmas 2, 5, 6, 7, 9 can be found in [15]. We now present examples to show how the above lemmas are used in exact computation sequences.

3. Examples

Example 1

Let us consider the example of a free group with two generators a and b. The set of terms L_0 is the language generated by the grammar G_0 = <{S},{+,–,a,b,0},P,S>, where P is the set of productions shown below:

(1) $S\to 0|a|b|+SS|-S$.

The set of axioms is $A_0=\{+x+yz=++xyz,+x-x=0,+-xx=0,+x0=x,+0x=x\}$.

Our first goal is to find if any operation satisfies the conditions of lemma 10, since in this case the output of the step function is also an initial algebra. We obtain the set of theorems $T_1=\{-+xy=+-y-x,--x=x,-0=0\}$.

In [15] we presented two methods for finding theorems in Th (A_0). Both start out by transforming the equations into reductions by using a complexity function. This way, we are sure that the system is terminating. In the first method we use the critical pairs method of Knuth and Bendix [11] and Huet [4]. The resulting equation is not transformed into a rule but added to the set of theorems. In the second we allow a limited number of applications of anti-reductions, that is, rules in which the equation is applied in the direction which increases the complexity.

By lemma 10 the system of reductions $R_1'=\{-+xy \rightarrow +-y-x\}$ is locally confluent and terminating. The construction from lemma 10 gives us the standard grammar $G_1' = <\{A,S\},\{-,+,a,b,0\},P'_1,\{S\}>$ for $R'_1(L_0)$, P'_1 being the set of productions shown in (2).

(2)

$A \rightarrow -A|a|b|0$

$S \rightarrow -A|+SS$

The atoms in the standard language are the terms generated by the variable A. The set of axioms A_0 becomes, by applying lemmas 5 and 8:

$A'_1=\{+x+yz=++xyz,+xR'_1(-x)=0, R'_1(-x)x=0, +x0=x,+0x=x\}$.

Similarly, $R'_1(\{--x=x,-0=0\})=\{--x=x,-0=0\}$.

Now we can use the set of reductions $R_1''=\{--x \rightarrow x,-0 \rightarrow 0\}$. This transforms the standard language L_1' into the standard language L_1'' generated by the grammar $G_1 = <\{A,S\},\{a,b,0,-a,-b,+\},P_1,S>$ where P_1 is shown in (3).

(3)

$A \rightarrow 0|a|b|-a|-b$

$S \rightarrow A|+SS$

The set of equations A'_1 becomes $A_1 = R_1''(A'_1) = \{+x+yz=++xyz, +xR_1(-x)=0,+R_1(-x)x=0, +x0=x,=+0x=x\}$. By R_1, we denote $R_1''\text{o}R'_1$.

It is interesting to note that we can start with the set of reductions $R_1=\{-+xy \rightarrow +-y-x,--x \rightarrow x,-0 \rightarrow 0\}$, and we see that R_1,R_1',R_1'' are confluent and terminating, and $R_1''(R'_1(L_0)) \subseteq R_1'(L_0)$. By the splitting lemma, $R_1=R_1''\text{o}R_1'$.

For the set of theorems E_2 we choose $\{+x+yz=++xyz\}$. We choose it since it is linear and generates a nice output language. We choose $R_2 = \{+x+yz \rightarrow ++xyz\}$ though either orientation is confluent and terminating. $L_2=R_2(L_1)$ is given by the grammar $G_2=<\{S\},\{a,b,0,+\},P_2,S>$, where P_2 is given by the set of productions (4).

(4) $S \rightarrow a|b|0|-a|-b|+Sa|+Sb|+S0|+S-a|+S-b$.

The set of equations A_1 becomes $A_2=R_2(A_1) = \{+x0=x, R_2(+0x)=x, R_2(+xR_1(-x))=0, R_2(+R_1(-x)x)=0\}$.

In the last two equations, x is a variable in L_1, while in the first two, x is a variable in L_2. By lemmas 6 and 8, $R_2\ (+xR_1(-x))=0$ becomes $R_2(+xR_2R_1(-x))=0$, where $x \in L_2$, and we can do the same with the last equation. We are now looking for a set of equations E_3. The equation 1. $+x0=x$ is good but the rest are sets of equations which are difficult to handle. In equations 2, 3, and 4 of A_2, we can consider the sets of equations obtained by considering x to be an atom in L_2, and we get the set of equations shown in (5).

(5)

2. $+a-a=0$	6. $+-bb=0$
3. $+b-b=0$	7. $+0a=a$
4. $+00=0$	8. $+0-a=-a$
5. $+-aa=0$	9. $+0b=b$
10. $+0-b=-b$	

By using the forced confluence method in (L_0,A_0), we get: $+x+y-y \rightarrow ++xy-y = +x0 \leftarrow +x+y-y$. From the equation $+x0=x=++xy-y$ applied to the atoms of L_2, we get the equations:

11. $++x00=x$	14. $++xa-a=x$
12. $++x-aa=x$	15. $++x-bb=x$
13. $++x-bb=x$	

We take E_3 to be the set of equations 1. - 15. R_3 is easily obtained from E_3 by orienting the equations in the decreasing order of length.

Then $L_3 = R_3(L_2)$ is the language whose grammar is $G_3 = \langle\{A,A',B,B',S\},\ \{a,-a,0,b,-b,+\},\ P_3,S\rangle$ whose set of productions is given in (6):

(6)

$A \rightarrow +Aa|+Ba|+B'a$

$A' \rightarrow +A'-a|+B-a|+B'-a$

$B \rightarrow +AB|+A'b|+B'-a$

$B' \rightarrow +A-b|+A'-b|+B'-b$

$S \rightarrow 0|A|A'|B|B'$

It can also be shown, by induction on the height of the trees in L_2 that $R_3(A_3)=\emptyset$. A method by which one can obtain a grammar for L_3 from R_3 and the grammar for L_2 is presented in the next section.

By Lemma 1, $R_3 o R_2 o R_1$ is a representation function for $\langle L_0,A_0\rangle$.

Example 2

Let us consider the case of a boolean algebra generated by the a finite number of elements $a_1,a_2,...,a_n$. The set of terms L_0 is the language generated by the grammar G_0=<$\{S,A\},\{\neg,\vee,\wedge,T,F,a_1,...,a_n\},P_0,S$>, where P_0 is the set of productions shown in (7):

(7)

$A\rightarrow a_1|a_2|...|a_n|T|F$

$S\rightarrow A|\neg S|\vee SS|\wedge SS$

The set of axioms is

$A_0=\{\neg\neg x=x, \neg\wedge xy=\vee\neg x\neg y, \wedge xx=x, \vee xy=\vee yx, \vee x\vee yz=\vee\vee xyz, \vee xx=x, \vee x\wedge xy=x,$ $\wedge x\vee xy=x, \wedge x\vee yz=\vee\wedge xy\wedge xz, \vee x\wedge yz=\wedge\vee xy\vee xz,$ $\wedge x\neg x=F, \vee x\neg x=T, \neg T=F, \neg F=T, \wedge xT=x,$ $\wedge xF=F, \vee xT=T, \vee xF=x.\}$

By lemma 10, the easiest thing to do is to take the set of equations

$E_1=\{\neg\wedge xy=\vee\neg x\neg y, \neg\vee xy=\wedge\neg x\neg y\}$ and transform them into the set of reductions

$R_1=\{\neg\wedge xy\rightarrow\vee\neg x\neg y, \neg\vee xy\rightarrow\wedge\neg x\neg y\}$.

The grammar for $L_1=R_1(L_0)$ is obtained by applying the construction from the proof of lemma 10 and it is G_1=<$\{A,S\},\{a_1,...,a_n,T,F,\wedge,\vee,\neg\},P_1,S$>, where P_1 is given in (9)

(9)

$A\rightarrow a_1|...|a_n|T|F|\neg A$

$S\rightarrow A|\wedge SS|\vee SS$

The set of axioms is

$A_1=R_1(A_0)=\{\neg\neg x=x, \wedge x=\wedge yx, \wedge x\wedge yz=\wedge\wedge xyz, \wedge xx=x, \vee xy, \vee yx, \vee x\vee yz=\vee\vee xyz, \vee xx=x, \vee x\wedge yz=x,$ $\wedge x\vee xy=x, \wedge x\vee xy\wedge xz, \vee x\wedge yz=\wedge\vee xy\vee xz, \wedge xR_1(\neg x)=F, \vee xR_1(\neg x)=T, \neg T=F, \neg F=T, \wedge xT=x, \wedge xF=F, \vee$ $xT=T, \vee xF=x.\}$

We can now choose $E_2=\{\neg\neg x=x, \neg F=T, \neg T=F\}$

We get the system of reductions $R_2=\{\neg\neg x\rightarrow, \neg F\rightarrow T, \neg T\rightarrow F\}$, which is confluent and terminating. The language L_2 has the standard grammar G_2= <$\{A,S\},\{a_1,...,a_n,\neg a_1,...,\neg a_n, T,F,\wedge,\vee\},P_2,S$>, where P_2 is shown in (10):

(10)

$A\rightarrow a_1|a_2|a_n|\neg a_1|...|\neg a_n|T|F$

$S \to A | \wedge SS | \vee SS$

The set of equations A_2 is

$\{\wedge xy = \wedge yx, \wedge x \wedge yz = \wedge \wedge xyz, \wedge xx = x, \vee xy = \vee yx, \vee x \vee yz = \vee \vee xy, \vee x \wedge yz = \wedge \vee xy \vee xz, \quad \wedge x R_2(\neg x) = T, \wedge xT = x, \wedge xF = F, \vee xT = T, \vee xF = x\}$.

Since $\wedge$ quasidistributes over $\vee$, we can take $E_3 = \{\wedge x \vee yz = \vee \wedge xy \wedge xz, \wedge \vee yzx = \wedge \vee yx \wedge zx\}$. The resulting set of reductions R_3 obtained by orienting the equations in E_3 from left to right is confluent and terminating and L_3 is the language given by the standard grammar G_3 = $<\{S,A\},\{a_1,\dots,a_n,\neg a_1,\dots,\neg a_n,T,F,\wedge,\vee\},P_3,S>$, where P_3 is the set of productions shown in (11):

(11)

$A \to a_1 | a_2 | \dots | a_n | \neg a_1 | \dots \neg a_n | T | F | \wedge AA$

$S \to A | \vee SS$

Next we can apply a set of reductions to the set of atoms of L_3 by using the commutativity, associativity, idempotence, and the equations $\wedge xT = x$ and $\wedge xF = F$.

Due to space constraints, the derivation will not be carried further.

4. Four Problems Encountered in Dealing With Exact Computation Sequences

In this section we discuss the four problems mentioned at the end of the Introduction.

Problem 1

Recall that the first problem is *to obtain characterizations of* L_i from the properties of the set of reductions R_i and the input set of terms L_{i-1}. Assuming that we are given a characterization of the set L_{i-1} and we know the set of meta-reductions S_i, we want to determine L_i. By lemma 1, we know that if S_i is finite and L_{i-1} is recursive, then L_i is a recursive set.

Lemma 7 tells us that if S_i consists of a finite set of left linear rules then it preserves context-freeness and deterministic context- freeness.

Lemma 9 gives a sufficient condition for transforming a standard grammar into a standard grammar.

We note that in this problem we are interested in the description of L_{i-1} and the part $(C) \Rightarrow l$ of the meta-reductions $(C) \Rightarrow l \to r$. Properties such as the form of the term r in the reductions $(C) \Rightarrow l \to r$ are not of interest in this problem.

We have methods to handle the case when the rules in S are left-linear and the condition C has the value true. One of these methods is the *production expansion*. Similar techniques have been used by

Barros and Remy [1]. This method requires that we have a context-free grammar for L_{i-1}. We shall illustrate how this method works by giving an example.

Let G=<{N},{+,0,S},P,N> be the grammar for L_{i-1} whose productions are shown below:

$N \rightarrow 0|SN|+NN$

Assume that the set of rules is $R=\{+x0 \rightarrow x, +xSy \rightarrow S+xy\}$. Then we expand the second N in the production $N \rightarrow +NN$. We get the grammar $N \rightarrow 0|SN|+N0|+NSN|+N+NN$. We say that a set of productions *is safe* if they cannot produce an instance of the left hand side of a rule. Productions $N \rightarrow +NO$ and $N \rightarrow +NSN$ are unsafe and are crossed out. The production $N \rightarrow +N+NN$ is also crossed out since it fails to produce a term in an expansion.

We are left with productions $N \rightarrow 0|SN$. Both are safe for R. This method was used in lemma 9 to obtain a standard grammar for L_i.

It is important to obtain a good description of the language L_i, since this description can be used for inductive proofs on L_i.

In particular, it is desirable to use context-free grammars. Boyer and Moore [2] use induction on terms to carry out proofs as well as Huet and Hullot [5], Fribourg [3] and Musser [12], Kirchner [10], Paul [13], and Remy [16].

We would like to have a better characterization of the relation between the input language L_{i-1} and the set of rules in S_i on one side, and the output language L_i. We need methods and results for handling non-linear pure rules and for dealing with conditional reductions. Results in that direction are presented in Remy and Zang [17]. For example, the commutative axiom is handled nicely by conditional meta-reductions. For example, if L_0 is the language generated by the grammar G=<{S},{a_1,...,a_n,+},P,S>, where P is the set of productions $\{S \rightarrow a_1|...|a_n|+SS\}$, and the set of axioms is $A_0=\{+xy=+yx\}$, we can define a function $h{:}L_0 \rightarrow N$, or into any other well order (N^k,>) such that h is injective. Then the conditional meta-reduction $(h(x)>h(y))$=>$+xy \rightarrow +yx$ produces a representation function for <L_0,A_0>. We would like to see theorems dealing with languages that do not have the subterm property.

Problem 2

In this problem, we are given L_{i-1}, the set of meta-reductions S_i and a set of axioms A_{i-1}. We want to characterize the set of axioms $A_i=S_i(A_{i-1})$. So far, we have weak results like lemmas 5 and 8.

As observed in the first step of the exact sequence in Example 1 of Section 2, lemma 5 is good in handling equations which contain operations that cannot be Ø-unified with the left hand sides of the reductions (C)=>$l \rightarrow r$. If a Ø-unification is possible, then an equation in A_{i-1} may be transformed into many equations in A_i. The equation $-+xx=0$ in the example mentioned above was transformed into the set of equations $+xR_1(-x)=0$. In getting this equation we used the fact that L_1 has the subterm property and we applied lemma 8. However, as we add more steps, the equations will become more and more complicated containing may occurrences of the step function symbols in them.

For example, let us take the case of free semigroup with two generators to which we add the commutativity axiom. The language L_0 is the set generated by the standard grammar $G_0=<\{S\},\{a,b,+\},P_0>$, where P_0 is the set of productions $S \rightarrow a|b|+SS$. We take A_0 to be $\{++xyz=+x+yz,+xy=+yx\}$. As the initial set E_0, we take the associativity axiom. Then $A_1=R_1(A_0)=\{R_1(+xy)=R_1(+yx)\}$, and there is not much that we can do with this equation. However, if we take x and y to be atoms in L_1 we get the theorem $+ba=+ab$. By using the forced confluence method, we obtain the theorem $++xab=++xba$. It can be shown that the set of equations $\{+ba=+ab,++xab=++xab\}$ is equivalent over L_1 to the set of axioms A_1

It can also be shown that the set of equations 1 - 15 shown in step 3 of example 1 of section 2 is equivalent to the set equations A_2 over L_2. We would like to obtain theorems which identify cases for which a system of equations E is a consequence of a system of equations E' when both E and E' are equations over a language L.

Reductions modulo a set of equations have been studied by Huet [4], Jouannaud and Munoz [9], Jouannaud [7], and Jouannaud and Kirchner [8].

It would be interesting to study how those results behave when we put certain conditions on the set of terms L.

Problem 3

In problem 3 (which is closely related to problem 2), we are given a set of equations E and a set of rules R and we want to know under what conditions an equation in E vanishes under the rules R. For example, the set of axioms in A_2 in example 1 from the second section vanishes under the set of rules R_3. At this point, we are looking for two things:

- A good induction method which will verify that a set of reductions satisfies an equation in A_{i-1} which contains step function symbols. We would like to get efficient methods for handling specific classes of axioms such as commutativity, idempotence, absorption.
- Theorems which give us sufficient conditions for a set of reductions to satisfy an equation. Induction methods have been presented by Huet and Hullot [5], Musser [12], and other references given above.

We are interested to see how one can use those results to get sufficient conditions for the equations to vanish. One may try to change the set of equations into an equivalent set by using results from Problem 2.

An interesting subproblem is to see under what conditions conditional reductions reduce to pure reductions. For example, conditional reductions are used to handle commutativity but if an operation is both commutative and associative, then the conditional reduce to pure reductions.

Problem 4

This problem is to give criteria for choosing the set of equations E_i that are to be eliminated. As a rule of thumb, we choose those equations which simplify both the form of L_i and of the set of equations A_i. For example, if an operation has the quasidistributivity property, then it carries a standard language into a standard language and the set of reductions is terminating and confluent. We have methods for dealing with specific axioms such as commutativity and associativity.

Another criteria is to reduce the problem to a problem for which we already know the answer. For example, if we have an exact sequence which handles a pair $\langle L_1,A_1\rangle$ and we can find a confluent and terminating system of reductions S such that $S(L_0)=L_1$ and $S(A_0)=A_1$, then we apply that system of reduction by choosing the equations in S.

5. Conclusion

A step by step approach seems to be more natural than a vertical approach as used by Knuth and Bendix [11]. This approach is more structured and allows a step by step construction. Note that H. Kirchner has exploited a similar idea. We can have packages for handling various kinds of axioms and we can construct an exact computation sequence by calling subprograms which deal with different classes of axioms. At the same time we must have a theorem proving mechanism since combinations of axioms yield various theorems which are in some cases easier to handle, as shown in the first step of example 1 from section 2.

We would like to relax the conditions imposed on exact computation sequences. For example let L_0 be the language generated the atoms $0,a,b,c$ under the operation + and A_0 a set of group axioms with identity 0. Let S be an exact computation sequence for $\langle L_0,A_0\rangle$, and let L' be the set of terms generated by the atoms $0,a,b$ under the operation +. Then $\langle L',A_0\rangle \to S$ (where $\langle L',A_0\rangle \to S$ is an inclusion), is a computation sequence, though it is not exact since the onto condition for the language is violated. We can apply this construct to all subalgebras. A subalgebra is defined as being closed under the operations and containing all the constants that occur in the set of equations A_0. It is interesting to study cases where $\langle L',A_0\rangle \to S$ is a computation sequence and L' is not a subalgebra. By a computation sequence, we mean a sequence $\langle L_0,A_0\rangle \xrightarrow{S_1} \langle L_1,A_1\rangle \to \ldots \xrightarrow{S_n} \langle L_n,\emptyset\rangle$, where $L_i \subseteq L_0$, and $S_i(L_{i-1}) \subseteq L_i$, such that the composition $S_n o S_{n-1} o \ldots o S_1$ is a representation function for $\langle L_0,A_0\rangle$. We keep the restriction that each S_i is a set of confluent and terminating reductions.

It would be worthwhile to study the algebraic properties of these computation sequences. One can define homomorphisms of such sequences, which we believe to be useful in many applications.

6. Bibliography

1. Barros L., and Remy J.L., Ecologiste: A system to make complete and consistent specifications easier, Proc. Workshop Rewrite Rule Laboratory, GE Research and Development, Schenectady, NY, 1983

2. Boyer R., Moore J.S., *Computational Logic*, Academic Press, 1979.

3. Fribourg L., A Narrowing Procedure for Theories with Constructors, 7th *International Conference on Automated Deduction*, R.E. Shostak, editor, Springer Verlag Lecture Notes in Computer Science, Vol. 170, 259-281.

4. Huet G., Confluent Reductions: Abstract Properties and Applications to Term Rewriting Systems, *JACM*, 1980, 767-821.

5. Huet G., Hullot J.M., Proofs by Induction in Equational Theories with Constructors, *JCSS*, Vol. 25, 1982, 239-266.

6. Huet G., Oppen D., Equations and Rewrite Rules: A Survey, *Formal Language Theory: Perspectives and Open Problems*, R. Book, editor, Academic Press, 1980.

7. Jouannaud J.P., Church Rosser Computations with Equational Term Rewriting Systems. Technical Report, CRIN, Nancy, France, January, 1983.

8. Jouannaud J.P., and Kirchner H., Completion of a set of rules modulo a set of equations, Technical Report SRI-CSL and CRIN 84-R-046, Nancy, 1984.

9. Jouannaud J.P., Munoz M., Termination of a set of Rules Model a Set of Equations, 7th *International Conference On Automated Deduction*, editor R.E. Shostak, Springer Verlag Lecture Notes in Computer Science, Vol. 170, 175-193.

10. Kirchner H, A general inductive completion algorithm and applications to abstract data types, Proc 7th *International Conference On Automated Deduction*, editor R.E. Shostak, Springer Verlag Lecture Notes in Computer Science, Vol. 170.

11. Knuth D., Bendix P., Simple Word Problems in Universal Algebras, in *Computational Problem in Abstract Algebra*, Ed. J. Leech, Pergamon Press, 1970, 263-297.

12. Musser D.L., On Proving Inductive Properties of Abstract Data Types, Proc. 7th *POPL*, Las Vegas, 1980.

13. Paul E., Preuve par induction dans les theories equationelles en presence de relations entre les constructeurs, Proc. 9th *Colloquium on trees in algebra and programming*, Bordeaux, France, 1984.

14. Pelin A., Gallier J.H., Solving Word Problems in Free Algebras Using Complexity Functions 7th *International Conference on Automated Deduction*, editor R.E. Shostak, Springer Verlag Lecture Notes in Computer Science, Vol. 170, 476-495.

15. Pelin A., Gallier J.H., Computing Normal Forms Using Complexity Functions Over N^k, Submitted to *TCS*, 1985.

16. Remy J.L., Etude des Systemes de reecriture conditionels et applications aux types abstraits algebriques, These d'Etat, Universite de Nancy, 1984.

17. Remy J.L., and Zhang H., REVEURS: a system for validating conditional algebraic specifications of algebraic abstract data types, Proc. 5th *ECAI*, Pisa, Italy, 1984.

AN ALGEBRAIC THEORY OF FLOWCHART SCHEMES

(extended abstract)

Gh. Ştefănescu

Department of Mathematics,
National Institute for Scientific and Technical Creation,
Bd. Păcii 220, 79622 Bucharest, Romania

1. INTRODUCTION

The first proofs of the main results in this paper may be found in [13]; revised proofs will appear elsewhere.

A. A motivation. This research is an attempt to develop Backus' ideas [2], namely to construct a mathematical theory in which program transformations are allowed and in which correctness can be obtained using simple algebraic computations. We use flowchart schemes to represent programs and we look for a calculus similar to the calculus of polynomials.

A framework for such a calculus is the following:

(F1) fix a support theory T (with some operations) and a set X of variables; give a formal representation for the basic elements of the calculus (called X-nomials over T), extend the operations from T to arbitrary X-nomials over T, and give the interpretation of these nomials when the variables have concrete values;

(F2) give a natural equivalence relation between these formal representations of nomials;

(F3) find (if it is possible!) a type struc for the support structure T such that:

- equivalent nomials have the same interpretation in all struc structures; and
- the classes of equivalent X-nomials over T form the struc structure freely generated by adding X to T.

(In a more categorial language this is the coproduct of a given struc structure with a struc structure freely generated by a set.)

The main point of the calculus is **(F2)**. For the case of X-polynomials over a ring T, represented as finite formal sum of terms (monomials), the equivalence relation is the reduction of similar terms; this can be seen as the equivalence relation generated by the following basic rules,

(id-p) identify similar terms and compute the resulted coefficients, and

(ac-p) delete terms with zero coefficient;

moreover, two polynomials are equivalent iff in two steps (firstly, identifying similar terms and secondly, deleting terms with zero coefficient) they can be reduced to the same polynomial. With this equivalence, in **(F3)** we can take ring for struc.

B. A representation of flowchart schemes. This subsection considers the first step of the calculus (i.e. **(F1)** in Subsection A).

(a) The support structure T is suposed to be given by a family of sets $T(m,n)$, $m,n \geq 0$. An element $f \in T(m,n)$ (also written , or $f : m \to n$, when T is known) is considered as a known process of computation with m inputs and n outputs.

I think that the most natural operations on flowchart schemes are: composition (or sequential composition) $\cdot$, sum (or parallel composition) $+$, and feedback $\uparrow_m$, $m \geq 0$; in a picture they are as follows:

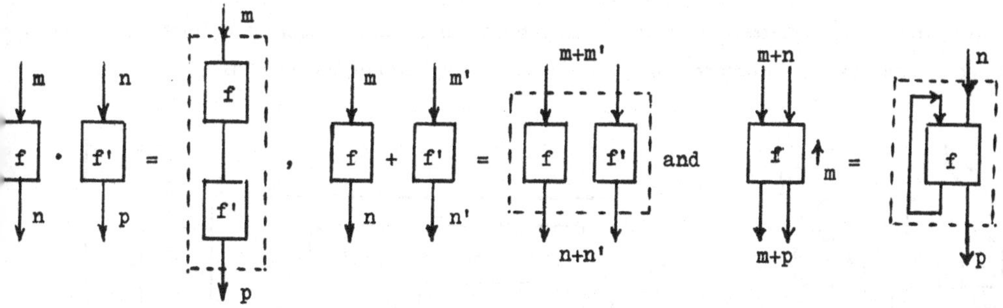

Moreover, if T contains the flowchart schemes of type g = 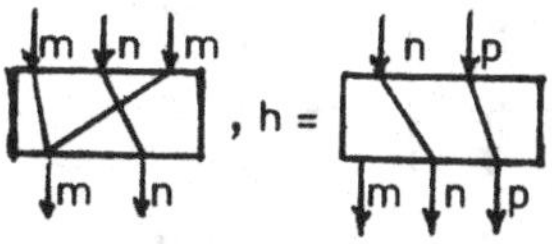, then instead of general composition it is enough to define left composition with flowchart schemes of type g above, and right composition with flowchart schemes of type h above, since

$f \cdot f' = (g\,(f + f')h)\uparrow_{m+n}$, for 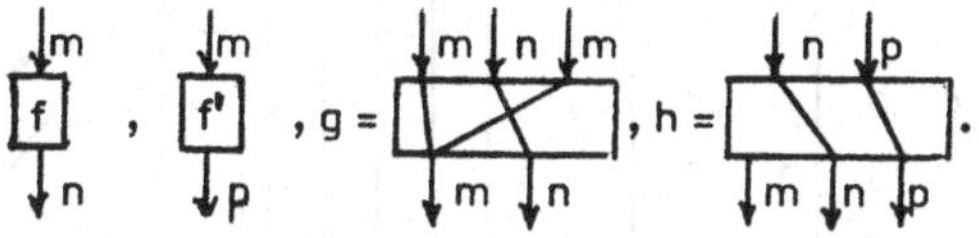.

The other operations can be defined if T contains some special elements. If T contains the elements $1_p \mathbin{\updownarrow} 1_p$ = 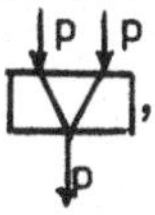, $p \geq 0$, then we can define:

tupling: $f \mathbin{\updownarrow} f' = (f + f')(1_p \mathbin{\updownarrow} 1_p)$, for $f : m \to p$, $f' : n \to p$; and

iteration: $f^{\dagger} = ((1_m \mathbin{\updownarrow} 1_m)f)\uparrow_m$, for $f : m \to m+n$.

If T contains the elements $1_p \mathbin{\Updownarrow} 1_p$ = 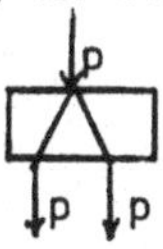, $p \geq 0$, then we can define:

> <u>cotupling</u>: $f \overset{\uparrow}{+} f' = (1_p \overset{\uparrow}{+} 1_p)(f + f')$, for $f : p \to m$, $f' : p \to n$; and
>
> <u>coiteration</u>: $f^{\dagger} = (f(1_m \overset{\uparrow}{+} 1_m))\uparrow_m$, for $f : m+n \to m$.

If T contains both $1_p \underset{\downarrow}{+} 1_p$ and $1_p \overset{\uparrow}{+} 1_p$, then we can define:

> <u>choice</u>: $f \overset{\uparrow}{\underset{\downarrow}{+}} f' = (1_m \overset{\uparrow}{+} 1_m)(f + f')(1_n \underset{\downarrow}{+} 1_n)$, for $f : m \to n$, $f' : m \to n$;
>
> <u>strict repetition</u>: $f^{+} = ((1_m \underset{\downarrow}{+} 1_m)\, f\, (1_m \overset{\uparrow}{+} 1_m))\uparrow_m$, for $f : m \to m$; and
>
> <u>repetition</u>: $f^{*} = ((1_m \underset{\downarrow}{+} 1_m)(1_m \overset{\uparrow}{+} 1_m)(f + 1_m))\uparrow_m$, for $f : m \to m$.

(b) The internal vertices in a flowchart scheme will be labeled with elements in a set X of double-ranked variables, i.e. every $x \in X$ has a number x_{in} of inputs,and another number x_{out} of outputs - another writing is $x \in X(x_{in}, x_{out})$ -; for a sequence e in the free monoid X^*, e_{in} (respectively e_{out}) denotes the sum of the inputs (respectively the sum of the outputs) of the letters of e.

(c) The usual nondeterministic flowchart scheme in Figure 1.a can be ordered as in Figure 1.b and can be briefly represented as in Figure 1.c (on the basis of connections).

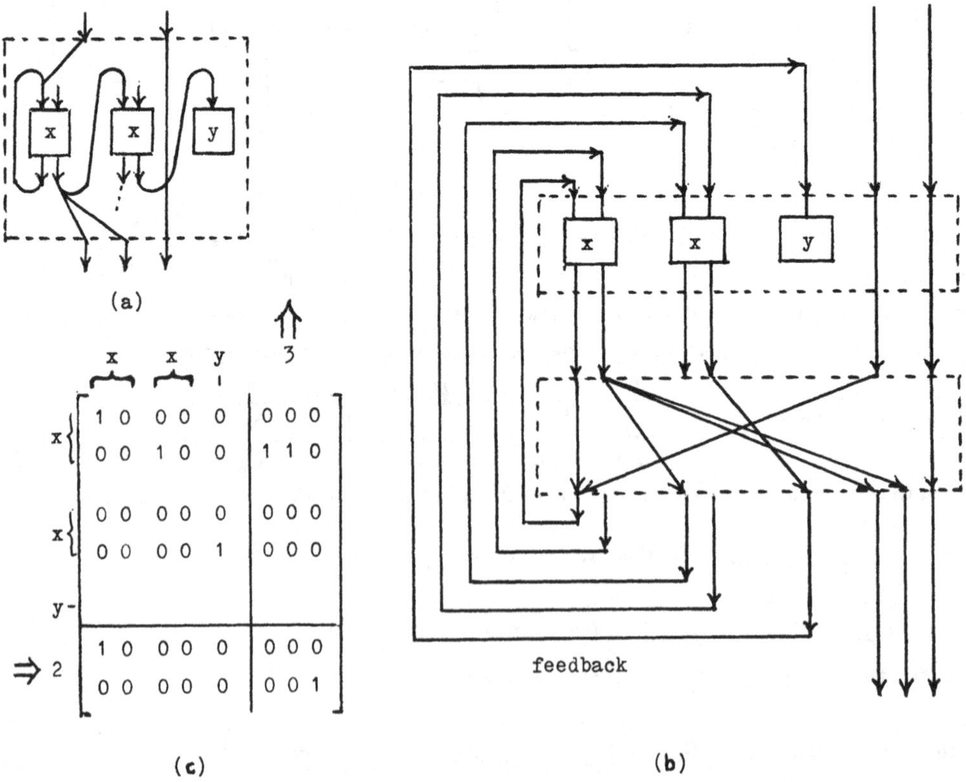

Figure 1.

Consequently, we can define an X-flownomial over T with m inputs and n outputs as

$$((x_1 + \dots + x_k + 1_m)\, l) \uparrow^{e_{in}}$$

where $x_1,\dots,x_k \in X$, $e = x_1 \dots x_k \in X^*$, and $l \in T(e_{out} + \mathbf{m}\ ,\ e_{in} + \mathbf{n}\ )$. Denote this set by $\mathbf{Fl}_{X,T}(m,n)$. A flow theory is $\mathbf{Fl}_{X,T}$ for some X, T. [Remark: since there is no charts in the above representation, we prefer to call the elements of the calculus flownomials instead of flowchart schemes.]

(d) If T contains the transpositios $1_m \nleftrightarrow 1_n$ = 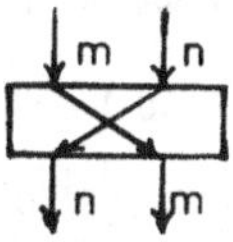

, then the basic operations in T can be extended to arbitrary flownomials. [Exercise: use pictures to deduce their natural formulas.] However, since we want to compare our algebraic structures (for the deterministic case and for the nondeterministic case) with other ones [4,5,6,7,9,10,12] we shall give their characterizations in terms of the old operations: composition, tupling, and iteration in the deterministic case, respectively composition, choice, and repetition in the nondeterministic case.

(e) When the variables in X and the morphisms in T have interpretations in a concrete computation theory Q (with composition, sum and feedback defined), namely $I_X(x) \in Q(x_{in}, x_{out})$, for $x \in X$, respectively $I_T(f) \in Q(m,n)$, for $f \in T(m,n)$, the interpretation of the X-flownomial over T $((x_1 + \dots + x_k + 1_m)\, l) \uparrow^{e_{in}}$ is obviously $((I_X(x_1) + \dots + I_X(x_k) + I_T(1_m))\, I_T(l)) \uparrow^{e_{in}}$.

2. DETERMINISTIC FLOWCHART SCHEMES

This deterministic case correspond to the physical world: physical processes have deterministic evolutions; two processes may be considered equivalent if they have the same step evolution.

A. More about the support structure. As it was seen the operations we work with are composition, tupling and iteration; it is known from [8] that all deterministic flowchart schemes can be obtained from the atomic flowchart schemes using these operations.

For the beginning we fix the standard model for the interpretation of flowchart schemes. This model, proposed in [7] by Elgot, consists in the following. Fix a set D representing the set of all value-vectors for the registers in a computing device. A program scheme F = 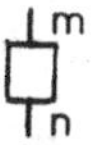

is interpreted as a partial function $f : [m] \times D \longrightarrow [n] \times D$ (here [m] denotes the set $\{1,\dots,m\}$) with the meaning that if program execution begins at line $k \in [m]$ of the program with initial value-vector d and if $f(k,d) = (j,d')$, then d' is the new value-vector when program halts at line $j \in [n]$. The set of all partial functions from $[m] \times D$ to $[n] \times D$ will be denoted by $\mathbf{Pfn}_D(m,n)$; when D has exactely one element we shall write $\mathbf{Pfn}_.$ instead of $\mathbf{Pfn}_D$.

The operations in $\mathbf{Pfn}_D$ have the following definitions.

Composition: for $f \in \mathbf{Pfn}_D(m,n)$ and $g \in \mathbf{Pfn}_D(n,p)$, we define $f \cdot_D g \in \mathbf{Pfn}_D(m,p)$ as

$$(f \cdot_D g)(j,d) = g(f(j,d)) \quad , \text{ for } (j,d) \in [m] \times D.$$

Tupling: for $f \in \mathbf{Pfn}_D(m,p)$ and $g \in \mathbf{Pfn}_D(n,p)$, we define $f \mathbin{\ddagger_D} g \in \mathbf{Pfn}_D(m+n,p)$ as

$$(f \mathbin{\ddagger_D} g)(j,d) = \text{"if } j \in [m] \text{ then } f(j,d) \text{ else } g(j-m,d)\text{"} , \quad \text{for } (j,d) \in [m+n] \times D.$$

Iteration: for $f \in \mathbf{Pfn}_D(m,m+n)$ we define $f^{\dagger_D} \in \mathbf{Pfn}_D(m,n)$ as

$$f^{\dagger_D}(j,d) = \begin{cases} (j_r - m, d_r), & \text{where } (j_r,d_r) \text{ is the last defined value (i.e. } j_r > m) \text{ in the following sequence: } (j_0,d_0) = (j,d) \text{ and for } k \geqslant 0, \ (j_{k+1},d_{k+1}) = f(j_k,d_k) ; \\ \text{undefined}, & \text{if such an } r \text{ does not exists,} \end{cases}$$

for $(j,d) \in [m] \times D$.

At the abstract level we shall define in turn algebraic theories, preiteration theories and paraiteration theories; moreover we agree to begin the analysis using for the support structure a paraiteration theory.

An <u>algebraic theory</u> in [7] is given by: a family of morphisms $T(m,n)$, $m, n \geqslant 0$; two binary operations, namely composition $\cdot : T(m,n) \times T(n,p) \rightarrow T(m,p)$ and tupling $\ddagger : T(m,p) \times T(n,p) \rightarrow T(m+n,p)$; and some distinguished morphisms $1_n \in T(n,n)$, $0_n \in T(0,n)$ and $\pi_i^n \in T(1,n)$, for $i \in [n]$. Moreover, the operations must fulfil the following axioms

- the composition is associative and 1_n are its neutral elements;
- the tupling is associative, 0_p are its neutral elements and the morphisms have a unique source-spliting into components, i.e. the following two axioms holds

(T1) $f_i = \pi_i^n (f_1 \ddagger \dots \ddagger f_n)$, for $f_1, \dots, f_n : 1 \rightarrow n$, and $i \in [n]$;

(T2) $(\pi_1^m f) \ddagger \dots \ddagger (\pi_m^m f) = f$, for $f : m \rightarrow n$.

A <u>preiteration theory</u> in [4] is an algebraic theory in which an iteration $\dagger : T(m,m+n) \rightarrow T(m,n)$ is defined.

<u>Remarks.</u> (i) The initial preiteration theory is $\mathbf{Pfn}_\bullet$, hence every partial function in $\mathbf{Pfn}_\bullet$ can be considered as a morphism in an arbitrary preiteration theory (the interpretation of the nowhere defined function $\perp_{m,n} : [m] \rightharpoonup [n]$ is $1_m^\dagger 0_n$).

(ii) A preiteration theory has a natural sum and a natural feedback, namely:

- the sum is: $1_m + 0_n = \pi_1^{m+n} \ddagger \dots \ddagger \pi_m^{m+n}$, $0_m + 1_n = \pi_{m+1}^{m+n} \ddagger \dots \ddagger \pi_{m+n}^{m+n}$, and $f + g = (f(1_m + 0_n)) \ddagger (g(0_m + 1_n))$, for $f \in T(p, m)$, $g \in T(q, n)$;
- the feedback is: $f \uparrow^m = (0_m + 1_n)(f(1_m + 0_n + 1_p))^\dagger$, for $f : m+n \rightarrow m+p$.

This is used to define flownomials over a preiteration theory.

Some axioms for iteration are listed below.

(I0) $(f(1_m + g))^{\dagger} = f^{\dagger}g$, for $f: m \to m+n$, $g: n \to p$;

(I1) $f(f^{\dagger} \ddagger 1_n) = f^{\dagger}$, for $f: m \to m+n$;

(I2) $(f((1_m \ddagger 1_m) + 1_n))^{\dagger} = f^{\dagger\dagger}$, for $f: m \to m+m+n$;

(I3) $g(f(g + 1_p))^{\dagger} = (gf)^{\dagger}$, for $f: m \to n+p$, $g: n \to m$;

(I4) if $f(\varrho + 1_p) = \varrho g$ then $f^{\dagger} = \varrho g^{\dagger}$, for $f: m \to m+p$, $g: n \to n+p$, and a function $\varrho: [m] \to [n]$.

Comment. The axiom **(I0)** shows that the iteration $f^{\dagger}$ has a "uniform behaviour" with respect to the last n variables of the system $f: m \to m+n$, hence these last variables can be considered as parameters; the axiom **(I1)** says that the iteration $f^{\dagger}$ gives a "canonical" solution for the Elgot recursive system $x = f(x \ddagger 1_m)$; in the presence of **(I0)** and **(I1)** the following two axioms **(I2)** + **(I3)** show that the canonical solution of a system can be expressed in terms of the canonical solution of its components (it follows that the "pairing" axiom

(P) $(f \ddagger g)^{\dagger} = (f^{\dagger}(h \ddagger 1_p)) \ddagger h$, where $h = (g(f^{\dagger} \ddagger 1_{n+p}))^{\dagger}$, for $f: m \to m+n+p$, $g: n \to m+n+p$

holds and this is similar to the Gauss substitution method for solving linear systems); the last axiom expresses the preservation of the canonical solution when we rename the variable, not necessarily in an injective way, but in a consistent way, i.e. if two variable have the same rename then the right side of their equations are equal after renaming (this axiom is a particular case of the functorial axiom in [1]).

A paraiteration theory is a preiteration theory in which the iteration fulfils the parameter axiom **(I0)**.

The extension of the operations in T to arbitrary flownomials over T may be given with the folowing definitions (we suppose that a generic morphism for connections $l: e_{out} + m \to e_{in} + n$ is splitted into transfer $t: e_{out} \to e_{in} + n$ and input $i: m \to e_{in} + n$, i.e. $l = t \ddagger i$).

Composition: for the flownomials $F = ((x_1 + \dots + x_k + 1_m)(t \ddagger i))\uparrow_{e_{in}} : m \to n$ and $F' = ((x'_1 + \dots + x'_{k'} + 1_m)(t' \ddagger i'))\uparrow_{e'_{in}} : n \to p$ we define

$$F \cdot F' = ((x_1 + \dots + x_k + x'_1 + \dots + x'_{k'} + 1_m)(t(1_{e_{in}} + i') \ddagger (0_{e_{in}} + t') \ddagger i(1_{e_{in}} + i')))\uparrow_{e_{in} + e'_{in}}.$$

Tupling: for the flownomials $F = ((x_1 + \dots + x_k + 1_m)(t \ddagger i))\uparrow_{e_{in}} : m \to p$ and $F' = ((x'_1 + \dots + x'_{k'} + 1_m)(t' \ddagger i'))\uparrow_{e'_{in}} : n \to p$ we define

$$F \ddagger F' = ((x_1 + \dots + x_k + x'_1 + \dots + x'_{k'} + 1_m)(ty \ddagger t'y' \ddagger iy \ddagger i'y'))\uparrow_{e_{in} + e'_{in}},$$

Iteration: for the flownomial $F = ((x_1 + \dots + x_k + 1_m)(t \ddagger i))\uparrow_{e_{in}} : m \to m + n$ we define

where $y = 1_{e_{in}} + 0_{e'_{in}} + 1_p$ and $y' = 0_{e_{in}} + 1_{e'_{in} + p}$.

$$F^{\dagger} = ((x_1+\ldots+x_k + 1_m)(t \underset{\vee}{+} i)((1_{e_{in}} + 0_n) \underset{\vee}{+} (i((1_{e_{in}} \underset{\leftrightarrow}{+} 1_m) + 1_n)^{\dagger} \underset{\vee}{+} (0_{e_{in}} + 1_n)))\uparrow_{e_{in}} .$$

Remarks: (i) if the iteration fulfils the axiom **(I1)** then the morphisms in T can be considered as scalar flownomials (i.e. $k = 0$) and the operations on flownomials extend the operations in T indeed;

(ii) if the iteration fulfils the axioms **(I1)** and **(I3)** then we can prove the identity

$$((x_1 + \ldots + x_k + 1_m)(t \underset{\vee}{+} i))\uparrow_{e_{in}} = i(((x_1 + \ldots + x_k)\, t)^{\dagger} \underset{\vee}{+} 1_n)$$

and this shows that the interpretation of flownomials agree with the known interpretation of flowchart schemes [3,6,7].

B. IO-equivalent flowchart schemes. Two flowchart schemes will be considered equivalent if the computation processes they denote have the same steps in execution. There is a simple way to introduce this equivalence relation, namely as the congruence relation generated by the following two basic rules (reductions):

(id) $F \xrightarrow{s} F'$ if F' can be obtained from F by identifying vertices which have the same label and whose output connections are equal after identification; and

(ac) $F \xleftarrow{i} F'$ if F' can be obtained from F by deleting non-accessible vertices.

Example.

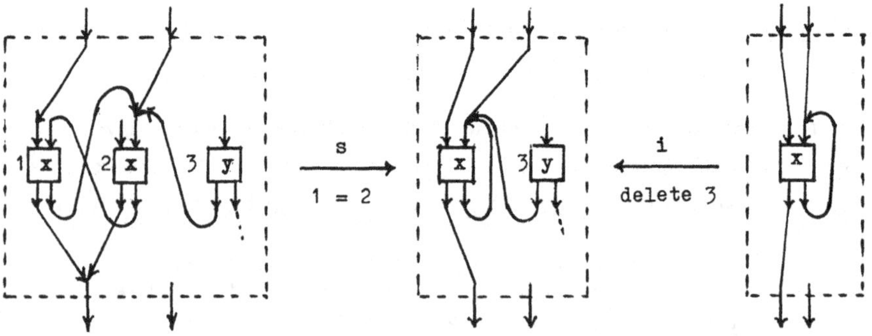

.

We can lift these informal relations to exact relations between X-flownomials over a paraiteration theory T. We say that two flownomials $F = ((x_1+\ldots+x_k + 1_m)\, \mathrm{l})\uparrow_{e_{in}} : m \to n$ and $F' = ((x'_1+\ldots+x'_{k'}+ 1_m)\, \mathrm{l}')\uparrow_{e'_{in}} : m \to n$ are in d-simulation (written $F \xrightarrow[d]{} F'$) if there exists a function $\varrho : [k] \to [k']$ such that

$$\varrho(i) = j \quad \text{implies} \quad x_i = x'_j$$

and its "block" extensions to inputs ϱ_{in} and to outputs ϱ_{out} fulfil

$$\mathrm{l}\,(\varrho_{in}+1_n) = (\varrho_{out}+1_m)\, \mathrm{l}' .$$

The relations $\xrightarrow{s}$, $\xrightarrow{i}$ (above $\xleftarrow{i}$ denotes the inverse relation $\xrightarrow{i}{}^{-1}$) are the restrictions of $\xrightarrow[d]{}$

to surjective functions ϱ , respectively to injective functions ϱ. Finally, we define our <u>IO-equivalence</u> $=_d$ as the congruence relation generated by $\xrightarrow[d]{}$ (or, equivalently, using $\xrightarrow[d]{} \subsetneq \xrightarrow{s} \cdot \xrightarrow{i}$, as the congruence relation generated by $\xrightarrow{s} \cup \xleftarrow{i}$).

The d-simulation is compatible with composition, tupling and iteration (for the compatibility with iteration we need the parameter axiom **(I0)**); by using some commutations between $\xrightarrow{s}, \xrightarrow{i}$ and their inverse relations we have the following characterization theorem for $=_d$.

<u>THEOREM.</u> If T is a paraiteration theory, then in $\mathbf{Fl}_{X,T}$ we have

$$=_d \; = \; \xrightarrow{s} \cdot \xleftarrow{i} \cdot \xrightarrow{i} \cdot \xleftarrow{s} .$$

This theorem shows that two flownomials are IO-equivalent iff in two steps (firstly, identifying similar vertices whose output connections are equal after identification and secondly, deleting non-accessible vertices) they can be reduced to the same flownomial.

<u>C. Strong iteration theories.</u> This subsection gives the technical part of the calculus (i.e. **(F3)** in 1.A). We say that a paraiteration theory is a <u>strong iteration theory</u> if the iteration fulfils the axioms **(I0)**, **(I1)**, **(I2)**, **(I3)**, and **(I4)** above.

<u>Remark.</u> Only the axiom **(I4)** is not equational, hence strong iteration theories form a quasi-variety; instead of **(I4)** iteration theories in [4,10] use a weaker equational axiom; theories with iterate in [5,6], still much weaker, require **(I4)** only for bijective functions ϱ. All these theories are generalizations of both pointed iterative theories in [4] and rational theories in [14], consequently all examples of pointed iterative theories and ω-continuous theories are examples of strong iteration theories.

<u>Basic example.</u> $\mathbf{Pfn}_D$ with the operations defined in Subsection **A** is a strong iteration theory.

The new axiom **(I4)** is <u>essential:</u> the support theory T must satisfy **(I4)** in order to IO-equivalent X-flownomials over T have the same interpretation in T. Moreover, two IO-equivalent flownomials have the same interpretation in all strong iteration theories; consequently we begin the analysis of the last problem in **(F3)** taking strong iteration theories for <u>struc</u>.

All the axioms of strong iteration theories, except **(T1)**, **(T2)** for tupling and **(I1)**, **(I4)** for iteration, hold even in $\mathbf{Fl}_{X,T}$; the left side of the equations in **(T1)**, **(T2)** and **(I1)** can be d-simulated in the right side, hence these axioms hold in the quotient structure; using the characterization theorem for $=_d$ we can prove that **(I4)** holds in the quotient structure; hence the classes of IO-equivalent flownomials over a strong iteration theory form also a strong iteration theory. Since the universality property can be proved as in [3,6] we have the following

MAIN THEOREM (deterministic case). If T is a strong iteration theory, then the classes of IO-equivalent X-flownomials over T form the strong iteration theory freely generated by adding X to T.

COROLLARY. The classes of IO-equivalent X-flownomials over **Pfn.** form the strong iteration theory freely generated by X.

It is known [11,10] that the theory of rational X-trees R_X form the iteration theory freely generated by X ; moreover it is easy to see that R_X is even the strong iteration theory freely generated by X; hence the interpretation of the classes of IO-equivalent X-flownomials over **Pfn.** in rational trees is an isomorphism; this show that

COROLLARY. In the simplest case T = **Pfn.** and $x_{in} = 1$, $\forall\, x \in X$, two flownomials are IO-equivalent iff the flowchart schemes they denote unfold the same tree.

This means that the IO-equivalence relation is a generalization of the strong equivalence relation in [8]. Consequently, we have a calculus for the classes of strongly equivalent deterministic flowchart schemes, essentially similar to the calculus of polynomials.

3. NONDETERMINISTIC FLOWCHART SCHEMES

This nondeterministic case correspond to the living world. Living processes are based on physical processes; their characteristic future consists in the ability of fixing aims. In view of its aim a living process can choose between several variants, or can stop if it feels that can not reach its aim and restart tring another way. Two living processes with the same starting state and the same aim may be considered equivalent if they reach the aim in the same way, namely they have the same successful paths.

A. More about the support structure. As we said before the operations we work with are composition, choice and repetition. A motivation as in the deterministic case leads to the basic model for nondeterministic programs, namely the theory of relations over D. The set of all relations between $[m] \times D$ and $[n] \times D$ will be denoted by $\mathbf{Rel}_D$; when D has exactly one element we shall write **Rel.** instead of $\mathbf{Rel}_D$. The operations in $\mathbf{Rel}_D$ are the following: composition is the usual composition of relations; choice is the union of relations; and repetition is Kleene's reflexive, transitive closure, namely $f^* = 1_m \cup f \cup f^2 \cup \ldots$, for $f \in \mathrm{Rel}_D(m,m)$.

At the abstract level we agree to begin the analysis using for the support theory a prereticulum, defined below.

A <u>matrix theory</u> in [9] is a theory T in which:

- $T(1,1)$ is a semiring with addition = choice and multiplication = composition;

- the morphisms in $T(m,n)$ are $m \times n$ matrices over $T(1,1)$ and the operations agree, i.e. choice = matrix addition and composition = matrix multiplication.

A <u>prereticulum</u> is an idempotent matrix theory (i.e. a matrix theory with $1_1 \overset{\wedge}{\underset{\vee}{+}} 1_1 = 1_1$) in which a repetition $^* : T(m,m) \longrightarrow T(m,m)$ is defined.

<u>Remark.</u> (i) The initial prereticulum is **Rel.** , hence every relation in **Rel.** (equivalently, every matrix over the Boolean semiring $\{0,1\}$) can be seen as a morphism in an arbitrary prereticulum.

(ii) A prereticulum T is naturally a paraiteration theory ($[f]+[g] = \begin{bmatrix} f \\ g \end{bmatrix}$ and $[f\ g] = [f]^*[g]$), hence we can define flownomials over a prereticulum. The extension of the operations from T to arbitrary X-flownomials over T , represented in the matricial form as in Figure 1.c, are:

$$\begin{array}{c|cc} & e & n \\ \hline e & D & C \\ m & B & A \end{array} \;\bullet\; \begin{array}{c|cc} & e' & p \\ \hline e' & D' & C' \\ n & B' & A' \end{array} \;=\; \begin{array}{c|cc|c} & e & e' & p \\ \hline e & D & CB' & CA' \\ e' & 0 & D' & C' \\ \hline m & B & AB' & AA' \end{array}\;; \qquad \begin{array}{c|cc} & e & m \\ \hline e & D & C \\ m & B & A \end{array} \;\overset{\wedge}{\underset{\vee}{+}}\; \begin{array}{c|cc} & e' & m \\ \hline e' & D' & C' \\ m & B' & A' \end{array} \;=$$

$$\begin{array}{c|cc|c} & e & e' & m \\ \hline e & D & 0 & C \\ e' & 0 & D' & C' \\ \hline m & B & B' & A \overset{\wedge}{\underset{\vee}{+}} A' \end{array}\;; \quad \text{and} \quad \left[\begin{array}{c|cc} & e & m \\ \hline e & D & C \\ m & B & A \end{array}\right]^* \;=\; \begin{array}{c|c|c} & e & m \\ \hline e & D \overset{\wedge}{\underset{\vee}{+}} CA^*B & CA^* \\ \hline m & A^*B & A^* \end{array}\;.$$

<u>B. IO-OI-equivalent flowchart schemes.</u> Two nondeterministic flowchart schemes will be considered equivalent if the computation processes they denote have the same successful paths. In order to define this equivalence relation, beside the basic rules $\overset{s}{\rightarrow}$, $\overset{i}{\leftarrow}$ in the deterministic case we shall use the following dual rules

(id^o) $F \overset{s^{-1}}{\longleftarrow} F'$ if F' can be obtained from F by identifying vertices which have the same label and whose input connections are equal after identification (conversely, F can be obtained from F' by multiplying vertices and taking their input connections accordingly to the input connexions in F'); and

(ac^o) $F \overset{i^{-1}}{\longrightarrow} F'$ if F' can be obtained from F by deleting non-coaccessible vertices.

Example.

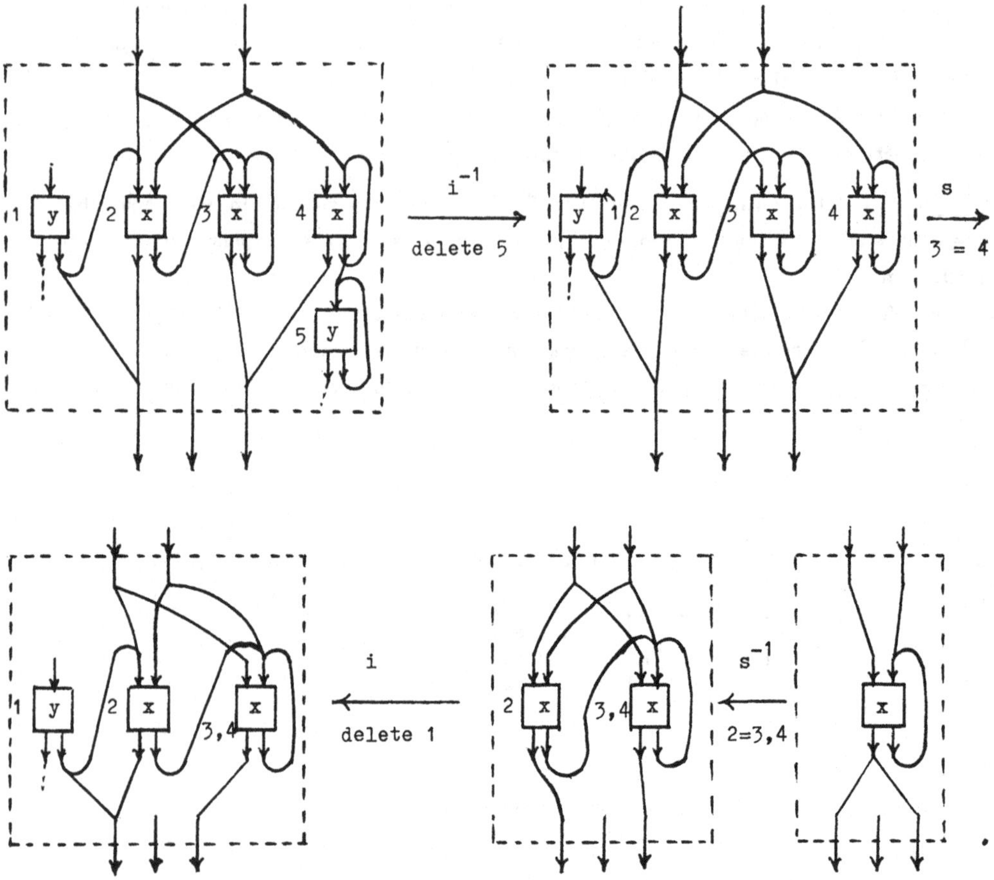

Mathematically, we say that two flownomials $F = ((x_1+\dots+x_k + 1_m)\, l)\uparrow_{e_{in}} : m \to n$ and $F' = ((x'_1+\dots+x'_{k'}+ 1_m)\, l')\uparrow_{e'_{in}} : m \to n$ are <u>in n-simulation</u> (written $F \xrightarrow[n]{} F'$) if there exists a relation $\varrho \subseteq [k] \times [k']$ such that

$$(i\, ,\, j) \in \varrho \qquad \text{implies} \qquad x_i = x'_j\, ,$$

and its "block" extensions to inputs ϱ_{in} and to outputs ϱ_{out} fulfil

$$l\,(\varrho_{in}+1_n) = (\varrho_{out}+1_m)\, l'.$$

The above relation $\xrightarrow{s^{-1}}$, $\xrightarrow{i^{-1}}$ correspond to ϱ^{-1} = surjective function, respectively to ϱ^{-1}= injective function. We define our <u>IO-OI-equivalence</u> $=_n$ as the congruence relation generated by $\xrightarrow[n]{}$ (when in $T(1,1)$ the equality $x\,(x \updownarrow y) = x$ holds, an equivalent definition is: the congruence generated by $\xrightarrow{s} \cup \xleftarrow{i} \cup \xleftarrow{s^{-1}} \cup \xrightarrow{i^{-1}}$).

A simple characterization theorem $=_n$ (like in the deterministic case) generally does not work. However, in the case of usual flowchart schemes ($T = \mathbf{Rel}_{\bullet}$) the n-simulation has the confluent property: $\leftarrow\cdot\rightarrow \subseteq \rightarrow\cdot\leftarrow$; hence

<u>PROPOSITION.</u> In $\mathbf{Fl}_{X,\mathbf{Rel}_{\bullet}}$ we have

$$=_n = \xrightarrow[n]{}\cdot\xleftarrow[n]{} = \xrightarrow{i^{-1}}\cdot\xrightarrow{s^{-1}}\cdot\xrightarrow{s}\cdot\xrightarrow{i}\cdot\xleftarrow{i}\cdot\xleftarrow{s}\cdot\xleftarrow{s^{-1}}\cdot\xleftarrow{i^{-1}} .$$

This shows that two flownomials over $\mathbf{Rel}_{\bullet}$ are IO-OI-equivalent iff in four steps (first, deleting non-coaccessible vertices; second, multiplying vertices and taking acordingly their input connexions; third, identifying similar vertices whose output connections are equal after identification; and fourth adding non-accessible vertices) they can be transformed into the same flownomial.

<u>C. Reticula.</u> This subsection considers the technical part of the calculus. We say that a prereticulum is a <u>reticulum</u> if the repetition fulfils the following four axioms:

(R1) $f f^* \updownarrow 1_n = f^*$, for $f : n \to n$;

(R2) $(f \updownarrow g)^* = f^* (g f^*)^*$, for $f, g : n \to n$;

(R3) $f (g f)^* = (f g)^* f$, for $f : m \to n$, $g : n \to m$; and

(R4) if $f \varrho = \varrho g$ then $f^* \varrho = \varrho g^*$, for $f : m \to m$, $g : n \to n$

and a relation $\varrho \subseteq [m] \times [n]$.

<u>Comments.</u> (i) The axioms **(R1)**, **(R2)**, **(R3)** and **(R4)** for functions ϱ rewrite respectively the axioms **(I1)**, **(I2)**, **(I3)** and **(I4)** in terms of $+$ and * ; consequently, a reticulum is a strong iteration theory; similarly, the dual category of a reticulum is also a strong iteration theory.

(ii) All these axioms can be proved in the axiomatic systems for the algebra of regular events in [12].

<u>Basic example.</u> $\mathbf{Rel}_D$ with the above operations is a reticulum.

<u>MAIN THEOREM (nondeterministic case).</u> (i) If T is a reticulum and the n-simulation in $\mathbf{Fl}_{X,T}$ has the confluent property, then the classes of IO-OI-equivalent X-flownomials over T form a reticulum.

(ii) If the classes of IO-OI-equivalent X-flownomials over T form a reticulum, then this is even the reticulum freely generated by adding X to T.

<u>COROLLARY.</u> The classes of IO-OI-equivalent X-flownomials over $\mathbf{Rel}_{\bullet}$ form the reticulum freely generated by X.

Consequently, in the nondeterministic case the main theorem above gives only a partial result.

Open problem. Does the classes of IO-OI-equivalent X-flownomials over an arbitrary reticulum T form a reticulum?

4. FINAL REMARKS

(a) While at the theoretical level we have a calculus for the classes of IO-equivalent (respectively, IO-OI-equivalent) flowchart schemes essentially similar to the calculus of polynomials, at the practical level the situation is much worse: we have not a good and powerful representation of finite partial functions in **Pfn**$_{\cdot}$ (respectively, relations in **Rel**$_{\cdot}$) similar to the Arabian representation of natural numbers.

(b) When the algebraic structure for support theories in the deterministic case will be unanimous accepted, I propose to be called Elgot theory; moreover I propose strong iteration theories as a candidate to this notion.

Acknowledgements: I would like to thank V. E. Căzănescu for many useful discussions at several stages of this research.

REFERENCES

1. M. A. ARBIB and E. G. MANES, Partially additive categories and flow-diagram semantics, J. Algebra **62**(1980), 203-227.

2. J. W. BACKUS, Can programming be liberated from the von Neumann style? A functional style and its algebra of program, Comm. ACM **21**(1978), 613-641; see also: From function level semantics to program transformation and optimization, in "Mathematical Foundations of Software Development," Lecture Notes in Computer Science 185, pp.60-91, Springer-Verlag, Berlin - Heidelberg - New York - Tokio, 1985.

3. S. L. BLOOM, C. C. ELGOT and R. TINDELL, On the algebraic structure of rooted trees, J. Comput. System Sci. **16**(1978), 228-242.

4. S. L. BLOOM, C. C. ELGOT and J. B. WRIGHT, Solutions of the iteration equations and extensions of the scalar iteration operation, SIAM J. Comput. **9**(1980), 25-45; Vector iteration in pointed iterative theories, SIAM J. Comput. **9**(1980), 525-540.

5. V. E. CĂZĂNESCU, On context-free trees, to appear in Theor. Comput. Sci. (1985).

6. V. E. CĂZĂNESCU and C. UNGUREANU, Again on advice on structuring compilers and proving them correct, INCREST Preprint Series in Mathematics, No.75/1982.

7. C. C. ELGOT, Monadic computation and iterative algebraic theories, in "Logic Colloquium" (H. E. Rose and J. C. Shepherdson, Eds.), pp.175-230, North-Holland, Amsterdam, 1975.

8. C. C. ELGOT, Structured programming with and without GO TO statements, IEEE Trans. Soft. Eng. **SE-2**(1976), 41-53.

9. C. C. ELGOT, Matricial theories, J. Algebra **42**(1976), 391-421.

10. Z.ESIK, Identities in iterative and rational theories, Comput. Linguistic and Comput. Language **XIV** (1980), 183-207.

11. S. GINALI, Regular trees and free iterative theory, J. Comput. System Sci. **18**(1979), 228-242.

12 A. SALOMAA, Two complete axiom systems for the algebra of regular events, J. Assoc. Comput. Mach. **13**(1966), 158-169.

13. Gh. ŞTEFĂNESCU, On flowchart theories I, INCREST Preprint Series in Mathematics, No.39/1984; A completion of "On flowchart theories I," idem No.7/1985; and On flowchart theories II (nondeterministic case), idem No.32/1985.

14. J. B. WRIGHT, J. W. THATCHER, E. G. WAGNER and J.A.GOGUEN, Rational algebraic theories and fixed-point solution, in "Proceedings 17th IEEE Symposium on Foundations of Computer Science, Houston, Texas (1976)," pp.147-158.

AN ALGEBRAIC FORMALISM FOR GRAPHS.

Michel BAUDERON and Bruno COURCELLE
Département de Mathématiques et d'Informatique
Formation associée au CNRS
Université de Bordeaux I, F-33405 Talence, France.

ABSTRACT : We provide an algebraic structure for the set of finite graphs, whence a notion of graph expression for defining them and a complete set of equational rules for manipulating expressions.

By working at the level of expressions, one derives from this algebraic formalism a notion of graph rewriting which is as powerful and conceptually simpler than the usual categorical approach of Ehrig and alii.

1. INTRODUCTION.

The theories of word and tree grammars basically rely upon two concepts :

(1) an element may be built from other elements through an algebraic process (concatenation of words or operation in the term algebra)

(2) an element may be substituted for a part of an other one, giving rise to the notion of production in a grammar.

This allows to give a systematic treatment of algebraic grammars in a quite general framework (see [Co85]). However it can not be applied to graphs and graph grammars in the form in which they have been studied till now.

The purpose of this work is to lay the foundations of an algebraic theory of graph grammars that would fit in the aforesaid framework. It will successively involve giving a convenient definition for graphs (in fact hypergraphs, see below and [BC85]), defining and studying algebraic operations on graphs and graph expressions as well as defining the substitution for graphs. Due to space limitation, all proofs will be omitted and we shall concentrate ourselves on the exposition of the foundations of this theory. A complete exposition may be found in [BC85].

Let us justify informally our main definitions and results. Although trees are graphs of a special type, proofs and definitions are usually written in different ways. For (finite) trees, structural induction is the natural way, but it does not work for graphs. For (finite) graphs, proofs and definitions are usually given in a global way.

This difference is at the core of the distinction between "structured programs" and flowcharts. The semantics of the former may be defined in a structured way, i.e. by induction on programs considered as expressions, whereas the semantics of the latter is only definable in a global way, from the program considered as a graph.

Structural induction in general is applicable to every set of objects M, equipped with an algebraic structure and generated from some of these objects, considered as elementary (they correspond to the base case of the induction), since expressions (or terms) can be defined for denoting the elements of M. A tree can be considered as an expression denoting itself.

In this paper, we define an algebraic structure on the set of finite graphs (actually on the set of finite oriented labelled hypergraphs with a sequence of distinguished vertices). The (hyper)-edges will play the role of elementary objects. This means that we shall consider a graph as a **set of (hyper)edges** "glued" together by means of vertices and not as a **set of vertices** linked together by means of edges.

Two expressions can define the same graph : we say then that they are equivalent. We exhibit a set R of equational laws which characterizes the equivalence of expressions (i.e. two expressions are equivalent if and only if they are equivalent with respect to the congruence relation generated by R).

This generalizes the situation of words where the three laws of monoids (associativity, left- and right-unity) play the role of the above mentioned set R. And exactly as words on an alphabet A form the free monoid generated by A, graphs form the free R-algebra generated by the set of (hyper)edge labels.

Applications to graph rewriting systems follow immediately. Let M be an algebra ; any term rewriting system (over the signature of M) defines a binary relation on M that we call a rewriting relation on M. This applies to graphs, since we have expressions for denoting them.

Graph rewritings have been defined and investigated using a somewhat involved categorical framework in various papers by Ehrig, Kreowski and alii (see for instance [Eh79], [EK79], [EK81]). Significant simplifications have been provided by Raoult [Ra84]. We show that the same rewritings may be obtained by means of our graph expressions in a much simpler way - at least conceptually, graph rewritings being intrinsically involved.

Concretely, graph rewritings coincide with term rewriting systems modulo the set R of equational relations. It is quite fortunate that the set R is both bilinear and balanced (in every equation of R, a variable either occurs once and only once in each hand side or it does not occur at all). See for instance Courcelle [Co85, Chap1,11 and 16].

Let us now explain the use hypergraphs in place of ordinary oriented labelled graphs.

In the case of words, it is an important fact that a word can be substituted for an occurrence of a letter (think of context-free grammars). We want to have a similar property for graphs, with labelled edges playing the role of occurrences of letters. Hence, we need graphs having the same "type" as the elementary objects they are built with.

Hence, we want to be able to subtitute for an ordinary edge, a graph with two distinguished vertices. The orientation of edges is useful to avoid ambiguity : in absence of orientation, a graph could be substituted for an edge in two different ways (clearly, two distinguished vertices must be distinguished from each other ; there is somehow an "entry" and an "exit" vertex).

Using a pair of vertices as an "interface" (cf [Eh79]) between a graph and a context where to put it is insufficient. Hence, we need graphs equipped with a **sequence** of distinguished vertices (and not only a pair of vertices). Accordingly, we use hypergraphs in order to insure the homogeneity of the theory.

The sequence of distinguished vertices of some graph G (we shall denote it by **src**) will be used as an **interface** between G and the outside world. Let us call **interior** the vertices of G which do not belong to **src**. No operation concerning G will modify the interior vertices. Only vertices from **src** can be affected. For example, two vertices of **src** can be made identical or linked by a new (binary) edge. But none of these operations can concern an interior node. This means that in a possible implementation, one could use a fixed sequence of registers for **src** (for pointers) and that the graph operations could be performed in constant time by means of manipulation of these pointers.

This situation generalizes the case of words : a word can be augmented by concatenation to the left or to the right, but not by insertion in the middle (if as usual one restricts to concatenation). In this case, **src** consists of the left and the right ends of the word. One could introduce operations like "insert u at the ith position", but due to the presence of i, this introduces infinitely many operations, and this is a difficulty for an algebraic treatment.

In the case of graphs, it would be even more difficult to **designate** an interior vertex at which for instance a new edge should be attached. We eliminate this difficulty by restricting graph operations to concern only source vertices. Of course, the length of the **src** sequence is not bounded but this multiplicity is handled by means of **sorts** (one for each length). And in any concrete "finite" situation (say when considering a graph rewriting system), only finitely many sorts and finitely many operators do appear.

2. THE CATEGORY OF GRAPHS WITH SOURCES.

We shall use a fixed given ranked alphabet A and we shall denote by $\tau : A \longrightarrow \mathbf{N} - \{0\}$ the rank function.

2.1. Concrete graphs : A **concrete graph** (over A) is a quadruple : G = <**VG**, **EG**, **labG**, **vertG**> where :

- **VG** is a (finite) non empty set whose elements are the vertices of the graph,
- **EG** is the (finite) set of **edges**
- **labG** : EG $\longrightarrow$ A assigns to each edge of G a label in the alphabet A,
- **vertG** : EG $\longrightarrow$ VG* associates with each edge of G, the word on VG consisting of the ordered sequence of vertices of the edge, i.e., **vertG**(e) = **vertG**(e,1)....**vertG**(e,l) with :

$$l = |\mathbf{vertG}(e)| = \tau(\mathbf{labG}(e))$$

2.2. Examples :

- For any integer $n \in \mathbb{N}$, the set $[n] = \{1,\ldots,n\}$ can be seen as the discrete graph having $VG = [n]$ and $EG = \emptyset$

- For any $a \in A$, of rank n, there is a concrete graph $\hat{a}$ defined by : $V\hat{a} = [n]$, $E\hat{a} = \{.\}$, $lab\hat{a}(.) = a$ and $vert\hat{a}(.) = 1\ldots n$.

- If we restrict the elements of the alphabet A to be of rank two, our concrete graphs are exactly the standard oriented edge-labelled graphs.

- Another example : Let $VG = \{v_1,\ldots,v_7\}$ and EG consist of ten edges, we let $\tau(a) = \tau(a') = 1$, $\tau(b) = \tau(b') = 2$, $\tau(c) = 3$.

In the drawing of this graph, we have used the following conventions :

- binary edges are represented as usual,
- unary edges are labels attached to vertices (a same vertex may be tagged with several occurrences of a same label),
- edges of rank greater than 2 are represented as binary edges with intermediate nodes. On the drawing below, there is one edge e such that : $\mathbf{vertG}(e) = v_4v_5v_6$ and another e' with $\mathbf{vertG}(e') = v_6v_7v_7$.

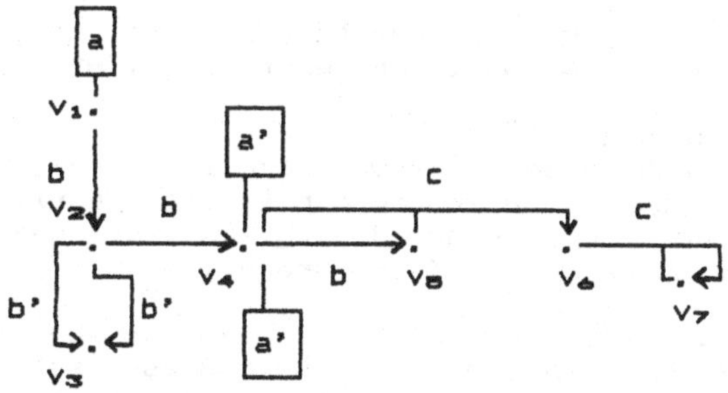

2.3. Graph morphisms : Let G and G' be two concrete graphs. A **graph morphism** (or simply, a morphism) $h : G \longrightarrow G'$ is a pair $h = (hV,hE)$ of maps $hV : VG \longrightarrow VG'$ and $hE : EG \longrightarrow EG'$ such that the following diagrams are commutative :

(hV^* denotes the canonical extension of hV to the monoid of words on VG and VG').

The composition of two graph morphisms is obviously a graph morphism itself and this makes the set of concrete graphs into a category. The identity morphism on the concrete graph G is $Id_G = (Id_{VG}, Id_{EG})$. Clearly, a morphism is an isomorphism whenever hV and hE are bijections.

2.4. Graphs : In many cases, the specific sets **EG** and **VG** chosen for defining a graph **G** are irrelevant. In other words, if there exists an isomorphism $h : G \longrightarrow G'$, we can consider G and G' as the "same graph". Actually, we define an (abstract) **graph** as the collection of all concrete graphs isomorphic to some given graph G. And we shall say in a loose way :

" let G be the graph <VG,EG,**labG**,**vertG**> ", which means the isomorphism class of G = <VG,EG,**labG**,**vertG**>.

Similarly, a morphism of (abstract) graphs will be a class of morphisms of concrete graphs. It will be defined up to isomorphisms on its source and target graphs. We shall denote by the same letter $\mathbb{G}_o$ the category of graphs and the set of graphs.

The most important consequence of this two-level definition of graphs is that most proofs will have to be done in two steps : first we shall make the proof in the category of concrete graphs, then we shall prove that it holds for abstract graphs (see [BC85]).

2.5. Concrete graphs with sources : A **concrete graph with n sources** (over A) or a concrete **n-graph** (over A) is a pair H = (G,**src**) where G is a concrete graph and **src** is a graph morphism from [n] to G.

Actually, **src** is nothing but a mapping from the set [n] to the set VG of vertices of G. The elements of **src**([n]) will be refered to as the **sources** of G. Note that a concrete 0-graph is nothing but a concrete graph.

If H = <G,**src**> we say that G is the underlying concrete 0-graph of H. We shall sometimes denote it by H^o.

2.6. Examples :

- For each integer $n \geq 0$, there is a n-graph whose underlying graph is [n] and whose source map is the identity map on [n]. This graph will be simply denoted by $\underline{n}$. We shall make intensive use of these graphs and overall of the graph $\underline{1}$.

- In the same way, with each basic graph $\mathbf{a}$ (corresponding to the label a of rank n in A), we can associate a n-graph $\underline{a}$ whose underlying graph is $\mathbf{a}$ and whose source map is the identity of [n] :

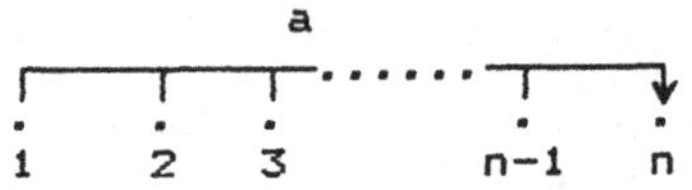

2.7. Morphisms of concrete n-graphs : Let H = (G,**src**) and H'= (G',**src'**) be two concrete graphs with n and n' sources respectively. A morphism from H to H' is a pair (h,σ), where h is a graph morphism from G to G' and σ a map from [n] to [n'] such that the following diagram is commutative :

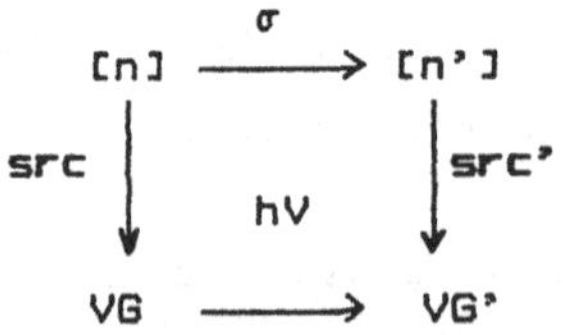

When n = n' and when the map σ is the identity on [n] , we shall say that the morphism is **source preserving** (and we shall call it an s.p-morphism).

An **n-graph** is the collection of all concrete graphs which are isomorphic to some concrete n-graph by some source preserving isomorphism. The category of n-graphs will be denoted by $\mathbb{G}_n$, and the category of graphs of all types will be denoted by $\mathbb{G}$. Clearly, one must keep in mind the remark of section 2.4., which remains valid in this context.

or $v = \mathbf{src}(i)$ and $v' = \mathbf{src}(j)$ for $(i,j) \in \delta$
with canonical mapping $f : VG \longrightarrow VG'$

- $EG' = EG$
- $\mathbf{vert}G' = f \circ \mathbf{vert}G$
- $\mathbf{lab}G' = \mathbf{lab}G$
- $\mathbf{src}' = f \circ \mathbf{src}$

If δ is the relation generated by a single pair (i,j) then we denote θ_δ by $\theta_{i,j}$.

If Δ is the trivial equivalence $\{(i,i) / i \in [n]\}$) then $\theta\Delta$ is the identity.

It is clear that if δ is the equivalence relation generated by a set of pairs $K = \{(i_1,j_1),\ldots,(i_k,j_k)\}$ then :

$$\theta_\delta = \theta_{i_1,j_1} \circ \ldots \circ \theta_{i_k,j_k}$$

If δ and δ' are two equivalence relations on $[n]$, then we denote by $\delta \vee \delta'$ the smallest equivalence relation which contains both δ and δ'. From the above remark, $\theta_{\delta\vee\delta'} = \theta_\delta \circ \theta_{\delta'}$.

As before this definition applies to graphs as well.

3.4. Algebraic laws. Proposition : For all graphs H, H' and H" of respective types n, n' and n" for all α, β, δ, δ', etc... of appropriate types, the following equalities hold true :

(R1) the law $\oplus$ is associative :
$$H \oplus (H' \oplus H'') = (H \oplus H') \oplus H''$$

(R2) $\sigma_\beta(\sigma_\alpha(H)) = \sigma_{\alpha\circ\beta}(H)$

(R3) $\sigma_{Id}(H) = H$

(R4) $\theta_\delta(\theta_{\delta'}(H)) = \theta_{\delta\vee\delta'}(H)$

(R5) $\theta\Delta(H) = H$

(R6) $\sigma_\alpha(H) \oplus \sigma_{\alpha'}(H') = \sigma_\beta(H' \oplus H)$
with $\alpha : [p] \longrightarrow [n]$, $\alpha' : [p'] \longrightarrow [n']$
and $\beta : [p+p'] \longrightarrow [n'+n]$ such that :
$\beta(i) = n'+\alpha(i)$ for $1 \leq i \leq p$
$\beta(i+p) = \alpha(i)$ for $1 \leq i \leq p'$

(R7) $\theta_\delta(H) \oplus \theta_{\delta'}(H') = \theta_{\delta+\delta'}(H \oplus H')$
where $\delta+\delta'$ is the equivalence on $[n+n']$ generated by $\delta \cup \{(n+i,n+j)/(i,j) \in \delta'\}$ (where δ and δ' are equivalence relations on $[n]$ and $[n']$ respectively).

(R8) $\theta_\delta(H \oplus \underline{1}) = \sigma_\alpha(\theta_{\delta'}(H))$
if δ is such that $(i,n+1) \in \delta$ for some $i \in [n]$, δ' is the restriction of δ to $[n]$ and $\alpha : [n+1] \longrightarrow [n]$ maps j to j for $j \leq n$ and n+1 to i.

(R9) $\theta_\delta(\sigma_\alpha(H)) = \sigma_\alpha(\theta_{\alpha(\delta)}(H))$
where $\alpha(\delta)$ denotes the equivalence relation on $[n]$ generated by $\{(\alpha(i),\alpha(j)) / (i,j) \in \delta\}$

(R10) $\sigma_\alpha(\theta_\delta(H)) = \sigma_\beta(\theta_\delta(H))$
whenever α and β have the same domain $[p]$ and $(\alpha(i),\beta(i)) \in \delta$ for all i in $[p]$

It is important to note that H, H' and H" are graphs but not concrete graphs : Rules R1, R6, R7, R8 do not hold for concrete graphs ; they only hold up to s.p.-isomorphisms. This is due to the necessity of taking a disjoint copy in the sum.

We shall use (R1) to (R10) as rules for transforming expressions.

3.5. Algebraic expressions : Let **N** be considered as a set of sorts. For every n,m in **N**, let $\oplus_{n,m}$ be a binary function symbol of

laws (R1) to (R10) defines a finite set of equations. Furthermore, these equations are linear and balanced i.e. the same variables occur in both hand sides and each variable occurs at most once.

3.7. Example : Let a,b be of rank 1, let c,d be of rank 2 and e be of rank 3. The graph H (with five vertices and six edges) :

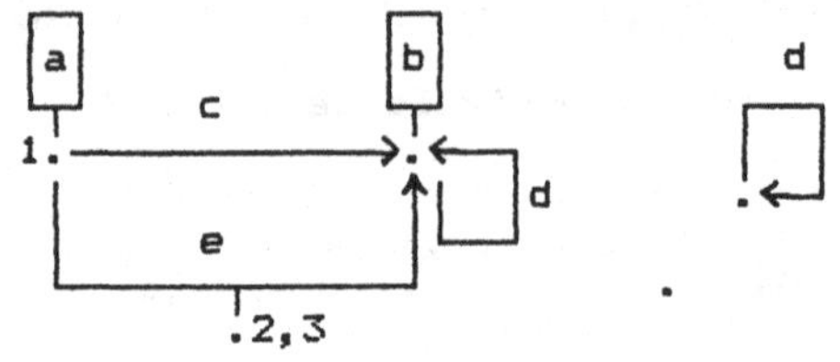

can be represented by the expression :

$$\bar{g} = \sigma_{\alpha}(\theta_{\delta}(\underline{a} \oplus \underline{c} \oplus \underline{b} \oplus \underline{d} \oplus \underline{e} \oplus \underline{d}) \oplus \underline{1})$$

where δ is the equivalence generated by
$\{(1,2),(3,4),(4,5),(5,6),(7,1),(9,4),(10,11)\}$
and $\alpha(1) = 1,\ \alpha(2) = \alpha(3) = 8.$

This expression is formed by source vertex fusion and source redefinition applied to :

$$\bar{g} = \underline{a} \oplus \underline{c} \oplus \underline{b} \oplus \underline{d} \oplus \underline{e} \oplus \underline{d} \oplus \underline{1}$$

which represents the graph H' :

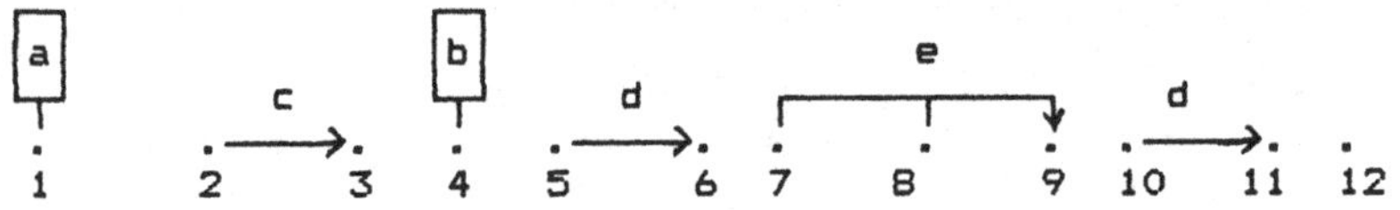

From this example, it is clear that an expression can be built to define any finite graph. Another expression for H is

$$\bar{g} = \sigma_{\beta}(\theta_{\delta''}(\underline{e} \oplus \sigma_{\tau}(\theta_{\delta'}(\underline{a} \oplus \underline{c} \oplus \underline{b} \oplus \underline{d}))))$$
$$+\ \sigma_{\emptyset}(\theta_{1,2}(\underline{d})) \oplus \sigma_{\emptyset}(1)$$

where δ' is generated by $\{(1,2),(3,4),(4,5),(5,6)\}$
and $\tau(1) = 1,\ \tau(2) = 2$
δ'' is generated by $\{(1,4),(3,5)\}$
and $\beta(1) = 1,\ \beta(2) = \beta(3) = 2.$

It is clear on this example that the expression $\bar{g}$ in the proposition of paragraph 3.5. is not unique. A most important tool for the proof of the main theorem is the existence for any graph of an expression which is said to be in "canonical form" (see [BC85]). In this example the first expression given is in canonical form.

4. SUBSTITUTIONS AND CONTEXT-FREE GRAPH GRAMMARS.

4.1. Substituting a graph for an edge in a graph : Let H = (G,src) be a concrete graph. Let e be an edge labelled by a of rank p. Let H' = (G',src') be a concrete p-graph. Our purpose is to define the concrete graph H[H'/e] i.e. the result of the substitu-

type $(n,m \longrightarrow n+m)$.

For every integer $n \in \mathbf{N}$ and every equivalence relation δ on $[n]$, we let $\theta_{\delta,n}$ be a unary function symbol of type $(n \longrightarrow n)$.

For every $p,n \in \mathbf{N}$ and every mapping $\alpha : [p] \longrightarrow [n]$ we let $\sigma_{\alpha,p,n}$ be a unary function symbol of type $(n \longrightarrow p)$.

If a belongs to A and is of rank n, then $\underline{a}$ denotes an n-graph and $\underline{1}$ denotes a 1-graph (see 2.6).

Hence one can form a many sorted set of (well-typed) expressions by using all these symbols. Each expression of type n denotes an n-graph in an obvious way ($\oplus_{n,m}$ denotes $\oplus$). Actually, when writing expressions, we shall omit the subscripts n, m, p. The sort of an expression can be computed bottom-up since the sorts of the basic graphs ($\underline{a}$ and $\underline{1}$) are fixed.

Let $\mathbb{E}_n$ (resp. $\mathbb{E}$) be the set of well formed expressions of type n (resp. of any type). Then every expression of type n denotes two things :

- a concrete graph **cval**(ξ)
- a graph **val**(ξ) which is the s.p.-isomorphism class of **cval**(ξ).

The mappings **cval** and **val** can be defined in a standard way by induction on the structure of ξ. In the case of **cval**, one needs a canonical way to take disjoint copies of concrete graphs for the definition of H $\oplus$ H'. More details on these mappings may be found in [BC85].

We shall say that two expressions ξ and ξ' of type n are **equivalent**, if they denote the same abstract graph i.e.,

$$\xi \equiv \xi' \quad \Longleftrightarrow \quad \mathbf{val}(\xi) = \mathbf{val}(\xi).$$

Proposition : For every finite n-graph H, there exists an expression ξ, such that H = **val**(ξ).

3.6. Manipulations of expressions : The main result of this section is the :

Theorem : Two expressions ξ and ξ' are equivalent if and only if

$$\xi \overset{*}{\underset{R}{\longleftrightarrow}} \xi' \quad \text{where} \quad \overset{*}{\underset{R}{\longleftrightarrow}} \text{ denotes the congruence}$$

on $\mathbb{E}$ associated with R, i.e. with the set of equational laws (R1) to (R10).

The complete proof of this theorem, which is rather technical may be found in [BC85].

Corollary : $\mathbb{G}$ is isomorphic to $\mathbb{E} / \overset{*}{\underset{R}{\longleftrightarrow}}$.

We can consider this statement as an extension of the classical fact that the monoid of words over X is isomorphic to the free monoid generated by X i.e. to the quotient of the free algebra by the congruence associated with the laws of monoids.

Remark : Each law (R1) to (R10) can be considered as an equation in the variables H, H' and H". Actually, these laws are equation schemes rather than single equations. But, whenever one limits the sorts n, m, p to be less than some fixed integer k, each of the

tion of H' for e in H.

Definition : We may assume that G' is disjoint from G and we set H[H'/e] = H" = (G",src") with :

G" = (VG", EG",**vertG"**,**labG"**) and
VG" = VG U VG' − {src'(1),...,src'(p)}
EG" = EG U EG' − {e}
vertG"(e_i) = **vertG**(e_i) or **vertG'**(e_i)
labG"(e_i) = **labG**(e_i) or **labG'**(e_i)
depending on whether e_i belongs to EG or EG'.

If an element a of A labels at most one edge of H we can write H[H'/a] for H[H'/e] where e is the unique edge labelled by a. If there is no such edge, then H[H'/a] = H.

Proposition : Let ξ be a graph expression, let w be an occurrence of $\underline{a}$ in ξ where a belongs to A. Let H = **cval**(ξ) and e be the edge of H corresponding to w. Let ξ' be an expression of type $\tau(a)$. Then **cval**(ξ)[**cval**(ξ')/e] = **cval**$(\xi[\xi'/w])$.

In this statement $\xi[\xi'/w]$ denotes the textual substitution of ξ' for $\underline{a}$ (at its occurrence w). This statement extends to abstract graphs (see [BC85]).

4.2. Graph expressions with parameters : Let now U be a new ranked alphabet over **N**−{0} whose elements u of rank $\tau(u) = m$ will be called parameters of type m, and let $\mathbb{E}(A,U)$ denote the set of well formed expressions formed over the ranked alphabet $A \cup U$, whose elements will obviously be called graph expressions with parameters. Each expression ξ in $\mathbb{E}(A,U)$ denotes a graph **val**(ξ) some edges of which are labelled by elements of U.

If ξ is an expression of type m in $\mathbb{E}(A,U)$ and u a parameter of type n occurring in ξ, it is clear that there exists an expression C of type n+m in $\mathbb{E}(A,U)$ such that $\xi = \sigma_\alpha(\theta_\delta(C \oplus \underline{u}))$ for some α and δ. The expression ξ will then be denoted by C[u].

4.3. Context-free graph grammars : Let us now define a **context free graph grammar** as a triple $\Gamma = \langle A,U,P \rangle$ where A and U are as above. A is the set of **terminal** symbols, U is the set of **non terminal** symbols, P is a finite set of pairs of the form (u,H) such that $u \in U$ and $H \in \mathbb{G}_{\tau(u)}(A \cup U)$.

For any two graphs H', H" in $\mathbb{G}(A \cup U)$ we write :

$$H' \xrightarrow[\Gamma]{} H"$$

if and only if H" = H[H'/e] for some pair (u,H) in P and some edge e of H' labelled by u.

We define the set of graphs generated by some H in $\mathbb{G}(A \cup U)$

as $$L(\Gamma,H) = \{ H' \in \mathbb{G}(A) \ / \ H \xrightarrow[\Gamma]{*} H' \} .$$

4.4. Algebraic graph grammars : An algebraic graph grammar is a polynomial system $S = \langle u_1 = p_1,..., u_n = p_n \rangle$ where $U = \{ u_1 ,..., u_n \}$ is a set of unknown and each p_i is of the form
$\xi_{i,1} +...+ \xi_{i,n_i}$.

We assume that for all $j = i,\ldots,ni$ the rank of u_i and the type of $\mathfrak{Z}_{i,j}$ are equal. Such a system has a least solution in $P(\mathbb{G})$ which is an n-tuple of sets of graphs $\langle A_1,\ldots,A_n\rangle$ (see [Co85] and [BC85]).

With every algebraic graph grammar S as above, we can associate a context-free graph grammar $\Gamma = \langle A,U,S\rangle$ where

$$S = \{ (u_i, \mathbf{val}(\mathfrak{Z}_{i,j})) \ / \ 1\leq i\leq n, \ 1\leq j\leq ni \}$$

Note that for every context-free graph grammar $\Gamma = \langle A,U,P\rangle$, one can find an algebraic graph grammar S such that S = P since every graph can be denoted by some graph expression. Actually, many different algebraic graph grammars S can be found for some given Γ. With these notations, we have :

Proposition : The least solution of S in $P(\mathbb{G})$ is the n-tuple of sets of graphs generated by Γ : $(L(\Gamma,u_1),\ldots,L(\Gamma,u_i))$.

Corollary : The algebraic sets of graphs and the context free sets of graphs coincide.

A similar result has been established by Habel and Kreowski [HK83] (for ordinary oriented graphs).

5. GRAPH REWRITINGS.

Our algebraic formalization of graphs yields a natural notion of graph rewriting, namely those coming from a ground rewriting system on graph expressions.

5.1. Productions and derivations : **A production** (of type n), is a pair $p = (G,G')$ of n-graphs. It can be represented as well by a pair $(\mathfrak{Z},\mathfrak{Z}')$ of graph expressions of the same type.

Let $p = (G,G')$ be such a production. We say that the m-graph **H' derives** from the m-graph H through p, and we write :

$$H \longrightarrow_p H'$$

if and only if there exists an expression $C[u]$ (of type m) with one occurrence of a parameter u of type n such that :

$$H = \mathbf{val}(C[\mathfrak{Z}]) \text{ and } H' = \mathbf{val}(C[\mathfrak{Z}']).$$

5.2. Double push-out : Let us keep the notations of the previous paragraph. The following lemma shows that the rewriting of an m-graph H into an m-graph H' through a production p of type n may be seen as the construction of a double pushout in the category of n-graphs.

It is important to notice that, although H and H' are m-graphs, the lemma asserts the existence of a push-out in the category of n-graphs. Indeed, each graph is turned into an n-graph through the appropriate source mapping so that all graphs are in the same category (see [BC85]).

Lemma : The rewriting of H into H' through the production p of type n implies the existence of the following double push-out in the category of n-graphs :

$$\begin{array}{ccccc} G & \xleftarrow{\text{src}} & \underline{n} & \xrightarrow{\text{src'}} & G' \\ {\scriptstyle h}\downarrow & & \downarrow{\scriptstyle k} & & \downarrow{\scriptstyle h'} \\ H & \xleftarrow[d]{} & K & \xrightarrow[d']{} & H' \end{array}$$

Definition : Let C be a category of graphs. A C-production is a 5-tuple (G,I,G',i,i') where i : I ⟶ G and i' : I ⟶ G' are two arrows in C. Given two graphs H and H' in the category C, one says that H $\longrightarrow_p$ H' (H rewrites into H' by applying p) iff there exists a commutative diagram :

$$\begin{array}{ccccc} G & \xleftarrow{i} & I & \xrightarrow{i'} & G' \\ \downarrow & & \downarrow & & \downarrow \\ H & \longleftarrow & J & \longrightarrow & H' \end{array}$$

where the two square are push-out in the category C (a "double push-out").

This definition is exactly that of Ehrig [Eh79]. We prove in [BC85] the equivalence of the two definitions (a converse of the previous lemma) in two specific situations :

(1) when C is the category of simple graphs i.e. of structures in the sense of [EK81]

(2) when C is the category of graphs with edge injective morphisms.

6. REFERENCES.

[BC85] BAUDERON M., COURCELLE B., Graph expressions and graph rewritings, (Report 8525, University Bordeaux I, Dept Maths Info 1985)

[Co85] COURCELLE B., Equivalences and transformations of regular systems. Applications to recursive program schemes and grammars, To appear in Theor. Comp. Sci. (Report 8521, University Bordeaux I, Dept Maths Info 1985)

[Eh79] EHRIG H., Introduction to the algebraic theory of graphs, Lect. Notes in Comp. Sci. 73 Springer 1979 1-69

[EK79] EHRIG H., KREOWSKI H-J., Push-out properties : An analysis of gluing constructions for graphs, Math. Nachr. 91 (1979) 135 -149

[EK81] EHRIG H., KREOWSKI H-J., MAGGIOLO-SCHETTINI A., ROSEN B R., WINKOWSKI J., Transformations of structures : An algebraic approach , Math. Systems Theory 14 (1981) 305-334

[HK83] HABEL A., KREOWSKI H-J., On context free graphs languages generated by edge replacement, Lect. Notes in Comp. Sci. 153 Springer 1983, 143-158

[Ra84] RAOULT J.C., On graphs rewritings, Theor. Comp. Sci. 32 (1984) 114-128

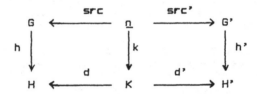

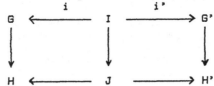

Membership for Growing Context Sensitive Grammars is Polynomial†

Elias Dahlhaus
Technische Universität Berlin
Fachbereich Mathematik
Strasse des 17. Juni 135
1000 Berlin - West 12
Germany

Manfred K. Warmuth
Department of Computer and Information Sciences
237 Applied Sciences
University of California
Santa Cruz, CA 95064
U.S.A.

1. Abstract: The membership problem for fixed context-sensitive languages is polynomial if the right hand side of every production is strictly longer than the left hand side.

2. *Introduction.*

Context-sensitive grammars (csgs) are one of the classical grammar families of formal language theory. They were introduced in [Ch59] and have been studied extensively since then (see [Bo73, Ha78] for an overview). Context-sensitive grammars are defined as rewriting systems, where the length of the right hand side of every production is at least as large as the length of the left hand side. This restriction on the productions is responsible for the fact that memberships for context-sensitive languages (csls) is equivalent to the problem of acceptance for nondeterministic linear bounded automaton [Ku64]. Therefore membership for csls is PSPACE complete [Ka72] and this is true even for certain fixed grammars. In this paper we show that if we restrict ourselves to "growing" productions, i.e. the right hand side of every production is strictly longer than the left hand side, then membership for fixed csls is polynomial.

This may appear surprising in view of the results obtained in [Bo78]. The growing csls are a subclass of $LINEAR_{CS}$ as defined in [Gl64, Bo71]. Languages of $LINEAR_{CS}$ are given by an arbitrary csg which has the property that every word w in the language has a derivation of length at most $c\,|w|$††, for some overall constant c which only depends on the grammar. In [Bo78] it was shown that there are NP-complete languages in $LINEAR_{CS}$. Thus our result that the family of growing csls is in P deserves an explanation.

Observe that in $LINEAR_{CS}$ "complex derivations" are allowed using non-growing productions; then the final word may be padded such that the length of the overall derivation is

† This research was done while the authors were visiting the Hebrew University of Jerusalem. The first author was supported by the Minerva Foundation and the second author by the United States-Israel Binational Foundation, grant no. 2439-82.

†† $|w|$ denotes the length of w.

linear in the length of the final word. In fact, for every language in P there is a polynomially padded version of this language which is in $LINEAR_{CS}$ [Bo78].

An arbitrary csg may be converted into a growing csg by adding a dummy symbol to the right hand side of every non-growing production. The grammar needs to be changed slightly so that the dummy symbols are "ignored." But now padding increases the length of the word exponentially. Each time a "signal" runs from one end of a sentential form to the other, the length increases by a constant factor.

Note that the question of emptiness for csls is undecidable [BPS61]. By padding a csg with dummy symbols a related growing csg is constructed. Clearly, emptiness for the corresponding growing csls is also undecidable. For the question of emptiness, the "exponential padding" is redundant.

The paper is outlined as follows. In Section 3 the basic notations are developed. Given a word w, we want to decide membership for a language defined by some fixed growing csg. A planar directed acyclic graph is associated with every derivation of w. In Section 4 we show that since all productions are growing there is a path of length $O(log(|w|)$ to some sink from each vertex in the graph. The sinks of the graph are labeled with the word w to be tested for membership. These short paths are then used in Section 5 in a polynomial cut-and-paste algorithm for deciding membership for a growing csl.

In the cut-and-paste algorithm each "piece" of a derivation graph is characterized by a tuple (called a frame) which contains all the essential information about the piece. Because of the short paths there is only a polynomial number of different frames which need to be considered. Frames were used extensively in [GS85, GW85a, GW85b] for studying polynomial cases of k-parallel rewriting.

In Section 6 the polynomiality of the membership problem for fixed, growing csls is contrasted with the fact that there are NP-complete languages defined by fixed, growing scattered grammars. In scattered grammars (scgs) [GH68, GW85c] the symbols to be rewritten in parallel are not required to be adjacent. In every production each symbol on the left hand side is rewritten into a string of length at least one. In growing scgs each symbol must be rewritten into a string of length strictly bigger than one. It is easy to see that in the derivation trees of growing scgs each node has an $O(log(|w|))$ path to a leaf where w is the word to be parsed. But since for scattered grammars the rewritten symbols don't need to be adjacent, we cannot cut and paste the derivation trees along the short paths.

The parallel complexity and the space complexity of the membership problem for fixed growing csls is discussed in the conclusion section. The main open problem is to determine the complexity of membership for ‘‘variable’’ growing csls, i.e. not only the word to be tested but also the growing csg is a variable of the input. The question is whether this problem can be solved in polynomial time or whether it is NP-complete.

3. Preliminaries

A *context-sensitive grammar* (csg) G is a quadruple (V,Σ,P,S) where:

i) V is a finite set of symbols, Σ is the subset of V which are the terminal symbols, and S is the startsymbol in $V-\Sigma$.

ii) P is a finite set of productions of the form $\alpha \rightarrow \beta$, s.t. $\alpha,\beta \in V^+$ and $|\alpha| \leq |\beta|$.

For two words u and v in V^*, u *derives* v, denoted $u => v$, if there exist $x,y,\alpha,\beta \in V^*$ s.t. $u = x\alpha y$, $v = x\beta y$ and $\alpha \rightarrow \beta \in P$. Let $\overset{*}{=>}$ denote the reflexive and transitive closure of =>. Using this notation we are ready to describe the *context-sensitive language* (csl) defined by the csg G: $L(G) = \{w \mid S \overset{*}{=>} w \text{ and } w \in \Sigma^*\}$.

Now the *membership problem* for a csl $L(G)$ is defined as follows:

Input: a word $w \in \Sigma^*$, where Σ is the terminal alphabet of G;

Question: is $w \in L(G)$?

Note that G is fixed, i.e. it is not a variable of the input. There are fixed csgs for which this problem is PSPACE complete [Ku64, Ka72].

We restrict ourselves to a subclass of csgs for which the membership problem is in P (Section 5). A csg G is *growing* if for all productions $\alpha \to \beta$ of the grammar, $|\alpha| < |\beta|$.

Following [Lo70] each *derivation* is associated with a planar directed acyclic graph called a *(derivation) graph*. The vertices in such a graph will be labeled with the corresponding symbols and productions used in the derivation. Let $\omega(x)$ denote the label of vertex x, where ω is a function from the set of vertices of the derivation graph to $V \cup P$. Vertices labeled with symbols (respectively productions) are called *symbol* (respectively *production) vertices*. We inductively define the *derivation graph* $D_k=(V_k,E_k)$ which is associated with the derivation $\alpha_1 \Rightarrow \alpha_2 \cdots \Rightarrow \alpha_k$:

Case $k=1$: Let $\alpha_1 = a_1 a_2 \cdots a_p$, where $a_i \in V$. Then $D_1=(V_1,E_1)$ has the vertices $V_1=\{x_1,x_2,\ldots,x_p\}$ s.t. $\omega(x_i)=a_i$ and no edges, i.e. E_1 is empty.

Case $k>1$: Assume $\alpha_1 \Rightarrow \alpha_2 \cdots \Rightarrow \alpha_{k-1}$ corresponds to the graph $D_{k-1}=(V_{k-1},E_{k-1})$ and $\alpha_{k-1} = u\alpha v \Rightarrow u\beta v = \alpha_k$. From D_{k-1} and the production $\alpha \to \beta$ the graph D_k is constructed for $\alpha_1 \Rightarrow \alpha_2 \Rightarrow \cdots \Rightarrow \alpha_k$. From the word $\beta = b_1 b_2 \cdots b_q$ create the vertices $V_\beta = \{y_i \mid 1 \le i \le q\}$ and from the production $\alpha \to \beta$ create an additional vertex y. Choose the vertices s.t. V_{k-1}, V_β and $\{y\}$ are distinct. The vertices of V_β are labeled with the symbols of β, i.e. $\omega(y_i)=b_i$, and y is labeled with the production $\alpha \to \beta$. Let V_α be the sinks (symbol vertices) of D_{k-1} corresponding to α. Now $V_k = V_{k-1} \cup V_\beta \cup \{y\}$ and $E_k = E_{k-1} \cup V_\alpha \times \{y\} \cup \{y\} \times V_\beta$.

An example is given in Figure 1. The planarity of the derivation graphs follows from the fact that only sinks are connected to the new production vertex. The sources of the graph D_k correspond to α_1 and the sinks to α_k. We say that D_k *derives* α_k. Since the graph is planar there is a natural left to right order amongst the sources: let $\alpha_1 = a_1 a_2 \cdots a_p$, then for $1 \le i < j \le p$ the vertex corresponding to a_i is to the *left* of the vertex of a_j and the vertex of a_j is to the *right* of a_i. Similarly, there is a natural left to right order amongst the sinks of a derivation graph, and amongst the predecessors and successor of every production vertex. Two sources (sinks) are called *adjacent* if they are adjacent in the left to right order of the sources (sinks).

In a derivation graph D a *path* π is defined to be a sequence $x_1,x_2,\ldots,x_{e+1}$ of vertices of D. The path π *starts* at x_1, *finishes* x_{e+1} and has *length* e. Notice that a path which contains one vertex has length zero. In this paper we assume that non-empty paths always end at a sink of D. The *distance* $d_x(y)$ of y from x is the length of the shortest paths which start at x and finish at y. Note that $d_x(x)=0$.

In the following lemma we will to show that there exists a shortest paths from each vertex to some sink s.t. no pair of paths is "crossing." To construct such a set of paths we use the following definition of consistency. Two paths are *consistent* if they have no common vertices, or if starting from the first common vertex the paths are identical. A set of paths is consistent if each pair is. Note that since derivation graphs are planar two consistent paths cannot "cross."

Lemma 1. For any derivation graph there exists a set of shortest paths from all vertices to sinks such that this set of paths is consistent.

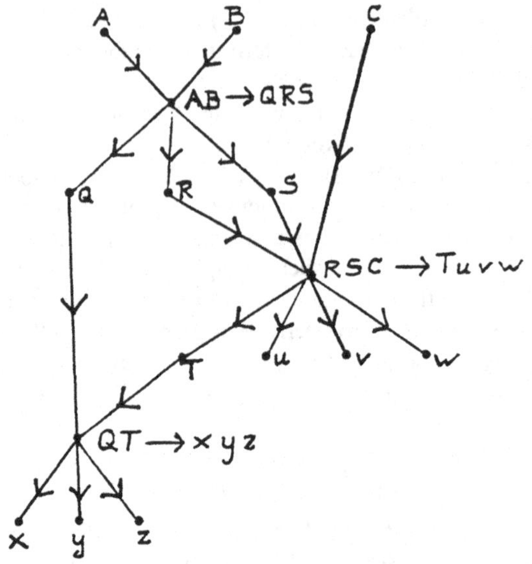

Figure 1. The derivation graph corresponding to the derivation
$\underline{AB}C \Rightarrow \underline{Q}R\underline{SC} \Rightarrow \underline{QT}uvw \Rightarrow xyzuvw$
(the rewritten symbols are underlined).

Proof: Let v_i, for $1 \le i \le m$ be the vertices of a derivation graph D and let π_i be a shortest path starting at v_i (and finishing at a sink of D). We now inductively construct paths π'_i (for $1 \le i \le m$), where π'_i starts at v_i, s.t. $\{\pi'_j \mid 1 \le j \le i\}$ is a consistent set of shortest paths.

Assume the set $\{\pi_j' \mid 1 \le j \le \overline{m} < m\} = \Pi'$ is consistent. If $\pi_{\overline{m}+1}$ is consistent with Π' we set $\pi'_{\overline{m}+1} = \pi_{\overline{m}+1}$ and there is nothing to show. Otherwise, let x be the first common vertex of $\pi_{\overline{m}+1}$ with some path π' of Π'. Since both π' and $\pi_{\overline{m}+1}$ are shortest paths, the suffixes of π' and $\pi_{\overline{m}+1}$ which start with x have the same length. Let $\pi_{\overline{m}+1}'$ be the path which agrees with $\pi_{\overline{m}+1}$ up until x and then follows π' to the sink. Clearly, $\pi_{\overline{m}+1}'$ has the same length as $\pi_{\overline{m}+1}$ and is consistent with Π'. This completes the description of the inductive construction. □

4. Short paths in derivation graphs

Consider derivation graphs for a growing csg which derive a word w. In this section we show that in such graphs there is a path of length $O(log(|w|))$ from each vertex to a sink. We prove this by assigning weights to the vertices, s.t. big weights will correspond to short paths.

Let us first discuss why there don't always exist short paths for derivation graphs of grammars which define languages in $LINEAR_{CS}$. In [Gl64] it was shown that $L = \{ucu^R cu : u \in \{a,b\}^*\}$ is not in $LINEAR_{CS}$.[1] Since growing csls are a subclass of $LINEAR_{CS}$

[1] The word u^R denotes the reverse of the word u.

[Bo73] the language L is not a growing csl. Intuitively, only $O(log(|w|))$ bits can be transmitted across paths of length $O(log(|w|))$. But in L, $O(|w|)$ bits need to be transmitted to synchronize the production of the words u, u^R and u in $w=ucu^Rcu$.

It is crucial that in the definition of L the word u is over a two symbol alphabet. Just producing three blocks of equal size as in the language $L' = \{a^n b^n c^n : n \geq 1\}$ is much easier. One can show that L' is a growing csl. In L' only $O(log(|w|))$ bits need to be transmitted. It is easy to see that $\hat{L} = \{a^{2^n} b^{2^n} c^{2^n} : n \geq 0\}$ is a growing csl. We let a special symbol scan the word. During each complete scan the number of symbols a, b, and c is doubled. From this it is easy to see that L' is also a growing csl. To produce the word $a^n b^n c^n$, $\lfloor logn \rfloor+1$ scans are used. Each scan corresponds to a bit in the bit representation of n. Again we double the number of symbols in each scan, but we also add an additional symbol if the corresponding bit is one.

We mentioned already in the introduction that for every language in P there is a padded version [Bo71] which is in $LINEAR_{CS}$. Thus even though $\{ucu^Rcu : u \in \{a,b\}^*\}$ is not in $LINEAR_{CS}$, the language $\{ucu^Rcud^{(|u|)^2} : u \in \{a,b\}^*\}$ is.

We proceed to prove the existence of short paths in derivation graphs of a growing csg. Let D be such a derivation graph deriving the word w. Each vertex x of D is associated with a subgraph of D. Let D_x be the subgraph induced by all vertices reachable from x.

The length of the paths will depend on the *growth ratios* of the productions in the grammar. The growth ratio of a production $\alpha \rightarrow \beta$ is the ratio $\frac{|\beta|}{|\alpha|}$. The minimum growth ratio of all productions of a grammar is the growth ratio of the grammar. Throughout the paper this minimum is denoted by g. Note that $g > 1$ for growing csgs.

The growth ratio of a production vertex is the ratio between the number of immediate successors over the number of immediate predecessors. Thus g is a lower bound on the growth ratios of the production vertices of D. Since each production vertex of D_x has at least as many immediate predecessors and the same number of immediate successors in D as in D_x, the growth ratios of the production vertices of D_x are also bounded by g.

We now assign weights to the vertices of D_x according to the following scheme:

i) $t_x(x) = 1$.

ii) For a production vertex p, the weight $t_x(p)$ is the sum of all the weights of the immediate predecessors of p.

iii) If p is a production vertex with k immediate successors, then each of these receives a weight of $\frac{t_x(p)}{k}$.

Note that $\sum_{s \text{ is sink of } D_x} t_x(s) = 1$. Since D_x has at most $|w|$ sinks, there is a sink in D_x of weight at least $\frac{1}{|w|}$.

The following lemma shows that big weights correspond to short paths.

Lemma 2: Let y be a symbol vertex of D_x and let d be a non-negative integer. If $t_x(y) \geq g^{-d}$ then $d_x(y) \leq 2d$.

Proof. We prove this by an induction on d. The base case of $d = 0$ is trivial. Assume the lemma holds for all $d' < d$ and let y be a symbol vertex of D_x s.t. $1 > t_x(y) \geq g^{-d}$. Let p be the production vertex which precedes y. Assume p has a immediate predecessors and b immediate successors. Since $t_x(p) = b \cdot t_x(y)$, p must have an immediate predecessor y' with weight at least $\frac{b}{a} t_x(y)$. By the above remarks $t_x(y') \geq g \cdot t_x(y) \geq g^{1-d}$. Applying the inductive hypothesis it follows that $d_x(y') \leq 2d-2$ and $d_x(y) \leq 2d$. □

Since there is a sink of weight at least $\frac{1}{|w|}$ in D_x, the lemma implies the existence of a path of length at most $2\left\lceil \log_g(|w|) \right\rceil$ from x to a sink. For fixed growing csgs this bound is $O(\log(|w|))$ since $g>1$ and since g only depends on the grammar. Paths are called *bounded* if they are of length at most $2\left\lceil \log_g(|w|) \right\rceil$. A derivation graph is *bounded* if there is a consistent set of bounded paths from all vertices to sinks. Combining the above remarks with Lemma 1, we get the basis for the polynomial algorithm:

Theorem 1: Every derivation graph of a growing csl which derives the word w is bounded.

5. Membership of growing csl is polynomial

In the last section we showed there are short paths from all vertices to sinks in derivation graphs. We now use these paths to "cut" derivation graphs into "pieces." Each piece is bordered by on the left and on the right by a path of length $O(log(|w|))$. There is an exponential number of derivation graphs and pieces. We therefore gather the essential information about a piece in a frame. Part of this information will be a description of the left and the right path. There will be only a polynomial number of valid frames, which are all found by the algorithm. The information gathered in the frames will be sufficient to decide membership. The same technique was used extensively in [GS85, GW85a, GW85b] to show that membership for various problems of k-parallel rewriting are polynomial. Also the classic Younger algorithm [Yo67] for context-free language membership can be described using a simple notion of frames: A frame parametrizes a possible derivation subtree by the label of the root and the boundaries of the subword that appears at the leaves of the tree.

For the algorithm we need to be able to describe paths in derivation graphs. One way of doing this is given in Figure 2. The productions are the labels of the production vertices on the path and the numbers specify which successors and predecessors are on the path. These numbers are necessary because for a given production $\alpha \rightarrow \beta$ in the grammar some symbols might have multiple occurrences in α or β. We could present the algorithm using the notation of Figure 2 which would be more efficient. But for the sake of simplicity of the presentation we assume that the grammar is in a special form.

A grammar is called a *one−grammar* if for each production $\alpha \rightarrow \beta$ in the grammar each symbol of the alphabet occurs at most once in α and at most once in β. Using standard methods of Formal Language Theory, it is easy to construct an equivalent one-grammar for a given grammar by increasing the size of the alphabet and by adding chain productions. For *chain* productions $|\alpha|=|\beta|=1$ must hold.

In the following we outline the construction of an equivalent one-grammar. Details are left to the reader. Assume there is a production $\alpha \rightarrow \beta$ in which some symbol A (terminal or non-terminal) appears twice in α. In this case the two occurrences of A in the production are replaced by two new non-terminals A_1 and A_2. Furthermore two new productions are added to the grammar: $A_1 \rightarrow A$ and $A_2 \rightarrow A$. By repeatedly applying the above, double ocurances of symbols in the left hand side of productions are eliminated. With a similar construction we can eliminate double occurrences from the right hand side of productions.

Since for the membership problem we assume that the grammar is fixed, the size of the equivalent one-grammar will also be independent of the input. Observe that the original grammar and the equivalent one-grammar define the same language. Furthermore derivation graphs for the original grammar translate into derivation graphs for the corresponding one-grammar and vice versa. A path in the derivation graph of the one-grammar is at most three times as long as the corresponding path in the "equivalent" derivation of the original grammar.

This motivates the following assumptions for the rest of this section. The fixed growing csg of the membership problem is given by its equivalent one-grammar. Paths in derivation graphs of the latter grammar are *bounded* if they are of length $6\lceil \log_g(|w|)\rceil$ instead of $2\lceil \log_g(|w|)\rceil$. Theorem 1 also holds for equivalent one-grammars with the new bound (Recall that g is the growth-ratio of the original growing csg.).

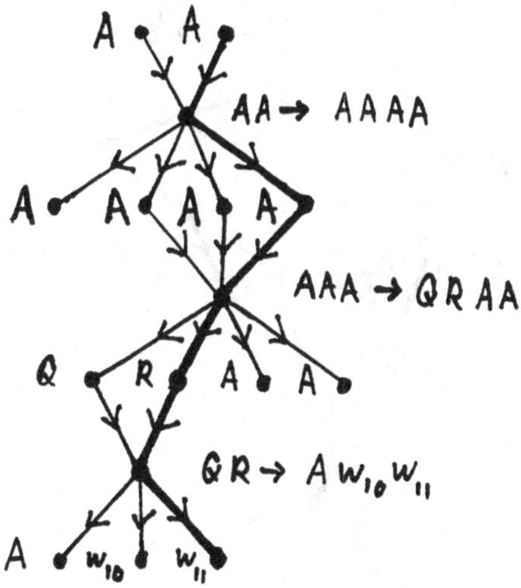

$2/AA \rightarrow AAAA/(4,3)/AAA \rightarrow QRAA/(2,2)/QR \rightarrow Aw_{10}w_{11}/3$

Figure 2. The description of a path (in boldface).

The input word w which is to be tested for membership is denoted as $w_1w_2 \cdots w_{|w|}$. To get a simple description of the algorithm we add dummy symbols to the beginning and end of w. Let [and] denote two symbols which are not in the alphabet of the grammar. Set w_0=[and $w_{|w|+1}$=].

To describe a path $\pi = x_1, x_2, \cdots, x_e$ in a derivation graph of a one-grammar, it is now sufficient to use the sequence $\omega(x_1)/\omega(x_2)/\cdots/\omega(x_e)$ which is called the *labeling sequence* of π and is denoted by $\Omega(\pi)$.

Similarly to paths, a labeling sequence is *bounded* if it is of length at most $6\lceil \log_g(|w|)\rceil$. A *frame* is a tuple (t,λ,ρ,l,r) s.t. $t \in V^2 \cup V$, λ and ρ are bounded labeling sequences and $0 \le l \le r \le |w|+1$.

Intuitively, the above frame specifies a "piece" of a planar derivation graph which might appear in the cut-and-paste process. This piece is bordered on the left (resp. right) by a path labeled with λ (resp. ρ). The piece derives $w_l, w_{l+1}, \cdots, w_r$, i.e. λ ends at a sink labeled with w_l, ρ ends at a sink labeled with w_r and the sinks in between are labeled accordingly. The word t specifies how the "piece" starts. If the left and right path start at the same vertex then t is the label of that vertex. In the case where the paths start at different vertices, t consists of the labels of both vertices. See Figure 3 for examples. The polynomial running time of the membership algorithm for fixed growing csls heavily relies on the fact that the number of frames is polynomial in $|w|$. Note that there is only a polynomial number of labeling sequences of bounded paths (length up to $6\lceil \log_g(|w|)\rceil$) since g is a positive constant.

Not every frame corresponds to a piece of a derivation graph, only valid frames do. A frame is *valid* if and only if it is a valid frame w.r.t. a bounded derivation graph D and a consistent set of bounded paths $\Pi=\{\pi_y \mid \pi_y$ starts with the symbol vertex y of D$\}$ from each symbol vertex of D to a sink.[2]

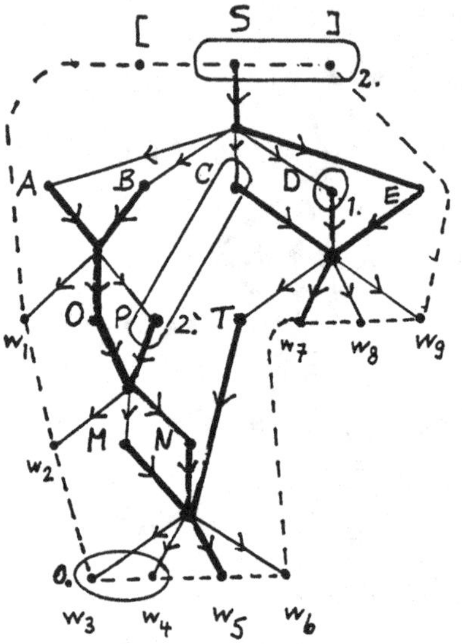

0. $(w_3w_4, w_3, w_4, 3, 4)$
1. $(D, D/CDE \rightarrow Tw_7w_8w_9/w_7, D/CDE \rightarrow Tw_7w_8w_9/w_7, 7, 7)$
2. $(S], S/S \rightarrow ABCDE/E/CDE \rightarrow Tw_7w_8w_9/w_7,], 7, 10)$
2'. $(PC, P/OP \rightarrow w_2MN/N/MNT \rightarrow w_3w_4w_5w_6/w_5,$
$C/CDE \rightarrow Tw_7w_8w_9/w_7, 5, 7)$

Figure 3. Some valid frames with respect to a derivation graph
and a set of bounded consistent paths (in boldface);
the symbols of the first component of each frame are encircled.

Definition: The valid frames of (D,Π) are given as follows:

1. The frame $(\omega(v),\Omega(\pi_v),\Omega(\pi_v),l,l)$ is valid if
   i) v is a symbol vertex of D;
   ii) π_v ends at a sink labeled with w_l.
2. The frame $(\omega(u)\omega(v),\Omega(\pi_u),\Omega(\pi_v),l,r)$ is valid if
   i) u and v are symbol vertices of D s.t. adding the edge (u,v) to D does not violate the planarity of D;
   ii) the edge (v,w) does not leave the planar circle which encloses all edges of D

[2] Note that D does not necessarily derive the whole word w, but in the case where w and the word derived by D have no subword in common then no *then no valid frames belong to D.*

and is defined by the edges between adjacent sources, the edge between the rightmost source and the rightmost sink, the edges between adjacent sinks, and the edge between the leftmost sink and the leftmost source (See dotted circle of Figure 3.).

iii) there is no path from u to v and vice versa;

iv) π_u ends at some sink s;

v) the $r-l+1$ sinks starting from s going to the right are labeled with $w_l, w_{l+1}, \ldots, w_r$;

vi) the $(r-l+1)^{st}$ such sink is the one at which π_v ends.

There are many valid frames belonging to (D,Π). For a particular frame we want to specify the subgraphs of derivation graphs which correspond to that frame. Let $F=(t,\lambda,\rho,l,r)$ be a valid frame of some tuple (D,Π). If t has one letter then $\rho=\lambda$ and the path of vertices in D which corresponds to λ is an *instance* of F. In the case where t has two letters then the subgraph I induced by the vertices v of D for which the following conditions hold is called an *instance* of the frame F:

i) v has a predecessor amongst the two vertices corresponding to t;

ii) π_v ends at a sink corresponding to w_m where $l \le m \le r$;

iii) if v is not on the path corresponding to λ but π_v and λ have some vertex x as their first common vertex, then the predecessor of x on λ is to the left of the predecessor of x on π_v;

iv) if v is not on the path ρ but π_v and ρ have some vertex x as their first common vertex, then the predecessor of x on ρ is to the right of the predecessor of x on π_v.

Intuitively I consists of all vertices of D "below" T, to the "right" of λ and to the "left" of ρ. Applying the above definition of valid frames gives the following equivalence.

Theorem 2: $S \overset{*}{=>} w$ if and only if there are two valid frames $([S,[,\mu,0,m)$ and $(S],\mu,],m,|w|+1)$.

Proof. Assume $S \overset{*}{=>} w$. Theorem 1 implies the existence of a bounded derivation graph for $S \overset{*}{=>} w$. By adding two vertices corresponding to the dummy symbols [and] one gets a bounded derivation graph D for $[S] \overset{*}{=>} [w]$. Let Π be some bounded set of consistent paths of D as defined above. Furthermore let v be the source of D which is labeled with S and assume π_v ends at the mth symbol of w. From the above Definition it follows that $([S,[,\Omega(\pi_v),0,m)$ and $(S],\Omega(\pi_v),],m,|w|+1)$ are valid frames of (D,Π).

To prove the reverse let I be an instance of $([S,[,\mu,0,m)$ and J be an instance of $(S],\mu,],m,|w|+1)$. Assume that I and J have distinct sets of vertices. By identifying the vertices on the rightmost path of I with the vertices on the leftmost path of J one can build a derivation graph for $[S] \overset{*}{=>} [w]$. Finally removing the two nodes labeled with the dummy symbols [and] leads to a derivation graph for $S \overset{*}{=>} w$. □

Theorem 3: The algorithm finds exactly all valid frames and can be implemented in polynomial time.

Proof: The first part of the theorem is proved in two inductions. In Induction 1 we show that all frames in the set *VAL* of the algorithm are valid according to the above Definition. Induction 2

Algorithm:

(* Constructs the set *VAL* of all valid frames. *):

(* We assume that $|w| \geq 3$. *)

0. Initialize *VAL* to $\{(w_i, w_i, w_i, i, i): 0 \leq i \leq |w|+1\} \cup$
$\{(w_i w_{i+1}, w_i, w_{i+1}, i, i+1): 0 \leq i \leq |w|\}$

Repeat

1. Add the frame $(A_i, A_i/P/\lambda, A_i/P/\lambda, l, l)$ to *VAL* if $P = A_1 A_2 \cdots A_k \rightarrow B_1 B_2 \cdots B_{k'}$ and $(B_j, \lambda, \lambda, l, l)$ is in *VAL*.

2.1. Add the frame $(A_i A_{i+1}, A_i/P/\lambda, A_{i+1}/P/\lambda, l, l)$ to *VAL* if $P = A_1 A_2 \cdots A_k \rightarrow B_1 B_2 \cdots B_{k'}$ and $(B_j, \lambda, \lambda, l, l)$ is in *VAL*.

2.2. Add the frame $(XA_1, \lambda, A_1/P/\rho_j, l, r_j)$ to *VAL* if $P = A_1 A_2 \cdots A_n \rightarrow B_1 B_2 \cdots B_{k'}$ and $(XB_1, \lambda, \rho_1, l, r_1)$ as well as $(B_i B_{i+1}, \rho_i, \rho_{i+1}, r_i, r_{i+1})$, for $1 \leq i < j$, are in *VAL*.

2.3. Add the frame $(A_k Y, A_k/P/\lambda_j, \rho, l_j, r)$ to *VAL* if $P = A_1 A_2 \cdots A_k \rightarrow B_1 B_2 \cdots B_{k'}$ and $(B_i B_{i+1}, \lambda_i, \lambda_{i+1}, l_i, l_{i+1})$, for $j \leq i < k'$, as well as $(B_{k'} Y, \lambda_{k'}, \rho, l_{k'}, r)$ are in *VAL*.

Until no new frame can be added to *VAL*.

shows that all valid frames are in the set *VAL* created by the algorithm. Notice that the above Definition and the Algorithm are outlined in the same way. A schematic description of the algorithm is given in Figure 4.

Induction 1: Let F be the first frame added to *VAL* by the algorithm which is not valid according to the above Definition. Let R be the set of frames of *VAL* which caused the algorithm to add F to *VAL*. Clearly the frames of R are valid. By combining instances for the frames of R one can build an instance for F (see proof of Theorem 2) and get a contradiction. For a complete proof we need to distinguish in which step F was added to *VAL* and reason in each case that F is valid. We only show this for Step 2.3. The remaining cases are similar.

Let I_i, for $j \leq i < k'$, be an instance of the frame $(B_i B_{i+1}, \lambda_i, \lambda_{i+1}, l_i, l_{i+1})$ (see Step 2.3) and $I_{k'}$ be an instance of $(B_{k'} Y, \lambda_{k'}, \rho, l_{k'}, r)$. Since these frames are valid the instances exist. Assume that the the vertex sets of the instances are disjoint. Let a, y and p be three new vertices s.t. $\omega(a) = A_k$, $\omega(y) = Y$ and $\omega(p) = A_1 \cdots A_k \rightarrow B_1 B_2 \cdots B_{k'}$. To build the instance for F we combine the instances by identifying the vertices on the rightmost path of I_i with the vertices on the leftmost path of I_{i+1}, for $j \leq i < k'$. Furthermore, we add the edges (a, p), (y, p) and the edges (p, v_i), for $j \leq i \leq k'$, where v_i is the vertex corresponding to B_i. Since there is an instance for F

this frame must be valid and we get a contradiction.

Similarly one can prove in a second induction that all valid frames are in the set *VAL* created by the algorithm. Assume $F=(t,\lambda,\rho,l,v)$ is a valid frame of (D,Π) (see the above Definition) which is not in *VAL*. Let T be the vertices of D which correspond to t. We choose F so that the number of vertices which have a predecessor in T is minimum. In Step 0 all valid frames are added to *VAL* for which the length of both λ and ρ is 0. Thus in the frame F either λ or ρ is of positive length. We distinguish the following cases.

1. $|T|=1$, $\lambda=\rho$ and λ has positive length;
2.1. $|T|=2$, the vertices of T have a common successor;
2.2. $|T|=2$, the vertices of T don't have a common successor, ρ has positive length;
2.3. $|T|=2$, the vertices of T don't have a common successor, λ has positive length.

We still need to show that in each case F is added to *VAL* by the algorithm which is a contradiction. We only show this for Case 2.2. The remaining cases are similar.

Let $T=\{u,v\}$, $\omega(u)=X$, $\omega(v)=A_1$, let the successor of v be labeled with $P=A_1A_2\cdots A_n \rightarrow B_1B_2\cdots B_{k'}$, and let v_i be the vertex of D corresponding to B_i. Because of the minimality of F the set *VAL* contains the frame $(XB_1,\Omega(\pi_u),\Omega(\pi_{v_1}),l,r_1)$ and the frames $(B_iB_{i+1},\Omega(\pi_{v_i}),\Omega(\pi_{v_{i+1}}),r_i,r_{i+1})$, for $1\le i<k'$, where w_{r_i} corresponds to the sink at which π_i ends, for $1\le i\le k'$. We conclude that F would have been added to *VAL* in Step 2.2 which is a contradiction.

The polynomiality of the algorithm follows from the fact that the number of different frames is polynomial and from the fact that only a constant number of different frames need to be considered to create a new valid frame. □

Combining Theorem 2 and Theorem 3 gives us the main result of this paper.

Theorem 4: The membership problem for fixed growing csgs is polynomial. □

In the conclusion section we discuss parallel algorithms for this problem.

6. NP-complete growing scattered languages

We will exhibit a fixed growing scattered language which is NP-complete. A *scattered grammar* (scg) G is a quadruple (V,Σ,P,S) where the components have the same meaning as for a csg except that the productions of P are defined differently [GH68]. The productions have the form $(A_1,A_2,\ldots,A_k)\rightarrow(\alpha_1,\alpha_2,\ldots,\alpha_k)$, s.t. $A_i\in V-\Sigma$, $\alpha_i\in V^*$ and $|\alpha_i|\ge 1$. In a *growing scg* the last condition is replaced by $|\alpha_i|>1$.

As in a derivation step for csgs, the left hand side of a production is replaced by the right hand side, but in the case of scgs the symbols A_i need not be adjacent. For two words u and v of V^*, $u \Rightarrow v$ if $u=u_1A_1u_2\cdots A_ku_{k+1}$, $v=v_1\alpha_1v_2\cdots\alpha_kv_{k+1}$ and $(A_1,\ldots,A_k)\rightarrow(\alpha_1,\ldots,\alpha_k)$ in P. The *(growing) scattered language* defined by the (growing) scg G is the set $L(G)=\{w \mid S \overset{*}{\Rightarrow} w \text{ and } w\in\Sigma^*\}$.

The main open problem concerning scattered languages (scls) is the question whether every csl is also a scl. This is rather unlikely, but it holds if productions of the type $(A_1,\ldots,A_k)\rightarrow(\alpha_1,\ldots,\alpha_k)$, s.t. $|\alpha_1\cdots\alpha_k|\ge k$ are allowed [GW85c].

It is easy to see that $L=\{ucu^Rcu: u\in\{a,b\}^*\}$ (see introduction of Section 4) is a growing scl. Since the symbols to be rewritten are not required to be adjacent, the derivations in different parts of the word can be synchronized.

In a growing scg it takes at most $|w|$ steps to derive a word w. Thus the growing

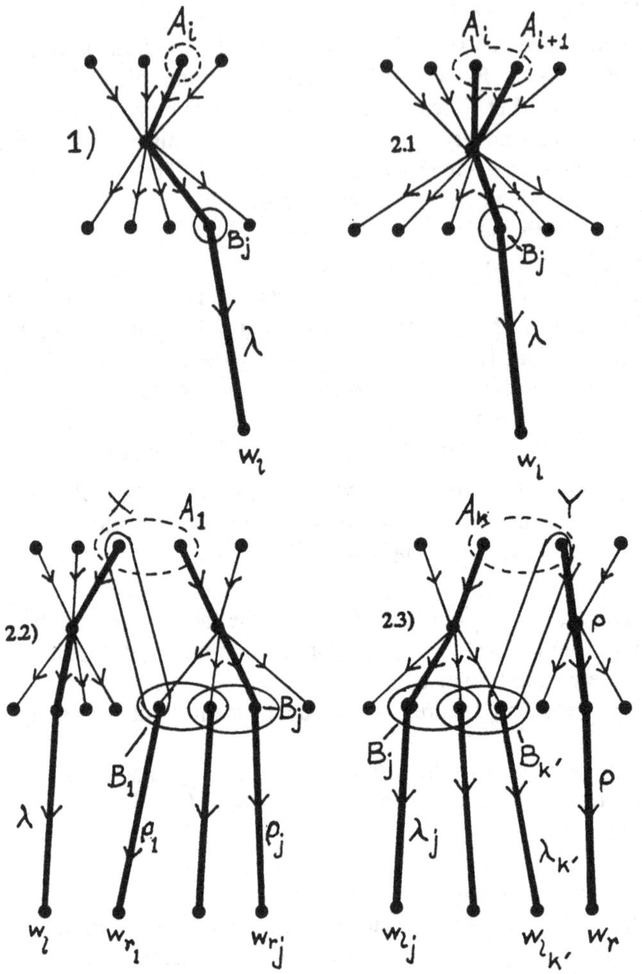

Figure 4: Schematic description of the cases 1-2.3 of the algorithm; labeled paths are indicated in boldface and the symbols of the first component of each frame are encircled.

scls are a subclass of NP. We will present a polynomial reduction of 3-partition to a growing scg.

3-Partition:
Instance: 3k numbers n_i and a bound B.
Question: Can the numbers be partitioned with k 3-element subsets each of which sums to B.

3-Partition was the first problem to be shown strongly NP-complete ,i.e. it remains NP-complete even if the n_i are encoded in unary [GJ78]. The language C for which we will provide a growing scg has the property that $<n_1, \ldots, n_{3k}, B>$ is an instance of 3-Partition if and only if the word $xa^{n_1}xa^{n_2}x \cdots xa^{n_{3k}} \cdots (yb^B)^k$ is in C.
Note that the word describes the corresponding instance and that its length is polynomial in the length of the unary encoding of the instance of 3-Partition. Thus the above equivalence implies that C is NP-complete.

Theorem 5: There are fixed, growing scls[3] which are NP-complete.

Proof: We will construct a growing scl $C = L(G)$ which fulfills the above equivalence. To simplify the construction, we assume that the numbers n_i are all at least three and $n > 1$.

$$G = (\{a,b,x,y,X,\bar{X},Y,\bar{Y},\hat{Y}\},\{a,b,x,y\},P,S), \textit{ where}$$

$$P = \{(S) \rightarrow (XYY\hat{Y}\hat{Y}), \quad (Y) \rightarrow (Y\hat{Y}\hat{Y}Y)$$

$$(X,Y) \rightarrow (xa\bar{X},yb\bar{Y}), \quad (X,\hat{Y}) \rightarrow (xa\bar{X},b\bar{Y}),$$

$$(\bar{X},\bar{Y}) \rightarrow (a\bar{X},b\bar{Y}),(aaX,bb),(aa,bb)\}.$$

To show the above equivalence observe that the grammar produces a sequence of blocks of a's followed by a sequence of blocks of b's. The sizes of the blocks of a's correspond to the numbers n_i. While X is deriving $xa^{n_i}X$ either some Y derives yb^{n_i} or some $\hat{Y}$ derives b^{n_i}. There is a block of b's for each n_i but the b-blocks are permuted and grouped in threes. We leave the details to the reader. □

7. Conclusions

We can express the membership problem for fixed growing csls as a membership problem for a variable context-free language. Given input word w and a fixed growing csg G, then we construct a context-free grammer G_w' from the frames (Section 5) of w and G. The frames form the non-terminals of G_w'. Note that the number of non-terminals is polynomial in w. The derivations of G_w' are defined using the recursions of steps 1-2.3. of the Algorithm of Section 5. The initial frames of Step 0 all derive the empty word ε. We still need to add a special start symbol S' which derives all combinations of two frames $([S,[,\mu,0,m)$ and $(S],\mu,],m,|w|+1)$ (See also Theorem 2.).

It is easy to see that $w \in L(G)$ iff $\varepsilon \in L(G_w')$. Also the derivation trees for ε in G_w' have only $O(|w|)$ nodes since the original grammar G is growing. We now sketch that growing csls are in the $LOG(CFL)$, the family of languages that are log-tape reducible to context-free languages [Su78]. $LOG(CFL)$ is exactly the family of languages recognized by a nondeterministic $log(n)$ tape bounded auxiliary pushdown automata within polynomial time [Su78]. n denotes the length of the input. To see that $L(G)$ is accepted by the latter type of automata, we simulate derivations of G_w' with a pushdown automata. The additional $log(|w|)$ tape is needed to store the nonterminals (frames) involved in the current production. Note that a frame requires at most $log(|w|)$ space.

It was further shown in [Ru80] that $LOG(CFL)$ are those languages accepted by an Alternating Turing Machine in $log(n)$ space and polynomial tree size. The space complexity

---

[3]The theorem also holds for the case were the language is *unordered* [Sa73] in addition to being growing and scattered.Note that the grammar used in the reduction is a growing unordered scattered grammar.

and parallel time complexity of $LOG(CFL)$ has been studied. Every language of $LOG(CFL)$ can be recognized in $(log(n))^2$ space by a deterministic Turing machine [Co70]. Thus growing csls can be recognized within the same space complexity as the lowest space complexity found for context-free languages [LSH65].

As for the parallel complexity [Ru80], $LOG(CFL)$ is contained in NC^2, the class of problems solved by uniform circuits of depth $O((log(n)^2)$ using a polynomial number of bounded fan-in gates. Explicit NC^2 circuits for the membership problem in a fixed context-free language are described in [UG85] and these circuits also solve the question $\varepsilon \in G_w'$. Furthermore PRAM algorithms are given for the same problems [UG85]. These algorithm run in $O((log(|w|)^2)$ parallel time and require a polynomial number of processors.

We showed that the membership problem for fixed growing csls can be solved in polynomial time and has reasonable space complexity and parallel time complexity. The main open problem is to determine the complexity of membership for "variable" growing csls, i.e. not only the word to be tested but also the growing csg is a variable of the input. The question is whether this problem can be solved in polynomial time or whether it is NP-complete.

Acknowledgements: We would like to thank Allen Goldberg and Habib Krit for helping to simplify the presentation of the results. Furthermore we are thankful to an anonymous referee who pointed out that growing csls are accepted by Alternating Turing Machines in $log(n)$ space with polynomial tree size and are thus contained in $LOG(CFL)$ (See conclusion section.).

References:

[BPS61] Y. M. Bar-Hillel, M. Perles, and E. Shamir, "On Formal Properties of Simple Phase Structure Grammars," *Z. Phonetik Sprachwiss. Kommunikationsforschung*, Vol. 14, pp. 143-172 (1961).

[Bo71] R. V. Book, "Time Bounded Grammars and their Languages," *Journal of Computer and Systems Sciences*, Vol. 5, pp. 397-429 (1971).

[Bo73] R. V. Book, "On the Structure of Context-Sensitive Grammar," *International Journal of Computer and Information Sciences*, Vol. 2, No. 2, pp. 129-139 (1973).

[Bo78] R. V. Book, "On the Complexity of Formal Grammars," *Acta Informatica*, Vol. 9, pp. 171-182 (1978).

[Ch59] N. Chomsky, "A Note on Phase-Structure Grammars," *Information and Control*, Vol. 2, pp. 137-167 (1959).

[Co70] S. A. Cook, "Path Systems and Language Recognition," *Proc. Second Annual ACM Symposium on Theory of Computing*, pp. 70-72 (1970).

[Gl64] A. Gladkii, "On the Complexity of Derivations in Phase-Structure Grammars," (in Russian), *Algebri i Logika, Sem. 3, Nr. 5-6, pp.29-44 (1964).*

[GS85] J. Gonczarowski and E. Shamir, "Pattern Selector Grammars and Several Parsing Algorithms in the Context-Free Style," Journal of Computer and System Sciences, Vol. 30, No. 3, pp. 249-273 (1985).

[GW85a] J. Gonczarowski and M. K. Warmuth, "Applications of Scheduling Theory to Formal Language Theory," Fundamental Studies Issue of Theoretical Computer Science, Vol. 37, No. 2, pp. 217-243 (1985).

[GW85b] J. Gonczarowski and M. K. Warmuth, "Manipulating Derivation Forests by Scheduling Techniques," Technical Report 85-3, Department of Computer Science, Hebrew University of Jerusalem.

[GW85c] J. Gonczarowski and M. K. Warmuth, "On the Complexity of Scattered Grammars," in preparation.

[GH68] S. Greibach and J. Hopcraft, "Scattered Context Grammars," Journal of Computer and Systems Sciences, Vol. 3, pp. 233-249 (1969).

[Ha78] M. A. Harrison, Introduction to Formal Language Theory, Addison-Wesley, Reading, Mass. (1978).

[Ka72] R. M. Karp, "Reducibility among Combinatorial Problems," in: R. E. Miller and J. W. Thatcher (eds.) Complexity of Computer Computations, Plenum Press, pp. 85-103, New York (1972).

[Ku64] S. Y. Kuroda, "Classes of Languages and Linear Bounded Automata," Information and Control, Vol. 7, pp. 207-223 (1964).

[Lo70] J. Loeckx, "The Parsing of General Phase-Structure Grammars," Information and Control, Vol. 16, pp. 443-464 (1970).

[Ru80] W. L. Ruzzo, "Tree-Size Bounded Alternation," Journal of Computer and System Sciences, Vol. 21, pp. 218-235 (1980).

[Sa73] A. Salomaa, Formal Languages, Academic Press, N.Y. (1973).

[LSH65] P. M. Lewis, R. E. Stearns and J. Hartmanis, "Memory Bounds for Recognition of Context-Free and Context-Sensitive Languages," Proc. Sixth Annual IEEE Symposium Switching Circuit Theory and Logical Design, pp. 191-212 (1965).

[Su78] H. Sudborough, "On the Tape Complexity of Deterministic Context-Free Languages," Journal of the ACM, Vol. 25, pp. 405-414 (1978).

[UG85] J. D. Ullman and A. V. Gelder, "Parallel Complexity of Logical Query Programs," Technical Report, Department of Computer Science, Stanford University (1985).

[Yo67] D. H. Younger, "Recognition and Parsing of Context-Free Languages in Time n^3," Information and Control, Vol. 10, pp. 189-208 (1967).

WEIGHTED GRAPHS :
A TOOL FOR LOGIC PROGRAMMING

Philippe DEVIENNE - EUDIL and LIFL (UA CNRS 369)
Patrick LEBEGUE - IUT A and LIFL (UA CNRS 369)

UNIVERSITY OF LILLE-I - Bât M3
59655 VILLENEUVE D'ASCQ Cédex FRANCE

ABSTRACT

Unfoldings of oriented graphs generate infinite trees that we generalize by weighting arrows of these graphs. Indexes along a branch are added during unfoldings and the result indexes variables. We study formal properties of these graphs (substitution, equivalence, unification,...). We use them to solve the halting problem of a recursive head-rewriting rule (as in PROLOG-like languages).

1.- INTRODUCTION

Usually, unification is recognized to act an important role in rewriting systems : for efficient unification algorithms, [4],[7],[12],[13] used tools such as directed acyclic graphs (dags), or oriented graphs (gos).

But the main problem is linked to the strategy of resolution, mainly in the case of recursive rules which may gives rise to infinite rewritings, i.e. an infinite resolution.

In general term rewriting systems, this problem has been often studied [2][10] by introducing termination orderings. But these results are limited to particular rules (left-side linear, recursive path ordering,..). Our approach is quite different, in the case of head-rewriting systems (as in Prolog-like languages) : We solve them for any single recursive rule (with or without occur-check), which can be written in Prolog in the following form :

$P(\beta) \rightarrow P(\tau)$ where β and τ are rational trees built on Σ and V ; and P is a predicate symbol, $P \notin \Sigma$.

The resolution in forward or backward strategy can be infinite for some goals or facts $P(\alpha)$ if the recursive rule can be infinitely applied.

To caracterize this infinite resolution, we introduce a new kind of graphs.

2 - WEIGHTED GRAPHS

Unformally, we consider graphs and their unfoldings ininfinite rational trees [7] (see fig. 1). A weighted graph (WG) is a graph with a top (as a tree) and paths which are weighted with relative integers (see fig. 2). In all the paper, we suppose that nodes are occurrences of letters of a finite graduate alphabet Σ in the usual sens [7] and, as in trees, we consider some leaves in a set V of variables; P, Q, ... can be interpreted in part 4 as predicate symbols.

$A(G)$ denotes the infinite tree deduced from G by :

- infinite unfolding of loops
- attribution to each occurrence of x of its index i, which is the algebraic sum of the weights along the branch leading from the top down to x.

Then we note x_i (see fig. 3) and all the variable leaves of $A(G)$ are of this type.

Furthermore, we associate to x a period j , j is a positive integer or is infinite. We note $x(j)$ and $x_i(j)$. If $J = +\infty$, we omit it. $x_i(j)$ can be interpreted as $x_{\text{"i modulo j"}}$.

graphs :

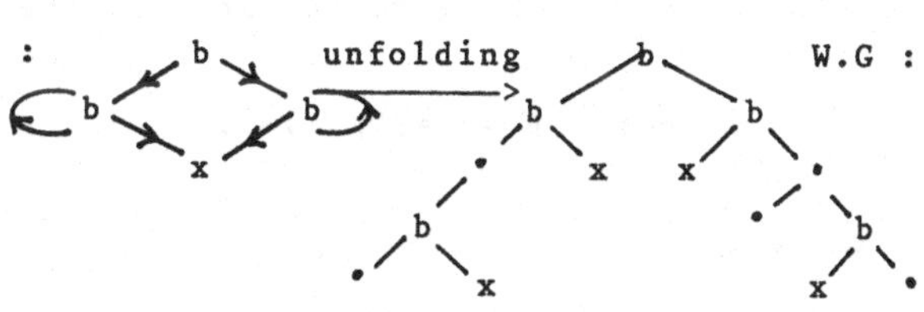

fig. 1

W.G :

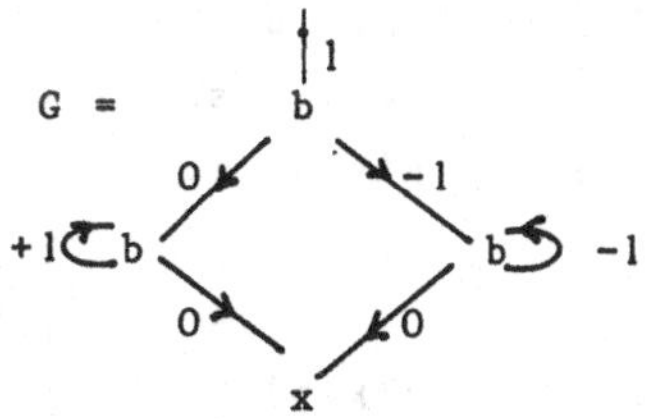

fig. 2

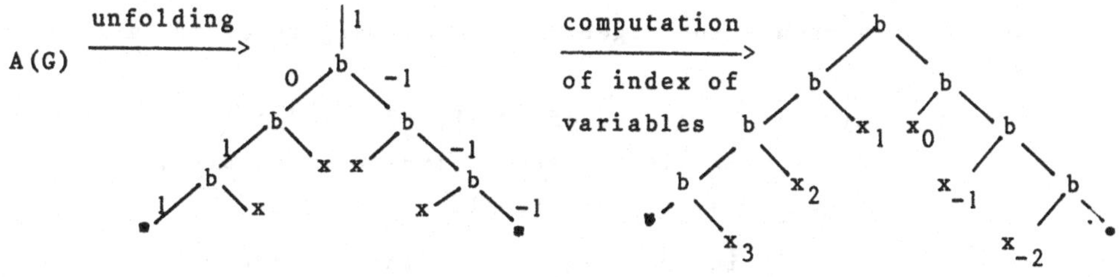

fig. 3

2.1.- Syntactical definition

DEFINITION 1 : We call WEIGHTED GRAPH (WG), built on Σ and V, the object $G = (X, LAB, SUCC)$ where :

> X is a finite set of nodes
>
> LAB is a mapping ; $X \to \Sigma \cup V$
>
> SUCC is a partial function : $X \times N \to X \times Z$
>
> $SUCC(s,i) = (s',q)$
>
> s' is the ith successor of s with a weight q. $(q \in Z)$

DEFINITION 2 : We call POINTED WEIGHTED GRAPH a weighted graph G wich is associated with :

> R a particular node called root $(R \in X)$
>
> W_R a weight associated to the root (head-weight). $(W_R \in Z)$
>
> Clas a function from V to N.

We note : $G/(R,W_R,Clas)$.

2.2.- Interpretation of weighted graphs :

We call $I(g)$ the interpretation of a pointed WG, $g = G/(R,W_R,Clas)$, the set of pairs $\{(k,A(g,k))\}$ where $k \in Z$, and $A(g,k)$ is a finite or infinite tree defined as follows :

Let $Dom(g)$ be the set, included in N^*, of valid paths in the graph g (in the classical way). [4].
For any given m in $Dom(g)$, we note $v = LAB[g(m)]$.

$$A(g,k)(m) = \begin{cases} v & \text{if } v \in \Sigma \\ (v,(q+k) \text{ modulo } Clas(v)) & \text{if } v \in V \end{cases}$$

q represents the algebraic sum of weights along m ;

Remarks :
- $A(g,k)$ is a tree of $M^\infty(\Sigma, V\times Z)$ ([1]) where $V\times Z$ is the set of indexed variables, $A(g,k)$ may be irrational. (fig. 3).
- Clas is interpreted as a period of index : For any i in Z x_i and $x_{(i \text{ modulo } Clas(x))}$ are equivalent

EXAMPLE 1 :

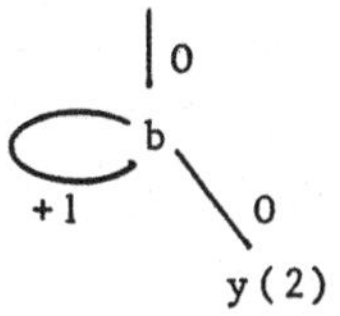

(i.e. Clas (y) = 2)

if k is even : A(g,k) = 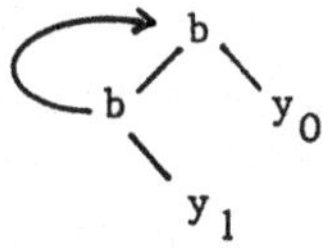

if k is odd : A(g,k) 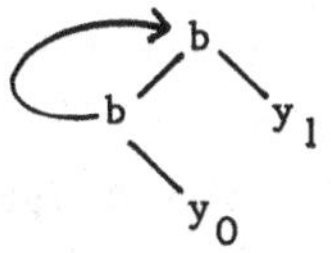

I(g) = {(2k , 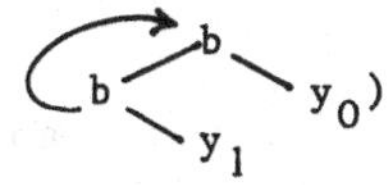

) ∀ k ∈ Z} {(2k+1 , 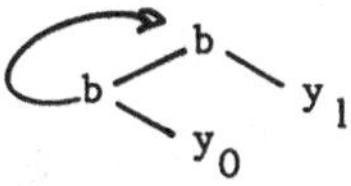

) / ∀ k ∈ Z}

EXAMPLE 2 :

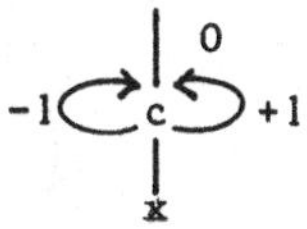

with Clas(x) = ∞

A(g,k) =

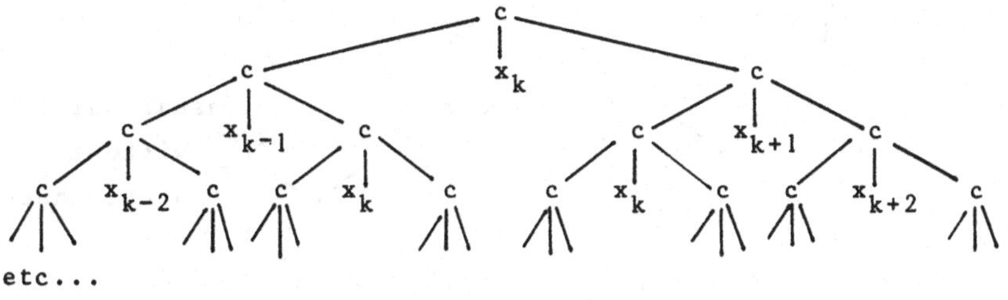

I(g) = {(k,A(g,k))/k ∈ Z}

3 - UNIFICATION

3.1.- Weighted substitutions (WS) :

We define on pointed WG's, an operation of substitution which associates to any variable (element of V) a pointed Weighted Graph. The interpretation of this object is deduced from the pointed WG one.

In classical conditions this operation is associative and coherent with the interpretation.

$$I(\sigma(g)) = I(\sigma)[I(g)] \ .$$

3.2.- Unification of W.G.

Two pointed WG g_1 and g_2 can be unified iff it exists a WS σ such that

$$\sigma(g_1) = \sigma(g_2) \ .$$

We describe a Z-unification algorithm in the set of pointed weighted graphs (over a graduate alphabet Σ). Our unification algorithm is exactly the classical one of two Graphs ([4]) , but we have to compute index and period of variables (Ex. 4) and weight of paths (fig. 5 and 6)

Ex. 4 : Let x_i, x_j variables of V, Unif$[x_i, x_j]$ can be written as :
Unif$[x_i, x_j]$ <=> for any $k \in Z$ $x_{i+k} = x_{j+k}$ (i.e. $x_k = x_{j-i+k}$)
$\Longrightarrow$ this unification leads to give a period abs(j-i) to x
(if x had a period then newperiod(x) = High Commun Factor [old period(x), abs(j-i)] .

Fig. 5

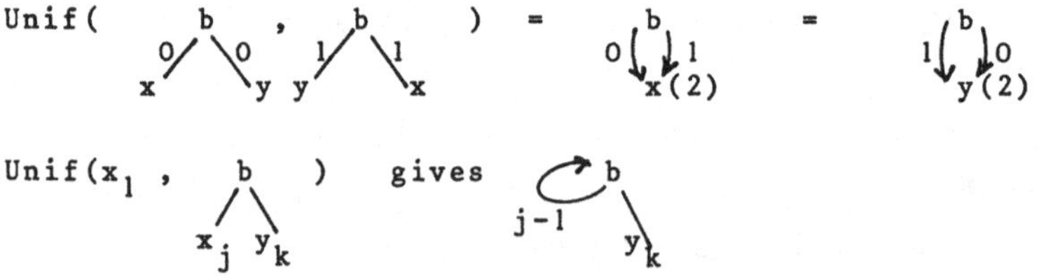

Fig. 6

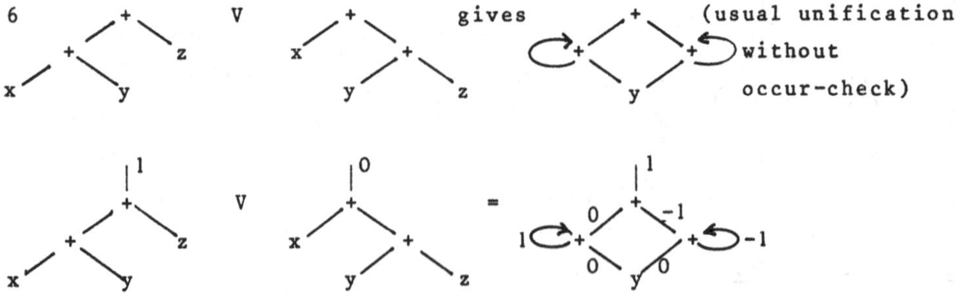

UNIFICATION ALGORITHM : Let g_1 and g_2 be two pointed Weighted Graphs constructed on the unique weighted graph G . g_1 and g_2 have for respective root r_1 and r_2 , and for respective head-weight w_1

and w_2 . $g_1 = G/(r_1,w_1,Cl_1)$, $g_2 = G/(r_2,w_2,Cl_2)$. We can previously compute the Clas : $Clas = Cl_1 \vee Cl_2$.

```
UNIFIED ((r1,w1),(r2,w2)) =
if (r1 = r2) then only the function clas is to be modified :
|                for all x in Var(g1) ∪ Var(g2) :
|                newclas(x) = high common factor (oldclas(x),|w1-w2|)
else if  Lab(r1) or Lab(r2)  is a variable (for instance Lab(r1)) then
|    |   we substitute to r1 the pointed WG of root r2 and
|    |   of head-weight (w1-w2)
|    else if  Lab(r1) = Lab(r2) then
|    |    |     - any arrow arriving to r1 , is to be redirected
|    |    |       to r2 with adding w1-w2 to the old weight.
|    |    |     - for i = 1 to number of sons of Lab(r1) do
|    |    |       |   r'1 = i th son of r1
|    |    |       |   r'2 = i th son of r2
|    |    |       |   w'1 = w1 + weight of the arrow between r'1
|    |    |       |                 and r'1
|    |    |       |   w'2 = w2 + weight of the arrow between r2
|    |    |       |                 and r'2
|    |    |       |   CALL UNIFIED ((r'1 ,w'1 ),(r'2 ,w'2 ))
|    |    |       endfor
|    |    else fail (The unified does nos exist)
|    |    endif
|    endif
endif.
```

Remark : It is important to note that :

- the existence of unified does not depend of weights
- when it exists, the unified of two trees or dags is a graph which generally contains loops.

The following property is our motivation of unification of pointed WG's. It permits to reduce the study of an infinite sequence of unifications "of the same pattern" to the study of an unification of two finite objects.

> $g_1 \vee g_2$ corresponds to the solution of the infinite system of equations $\{A(g_1,k) = A(g_2,k) \;/\; \forall k \in Z\}$.

4 - STUDY OF A SINGLE RECURSIVE RULE

We introduce here a property of decidability of the halting problem of a recursive rule, based on the unification in Weighted Graphs :

4.1.- Decidability criterion

THEOREM : The rule $P(\beta) \to P(\tau)$ gives for some $P(t)$ an infinite resolution iff,

- without occur-check (as in Prolog) : $\overset{|1}{\beta}$ and $\overset{|0}{\tau}$ can be unified or, equivalently, $\beta \vee \tau$ exists (in the same way without occur-check).
- with occur-check (β, τ DAGs) (as in the standard logic [9]) $\overset{|1}{\beta} \vee \overset{|0}{\tau}$ exists in W.G. and contains no looping path whose the algebraic sum of weights is null.

We call weighted loop restriction (WLR) this restriction on the loops of the weighted graphs.

SKETCH OF THE PROOF : We explicite the iterative application of the rule by renaming the variables of β and τ after every rewriting, i.e. by adding a suffix, the number of rewritings, to the variables of β and τ . With this notation, a Prolog recursive rule can be infinitely used iff we can give a solution to the infinite system of equations :

$$\{\beta_{i+1} = \tau_i \;/\; \forall\, i \in N\}$$
$$(\beta_0 \to \tau_0 = \beta_1 \to \tau_1 = \beta_2 \ldots)$$

where β_i and τ_i are the terms β and τ where we index all variables with the integer i .

We prove that such a system have a solution iff the following system $\{\beta_{i+1} = \tau_i \;/\; \forall\, i \in Z\}$ have also a solution.

We remark that $\beta_{i+1} = A(\overset{|1}{\beta}, i)$ in the previous notation and we know that the computation of $\beta \vee \tau$ is equivalent to solve the following system :

$$\{A(\overset{|1}{\beta}, i) = A(\overset{|0}{\tau}, i) \;/\; \forall\, i \in Z\} ,$$

which give the fundamental property.

COROLLARY 1 : Let us call "invariant" any tree which is rewritten with $\beta \to \tau$ in an other equivalent tree, and let g be $\overset{|1}{\beta} \vee \overset{|0}{\tau}$ then :

. the smallest invariant is $A(g)$ (i.e. $A(g,0)$)

. for any $k \in Z$ $A(g,k) \rightarrow A(g,k+1)$ by $\beta \rightarrow \tau$

$A(g,k) \rightarrow A(g,k-1)$ by $\tau \rightarrow \beta$

<u>COROLLARY 2</u> : Let the Prolog program that we call the WHILE Prolog :

$$\begin{cases} P(\alpha) \rightarrow ; & \text{is a fact} \\ P(\beta) \rightarrow P(\tau) & \text{if } P(\tau) \text{ than } P(\beta) \\ P(t)\ ? & \text{is the goal} \end{cases}$$

for any fact $P(\alpha)$ and any goal $P(t)$, the program π has a finite set of patterns of solutions (i.e. a solution is an instanciation of a pattern) if

let g be $\underset{\beta}{|1} \;\; V \;\; \underset{\tau}{|0}$

* g does not exist or
* g exists and :
  . without occur-check, $A(g)$ is a rational tree and on the looping paths of g the algebraic sum of weights is null,
  . with occur-check, $A(g)$ is finite.

4.2.- <u>Examples</u> : We will do here a complete study of a Prolog-like program rules to allow to the readers to understand the power of such an unification.

* for instance we study the rule :

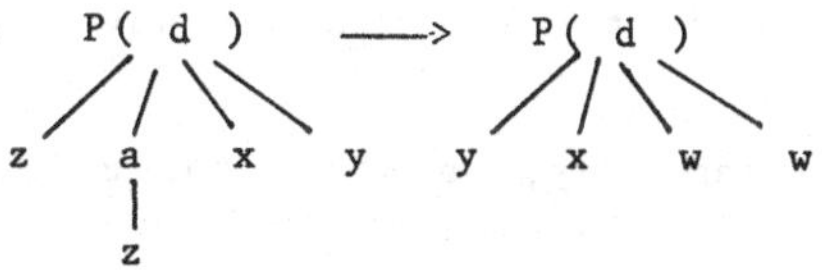

The behaviour of this rule cannot be seen without applying it, given as follows :

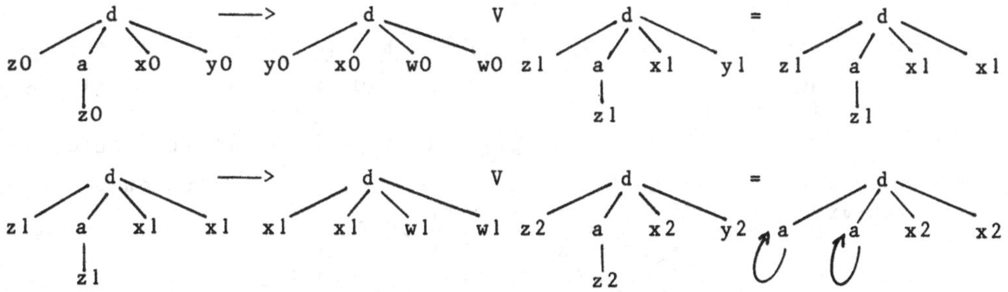

but in this last unification, x1 and x2 must be substitued by a^{∞} then :

- with occur-check, this rule stops always before the third step,
- without occur-check, this rule can give rise to an infinite resolution, we can see that the smallest invariant is here

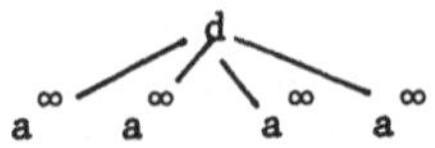

Now, we study the properties of this rule in the W.G. :

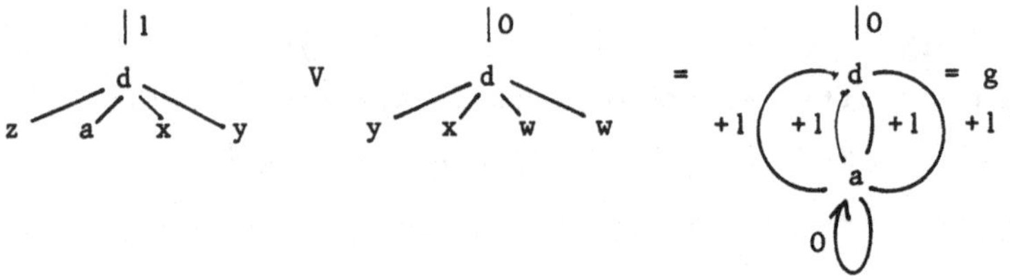

<u>THEOREM</u> : The rule can give an infinite resolution only without occur-check, because the WLR is not satisfied (the loop on a)

In the case of resolution without occur-check :

<u>COROLLARY 1</u> : The smallest invariant is A(g) =

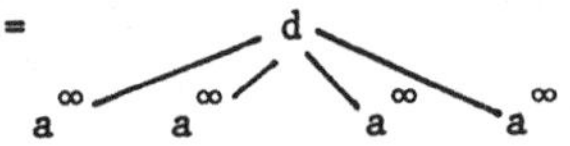

<u>COROLLARY 2</u> : For any fact or goal, the infinite resolution give an finite set of solutions, because A(g) is rational and the loop in g have a null weight.

* In the same way, we study the commutativity :

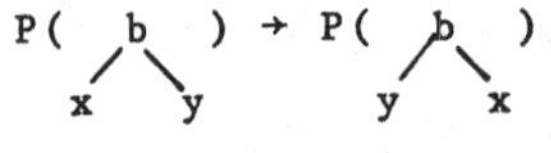

It is obvious that the rule can be used infinitely with or without occur-check, the smallest invariant is b(x,y) and for any fact or goal, the set of solutions is finite :

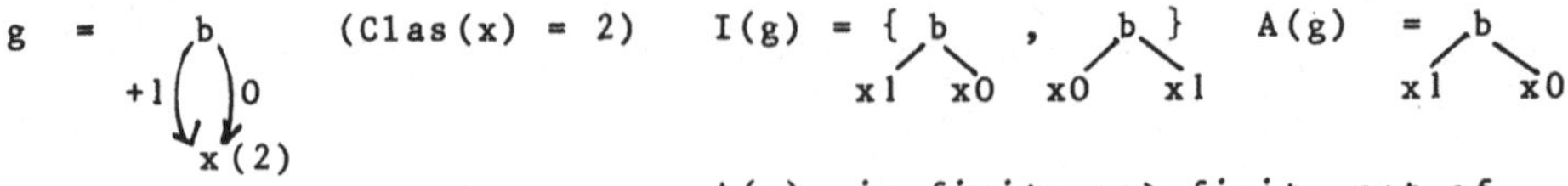

A(g) is finite ==> finite set of solutions.

* Associativity :

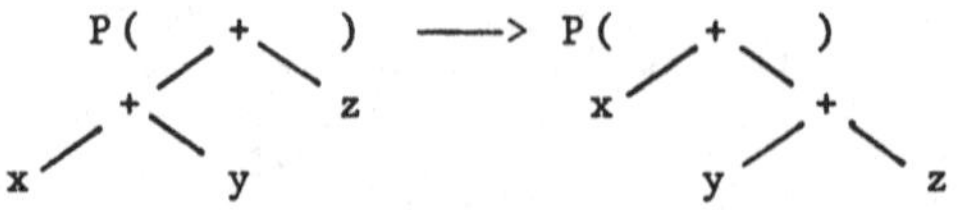

The rule is infinite for the resolution with or without occur-check :

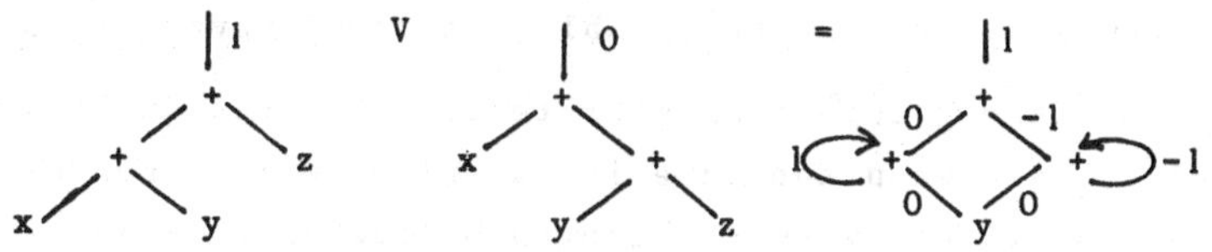

The smallest invariant is A(g) :

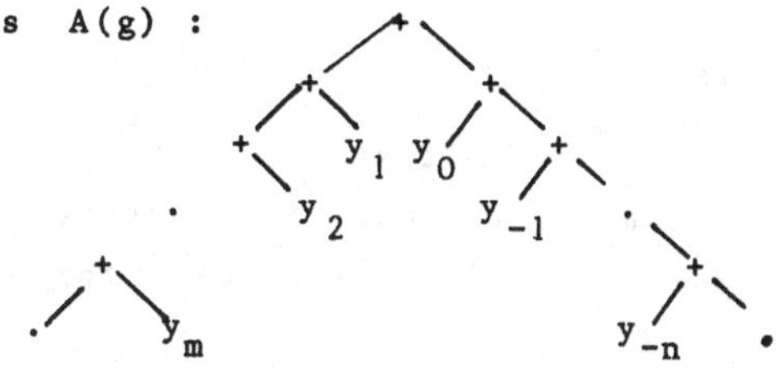

This rule can give rise to an infinite set of patterns of solutions :

α = 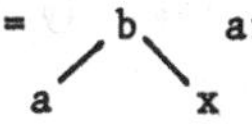 and t = 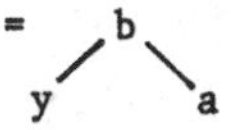 solutions : P(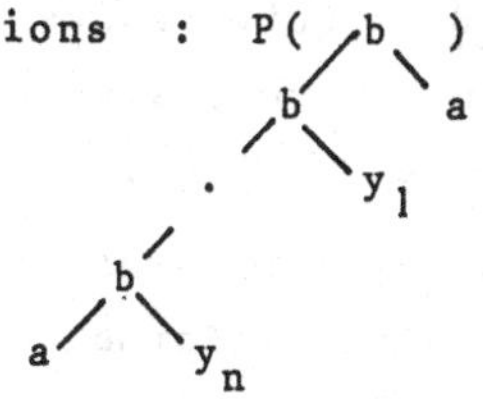) ($\forall n \in N$)

* Let be the other rule : P(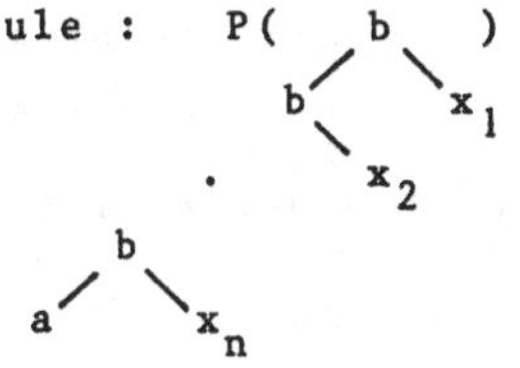) → P(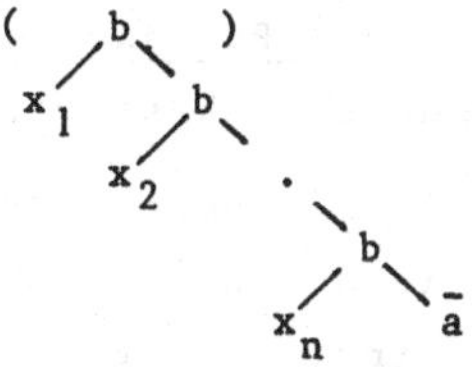)

For all t and α , this rule cannot give an infinite resolution, there is always a finite set of patterns of solutions, because $\beta \vee \tau$ does not exist.

There do not exist trees which could be rewritten more than 2^n times by $\beta \to \tau$ or $\tau \to \beta$

CONCLUSION

We have just given a simple criterion of halt decidability of a single recursive rule. This has been possible by the interpretation of the recursivity in the Weighted Graphs which contain the whole complexity of the resolution for this type of programs.

We have also defined for this new kind of graphs, weighted substitution, syntactical operations and an algorithm of Z-Unification, which give them a consistence apart from the problem which we have solved.

By similar methods, we are going to give criterions for the decision of halt of the WHILE, even when the rule is infinite, with a number of rewritings which depend linearly on the length of the fact $P(\alpha)$ and of the goal $P(t)$.

AKNOWLEDGEMENTS

The authors would like to thank Prof. Max DAUCHET, who suggested the subject of this paper, for many instructive discussions and encouragments. Our thanks to Prof. Gérard COMYN for helpful comments on this paper.

REFERENCES

[1] COURCELLE B., Equivalences and transformations of regular systems, applications to recursive program schemes and grammars. Université de BORDEAUX, n° 8430, Decembre 1984

[2] DERSHOWITZ B., A theoretical Basis for the Reduction of Polynomials to Canonical Forms. ACM-SIGSAM Bulletin, 39, August 1976, pp. 19-29

[3] DEVIENNE Ph. and LEBEGUE P., Rapport Technique sur les GO's pondérés, Thesis, to appear.

[4] FAGES F., Notes sur l'unification des termes de premier ordre finis ou infinis. Internal report INRIA-LITP

[5] GREIBACH., An infinite hierarchy of context-free languages. JACM, 16,1 (1969), pp. 91-106

[6] HEREL D. , KOZEN D., PARIKH R., Process Logic : Expressiveness, Decidability, Completeness. J. of Computer and System Sci., 25 (1982), pp. 144-170.

[7] HUET G., Confluent reductions : Abstract properties and applications to term rewriting systems JACM, 27 (1980), pp. 797-821

[8] KOZEN D., Indexings of subrecursive classes. Theoretical Computer Science, 11 (1980), pp. 227-301

[9] KOZEN D., Lower bounds for natural proof systems. 18th F.O.C.S., (1977), pp. 254-266.

[10] LESCANNE P., How to prove termination ? An approach to the implementation of a new recursive decomposition ordering. Proceedings of an NSF Workshop on the Rewrite Rule Laboratory Sept. 6-9 (1983), (Eds GUTTAG, KAPUR, MUSSER), General Electric Research and Development Center Report 84GEM008, April 1984, pp. 109-121.

[11] LLOYD J.W., Foundations of logic programming. Springer Verlag, (1984)

[12] MARTELLI A. and MONTANARI U., An efficient unification algorithm Transactions on Programming Languages and Systems 4, 2 (1982) pp. 256-282

[13] PATERSON M.S. and WEGMAN M.N., Linear unification. J. of Computer and Systems Sci., 16 (1978), pp. 158-167

CLASSICAL AND INCREMENTAL EVALUATORS FOR ATTRIBUTE GRAMMARS

by Gilberto Filé

Université de Bordeaux I
U.E.R. de Mathématiques et d'Informatique
351, Cours de la Libération
33405 Talence - Cedex - France

Introduction

The study of attribute grammars (AG's) has mainly concentrated on the problem of efficiently performing attribute evaluation. One class of AG's is particularly interesting in this respect : the class of the absolutely noncircular attribute grammars (ANC-AG's) [7] . This class is interesting because quite efficient evaluators can be (automatically) constructed for any ANC-AG and, moreover, testing whether any AG is ANC takes polynomial time. Because of these reasons ANC-AG's have been used in the construction of compiler writing systems [5] and studied in many theoretical papers [6,7,3,9,4].

In this article we present a unifying definition of ANC-AG's that includes as special cases the characterizations of this class given in [4,7,9] . Classically,[7], an AG is ANC if for every nonterminal X there exists a graph D(X) on its attributes having some special properties (see Section 1). Based on these graphs in [3,6] it is shown that recursive evaluators can be constructed easily for any ANC-AG. The evaluators of [3,6] are especially simple. Unfortunately, they may recompute some attributes several times. A second difect of these evaluators is that they "compute one synthesized attribute at the time". In [4] an improved model of evaluation is proposed : the new evaluator is still composed of a set of recursive procedures, but each procedure does not necessarily compute only one synthesized attribute. From this point of view the graph D(X) is transformed into a graph T(X) having blocks of attributes of X as nodes and where edges run from blocks of inherited attributes to blocks of synthesized attributes. Clearly the evaluators based on these graphs are, in general, subject to recompute attributes, as those based on the D(X)'s.

In [9] it is shown that for each nonterminal X of an ANC-AG one can always find a set A(X) of graphs having blocks of inherited or synthesized attributes of X as nodes (as T(X)), but such that for any 2 nodes there is a directed path from one of them to the other one. Hence,these graphs have, in addition to the arcs of T(X), also arcs running from blocks of synthesized attributes to blocks of inherited ones. Such graphs will be called totally ordered. From the A(X)'s it is easy to develop an evaluator that does never recompute attributes [9]. Unfortunately, such an evaluator risks to be very big : the number of graphs for each nonterminal may be exponential in the number of its attributes !

In this paper we show that an AG is ANC if to each nonterminal X one can associate a set ΠA(X) of partially ordered graphs (in contrast to the totally ordered of A(X),[9]). ΠA(X) lies between T(X) an A(X) in the following sense. T(X) and A(X) represent, respectively a minimal and a maximal level of the knowledge of the dependencies among the attributes of X that is necessary for constructing an evaluator. ΠA(X) is any intermediate level of this knowledge. One can show that from any ΠA(X) one can "throw away" knowledge for constructing T(X)and that one can "add knowledge" in order to construct an A(X). Clearly, the evaluators that are constructed from the ΠA(X)'s occupy such an intermediate place too and, in general, one should try to have in ΠA(X) the amount of knowledge that minimizes the recomputation and the size of the evaluator.

The ideas discussed above are also used for characterizing a class of AG's less known than ANC : the class of OPC-AG's that was introduced in [1] and that we prefer to call <u>doubly noncircular AG's</u> (DNC-AG). Using our characterization of DNC-AG we show how to construct (in a way essentially similar to that used for ANC-AG's) incremental evaluators for DNC-AG's. Again our approach allows the "tuning" of the evaluator in the attempt of finding a best compromise between recomputation and size.

The remainder of the article consists of 4 parts. In the first section we describe our approach, starting from the necessary definitions and arriving to the announced characterization of ANC- and DNC-AG's. Several examples are also given for clarifying the definitions. The 2nd and 3rd parts are dedicated to the development of the evaluators for ANC-AG's and DNC-AG's , respectively, and have an informal character. Finally, in the fourth part we point out some interesting problems that we intend to study in the future.

1 - <u>The characterization of ANC- and DNC-AG's</u>

The notion of <u>attribute grammar</u> (AG) is assumed to be known. Let us present below an AG that will be our running example throughout the paper.

<u>Example 1</u> : of an AG.

In this article AG's are always considered from a "schematic" viewpoint i.e., one is not interested in the interpretation of the functions used for computing the attributes, but only in the <u>dependency relation</u>

among the attributes. This relation is usefully represented by graphs. In Fig.1 an AG that we call GEX is shown by giving for each production $p \in [1,6]$ of GEX the <u>production graph of p</u>, denoted D(p), that represents the dependencies among the attributes of the nonterminals of p. Our purpose is to construct evaluators for AG's and, to this end, the production graphs are the necessary information. Inherited and synthesized attributes are shortened into i- and s-attributes, respectively. The left and right-hand side of a production are also shortened into lhs and rhs, respectively. In Fig.1, for $j \in [1,3]$, i_j is an i-attribute and s_j an s-attribute. Consider for instance production 2 of Fig.1 : we indicate the attribute i_1 of the lhs of 2 by $(i_1,0)$, whereas $(i_1,2)$ is the same attribute, but of the 2nd nonterminal of the rhs of 2.

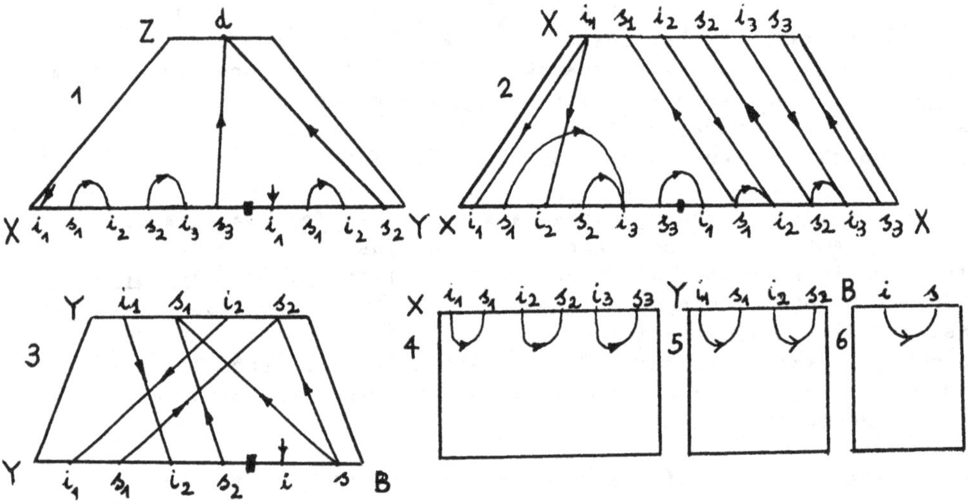

Figure 1 : The AG GEX.

Such couples are called the attributes of production 2. Thus, in general, for a production $p : X_0 \to X_1 \dots X_\gamma$ the <u>attributes of p</u> are all the couples (a,j) where $j \in [0,\gamma]$ and a is an attribute of the nonterminal X_j. Similarly, one defines the <u>attributes of a tree t</u> : for any node n of t with label Q and for any attribute a of Q, (a,n) is an attribute of n and t. □

<u>A partition of a set S</u> is, as usually, a set of nonempty, disjoint subsets of S whose union is S. <u>A partially ordered partition</u> (po-partition) of a set S is a couple $\Pi(S) = \langle\Pi,\to\rangle$, where Π is a partition of S and $\to$ is a relation on Π. $\Pi(S)^+$ is $\langle\Pi,\xrightarrow{+}\rangle$, where + indicates the transitive closure. If $\xrightarrow{+}$ is a total order on Π, i.e for any A and B in Π with $A \neq B$ either $A \xrightarrow{+} B$ or $B \xrightarrow{+} A$ (but not both), then $\Pi(X)$ is a <u>totally ordered partition of S</u> (to-partition). Clearly $\Pi(X)$ and $\Pi(X)^+$ can be viewed as graphs on Π. This point of

view will be often used in the sequel.

<u>Definition 1</u> : of assignment of po-partitions.

Given an AG G, an <u>assignment of po-partitions for G</u>, shortly, a pop-assignment for G, is a family of sets $\Pi A = \{\Pi A(X) / X$ is a nonterminal of $G\}$, where $\Pi A(X) = \{\Pi_1(X), \ldots, \Pi_n(X)\}$ and each $\Pi_i(X)$ is a po-partition of the set of attributes of X such that, if $\Pi_i(X) = <\Pi_i, \rightarrow>$, each element of Π_i contains either only i- or only s-attributes. If for every nonterminal X every $\Pi_i(X)$ is a to-partition then ΠA is a <u>top-assignment for G</u>.
There are 3 special types of pop-assignment :

1) a pop-assignment ΠA is <u>trivial</u> if, for every nonterminal X , $\Pi A(X) = \{\Pi(X)\}$ and $\Pi(X) = <\Pi, \rightarrow>$ such that $\Pi = \{\{a\} / a$ is an attribute of $X\}$;

2) a pop-assignment ΠA is <u>IS</u> if, for every nonterminal X and every $<\Pi, \rightarrow> \in \Pi A(X)$, for A and B in Π, $A \rightarrow B$ implies that A contains i-attributes and B s-attributes ;

3) a pop-assignment ΠA is <u>ISI</u> if for any $<\Pi, \rightarrow>$, A and B as above, $A \rightarrow B$ implies that A and B contain different types of attributes, i.e, one contains i- and the other s-attributes.

□

<u>Example 2</u> : of pop-assignment for GEX.
In what follows we give a pop-assignment $EX\Pi A$ for GEX :

1) $EX\Pi A(Z) = \{\Pi(Z)\}$, where $\Pi(Z) = <\{d\}, \phi>$.
2) $EX\Pi A(X) = \{\Pi_1(X), \Pi_2(X)\}$, where, representing the po-partitions with graphs,

$\Pi_1(X) =$ 

$\Pi_2(X) =$ 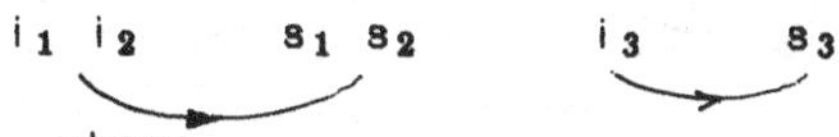

3) $EX\Pi A(Y) = \{\Pi(Y)\}$, where ,

$\Pi(Y) =$ 

4) $EX\Pi A(B) = \{\Pi(B)\}$, where $\Pi(B) =$ 

□

Pop-assignments are used for transforming the production graphs of an AG into new graphs that, if they satisfy some properties specified later, are useful for constructing the desired evaluators.

<u>Definition 2</u> : of augmented production graphs.

Consider an AG G, a production $p : X_0 \rightarrow X_1 \ldots X_\gamma$ of G and the sequence of po-partitions $\Pi(X_0) = \langle \Pi_0, \rightarrow_0 \rangle, \ldots, \Pi(X_\gamma) = \langle \Pi_\gamma, \rightarrow_\gamma \rangle$ of the attributes of $X_0, \ldots, X_\gamma$, respectively.
With <u>$D(p)[\Pi(X_0), \ldots, \Pi(X_\gamma)]$</u> we denote the graph defined as follows :

(i) the nodes are all the couples (A, i), where $A \in \Pi_i$, $i \in [0,\gamma]$,

(ii) there are 2 types of arcs :

(a) <u>edges that come from each $\Pi(X_i)$</u> : for each $i \in [0,\gamma]$, if $A \rightarrow_i B$, A and B in Π_i, then there is an edge running from (A, i) to (B, i) ,

(b) <u>edges that come from D(p)</u> : if an attribute (a, i) of p depends on attribute (b, j) and $a \in A \in \Pi_i$ and $b \in B \in \Pi_j$, then there is an edge running from (B, j) to (A, i).

The graph contains no other arc, such graphs are called augmented production graphs.

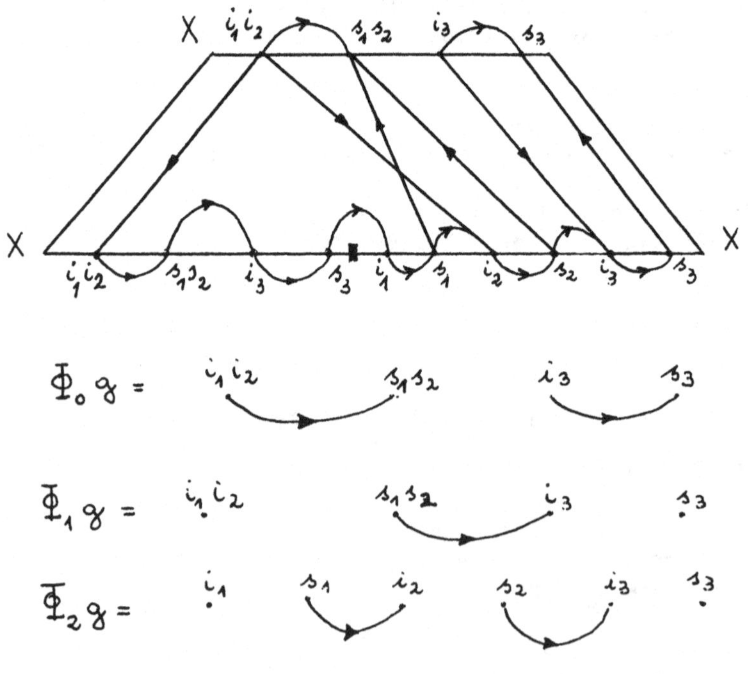

Figure 2.

Given an augmented production graph $g = D(p)[\Pi(X_0), \ldots, \Pi(X_\gamma)]$, for $i \in [0,\gamma]$, $\Phi_i g$ is the projection of the transitive closure of g, excluding the arcs that come from $\Pi(X_i)$, into the nodes $\{(A,i)/A\in\Pi_i\}$.

□

Example 3 : of augmented production graphs.

Fig.2 shows the graph $g = D(2)[\Pi_2(X),\Pi_2(X),\Pi_1(X)]$ and the projections $\Phi_i g$ for $i\in[0,2]$; production 2 of GEX is shown in Fig.1 and $\Pi_1(X)$ and $\Pi_2(X)$ are po-partitions of EXΠA, given in Example 2. □

Let us finally specify what sort of properties must be satisfied by the augmented production graphs of an AG in order to be useful for constructing an evaluator for it.

Definition 3 : Given an AG G and a pop-assignment ΠA for G

(1) ΠA is good for G if for any production $p : X_0 \rightarrow X_1 \ldots X_\gamma$ of G and po-partition $\Pi(X_0) \in \Pi A(X_0)$ the following holds : there exists at least one choice $\Pi(X_1) \in \Pi A(X_1),\ldots,\Pi(X_\gamma)\in \Pi A(X_\gamma)$, such that ,

(i) $g=D(p)[\Pi(X_0),\ldots,\Pi(X_\gamma)]$ is noncircular, and
(ii) $\Phi_0 g$ is a subgraph of $\Pi(X_0)^+$.

(2) ΠA is fine for G if for every production p as above, for every $i \in [0,\gamma]$, and for every po-partition $\Pi(X_i) \in \Pi A(X_i)$, there exists at least one choice of po-partitions $\Pi(X_0)\in\Pi A(X_0),\ldots,\Pi(X_{i-1})\in\Pi A(X_{i-1})$, $\Pi(X_{i+1})\in\Pi A(X_{i+1}),\ldots,\Pi(X_\gamma)\in\Pi A(X_\gamma)$, such that,

(i) $g = D(p)\ [\Pi(X_0),\ldots\Pi(X_\gamma)]$ is noncircular, and
(ii) $\Phi_i g$ is a subgraph of $\Pi(X_i)^+$. □

Example 4 : Of good and fine pop-assignments.

The pop-assignment EXΠA of Example 2 is good for GEX. Fig. 2 shows that for production 2, $\Pi_2(X)$ satisfies the conditions of Def. 3(1). It is easy to see that Def. 3(1) is satisfied in all other cases.
On the other hand EXΠA is not fine for GEX : in Fig.2 $\Phi_1 g$ and $\Phi_2 g$ are not contained in $\Pi_2(X)$ and $\Pi_1(X)$, respectively. Moreover, this fact is not reparable : there exists no fine pop-assignment for GEX. This is because of production 3, see Fig.1. Production 1 of GEX "forces" s_1 before i_2 and hence it forces the total order $H_1= i_1s_1i_2s_2$. On the other hand, production 3 of GEX exchanges the order for the nonterminals Y : if the order for the lhs Y is H_1, then the order for the rhs Y is $H_2= i_2s_2i_1s_1$, and, viceversa, if that of the lhs is H_2, that of the rhs is H_1. Thus, for respecting Def.3 (2), a pop-assignment ΠA' for GEX should contain 2 po-partitions $\Pi_1(Y)$ and $\Pi_2(Y)$ for Y forcing the orders H_1 and H_2, respectively ; but $\Pi_2(Y)$ contradicts the conditions of Def.3(2) for production 1 of GEX : any augmented production graph of production 1 using $\Pi_2(Y)$ contains a cycle.

One can easily modify GEX in such a way that it admits a fine pop-assignment : let production 3 have dependencies $(i_j,0) \rightarrow (i_j,1)$, $(s_j,1) \rightarrow (s_j,0)$, and $(s,2) \rightarrow (s_j,0)$, for $j \in [1,2]$, at the place of those of Fig.1. The new AG is called GEX1. EXΠA1, given below, is a fine pop-assignment for GEX1.

1) EXΠA1(X) = {Π(X)}, where

$$\Pi(X) = i_1 \rightarrow s_1 \rightarrow i_2 \rightarrow s_2 \rightarrow i_3 \rightarrow s_3$$

2) EXΠA1(Y) = {Π1(Y)}, where $\Pi 1(Y) = i_1 \rightarrow s_1 \rightarrow i_2 \rightarrow s_2$

3) EXΠA1(Z) = {Π(Z)} and EXΠA1(B) = {Π(B)}, see Example 2.

Observe that EXΠA1 is a top-assignment for GEX1, but that this is not necessary for an assignment to be fine. EXΠA1 is top only because our example is very simple. □

As announced before, the existence of a good or fine pop-assignment for an AG implies that one can easily construct an evaluator for it. We will examine how this is done in sections 2 and 3, but let us close this section showing that such pop-assignments exist for an AG iff the AG belongs to some special classes, viz., the classes of ANC- and DNC-AG's, defined originally in [7] and [1], respectively. The definition of these classes, reformulated in our terminology, is as follows.

(1) An AG G is <u>absolutely non circular</u> (ANC) if there exists a trivial IS pop-assignment that is good for G.

(2) An AG G is <u>doubly noncircular</u> (DNC) if there exists a trivial ISI pop-assignment that is fine for G.

<u>Main Theorem</u>. Let G be an AG.

(1) The following 4 points are equivalent,
- (a) G is ANC,
- (b) there is an IS good pop-assignment ΠA for G such that, for each nonterminal X of G, ΠA(X) is a singleton, [4],
- (c) there is a good pop-assignment for G,
- (d) there is a good <u>top</u>-assignment for G [9].

(2) The following 4 points are equivalent,
- (a) G is DNC,
- (b) there is an ISI fine pop-assignment ΠA for G such that, for each nonterminal X of G, ΠA(X) is a singleton,
- (c) there is a fine pop-assignment for G,
- (d) there is a fine <u>top</u>-assignment for G.

<u>Proof</u>. The proof of the theorem is omitted. We only remark that 1(c) ⇒ 1(d) is shown as follows. For any po-partition Π(X), let BΠ(X) be the set of the to-partitions corresponding to the topological

orderings of $\Pi(X)$. One can show that if ΠA is a good pop-assignment for an AG G then $\Pi A'$ is a good top-assignment for G, where $\Pi A'(X) = \cup \{B\Pi(X) / \Pi(X) \in \Pi A(X)\}$. □

2 - The ANC-evaluator

For the sake of conciseness we do not give the general construction, but only describe an ANC-evaluator for the AG GEX. From this description it should be easy to extract the general principles. The ANC-evaluator for the GEX is called P. P consists of a set of recursive procedures : one procedure, called eval$\Pi(T)B$ for each nonterminal $T \in \{Z,X,Y,B\}$ of GEX and block B of s-attributes in a po-partition $\Pi(T) \in EX\Pi A(T)$. Each procedure eval$\Pi(T)B$ has the form shown in Fig. 3.

```
proc evalΠ(T)B (n) ; node n ;
begin
      case production applied at n of
  p1 : CODE(p1,B)
  .
  .
  .
  pk : CODE(pk,B)
    end of case
end
```

Figure 3 : The structure of a recursive procedure of P.

Let Ω be the following function : $\Omega(1,\Pi(Z))=\langle\Pi_1(X),\Pi(Y)\rangle$, $\Omega(2,\Pi_1(X)) = \langle\Pi_2(X),\Pi_1(X)\rangle$, $\Omega(2,\Pi_2(X)) = \langle\Pi_2(X),\Pi_1(X)\rangle$, $\Omega(3,\Pi(Y))=\langle\Pi(Y),\Pi(B)\rangle$. Ω is useful for constructing P. Intuitively, it tells, given the po-partition of the lhs of a production, what po-partitions must be used for the rhs nonterminals. Observe that Ω is such that it gives for each production po-partitions that satisfy Def. 3(1) : for instance, $\Omega(2,\Pi_1(X))=\langle\Pi_2(X),\Pi_1(X)\rangle$ and $D(2)[\Pi_1(X),\Pi_2(X),\Pi_1(X)]$ satisfies Def. 3(1). Clearly Ω is not the unique function satisfying this property : $\Omega'(2,\Pi_1(X))=\langle\Pi_1(X),\Pi_1(X)\rangle$ and $\Omega'=\Omega$ otherwise, would also do.

The augmented production graphs determined by Ω are used for constructing P. Fig.4 contains the graph $g_1=D(1)[\Pi(Z),\Pi_1(X),\Pi(Y)]$ and the procedure eval$\Pi(Z)d$. It is easy to see that the procedure can be directly constructed from g_1 : as shown in Fig. 4 each node of g_1 corresponds to an instruction of the procedure.

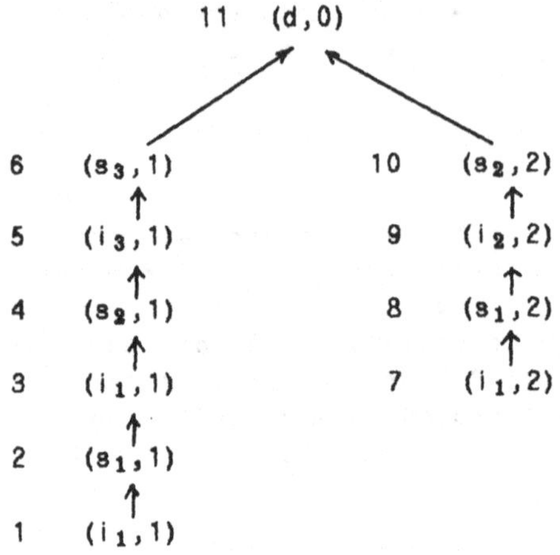


```
proc evalπ(Z)d (n) ; node n;
begin
      compute (i_1,n1) ;                 1
      call eval π_1(X)s_1 (n1);          2
      compute (i_2,n1) ;                 3
      call evalπ_1(X)s_2 (n1)            4
      compute (i_3,n1) ;                 5
      call evalπ_1(X)s_3 (n1) ;          6
      compute (i_1,n2) ;                 7
      call evalπ(Y)s_1 (n2)              8
      compute (i_2,n2) ;                 9
      call evalπ(Y)s_2 (n2)              10
      compute (d,n)                      11
end
```

Figure 4.

For representing trees we use the Dewey notation and hence, in the procedure of Fig. 4, n1 and n2 are the left and the right son of n respectively ; 0 is the root of a tree and n0=n for any node n.

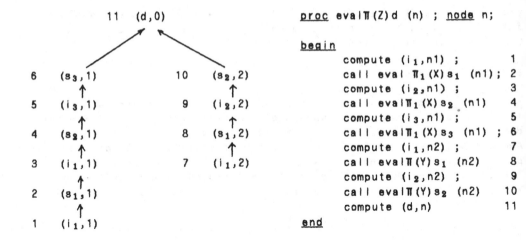

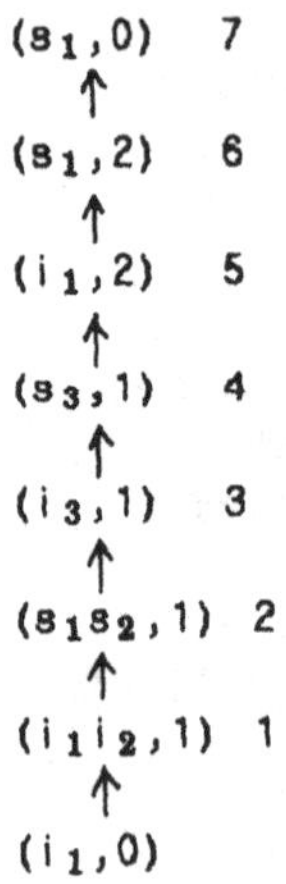


```
proc evalπ_1(X)s_1 (n) ; node n ;
begin
      case production applied et n of
   2: compute (i_1i_2,n1) ;              1
      call evalπ_2(X)s_1s_2 (n1) ;       2
      compute (i_3,n1) ;                 3
      call evalπ_2(X)s_3 (n1) ;          4
      compute (i_1,n2) ;                 5
      call evelπ_1(X)s_1 (n2) ;          6
      compute (s_1,n)                    7

   4 : compute (s_1,n)
      end of case
end
```

Figure 5.

Observe that evalπ(Z)d of Fig.4 is not the unique procedure one could construct from g_1 : what is important is that the order of the instructions of the procedure determines (under the given correspondence instutions/nodes of g_1) a topological ordering of the nodes of g_1. Let us now consider the procedure evalπ_1(X)s_1. For constructing the first part of it (corresponding to production 2) we have to consider the subgraph of g_2=D(2)[π_1(X),π_2(X),π_1(X)] consisting of all the nodes connected to $(s_1,0)$. Both this graph and the correspon-ding procedure are shown in Fig. 5.

Observe that for constructing eval$\pi_1(X)s_1$ we use g_2 because $\Omega(2,\pi_1(X)) = \langle\pi_2(X),\pi_1(X)\rangle$. At this point it should be easy to construct eval$\pi_1(X)s_2$ and eval$\pi_1(X)s_3$ from the corresponding subgraphs of g_2. Eval$\pi_2(X)s_1s_2$ and eval$\pi_2(X)s_3$ are constructed in a similar wey from the augmented graph $D(2)[\pi_2(X),\pi_2(X),\pi_1(X)]$. In Fig. 6 we give the procedure eval $\pi(Y)s_1$ together with the subgraph of $D(3)[\pi(Y),\pi(Y),\pi(B)]$ needed to construct it .

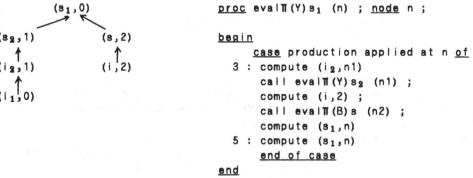

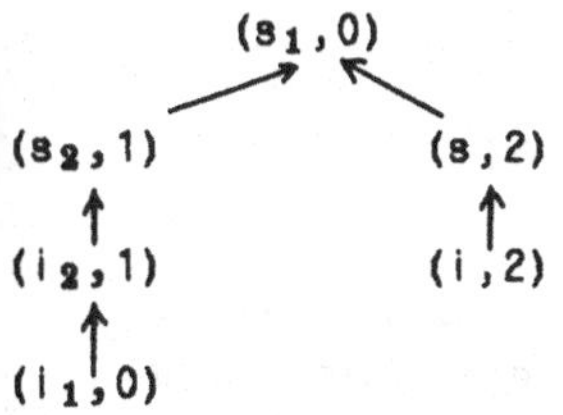


```
proc evalπ(Y)s1 (n) ; node n ;

begin
      case production applied at n of
   3 : compute (i2,n1)
       call evalπ(Y)s2 (n1) ;
       compute (i,2) ;
       call evalπ(B)s (n2) ;
       compute (s1,n)
   5 : compute (s1,n)
       end of case
end
```

Figure 6.

Eval$\pi(Y)s_2$ is going to be very similar to eval$\pi(Y)s_1$ and eval$\pi(B)s$ should not present any problem. For completing P one only needs to add to the procedures described so far a <u>main program</u> that starts the computation on any input tree : call eval$\pi(Z)d$ (0), 0 is the root of the tree. The evaluator P described above shows the following important points :

(1) The construction of an ANC-evaluator follows immediately from the augmented production graphs and thus it con be automatized easily.

(2) Using po-partitions with bigger blocks one obtains more efficient ANC-evaluators. For instance, there are good pop-assignments for GEX having only $\pi_1(X)$ as po-partition for X, but, clearly, the evaluator P described above, using also $\pi_2(X)$ performs less recursive calls in any subtree with root X (of an input tree) than any ANC-evaluator using $\pi_1(X)$ only. Observe that any pop-assignment πA for GEX having $\pi A(X) = \{\pi_2(X)\}$ could not be good for GEX. Thus, for fully exploiting the idea of having big blocks of attributes one needs to allow many po-partitions for the same nonterminal.

(3) <u>ANC-evaluators have a difect</u> : they may compute several times some attributes of their input trees. This is the case for the evaluator P described above. P evaluates 2 times the attributes of every node labeled by Y of an input tree : both procedures eval$\pi(Y)s_1$ and eval$\pi(Y)s_2$ must compute $(i,n2)$ and $(s,n2)$ because one does not know for any node n (labeled by Y and to which production 3 is applied) whether (s_1,n) is computed before (s_2,n) i.e., eval$\pi(Y)s_1(n)$

is executed before eval$\Pi(Y)s_2(n)$, or viceversa.

The recomputation of an ANC-evaluator can be reduced (or even eliminated) by adding new dependencies to some of the po-partitions of the pop-assignment ΠA used to construct the evaluator. Transforming ΠA into a good <u>top</u>-assignment (which, by the Main Theorem, is always possible) surely eliminates all recomputation, but is, in general, not necessary. Let us consider our evaluator P. For eliminating its recomputation it suffices to transform EXΠA so as to have 2 to-partition $\Pi_1(Y)$ and $\Pi_2(Y)$ for Y as follows :

$$\Pi_1(Y) = i_1 \to s_1 \to i_2 \to s_2$$

$$\Pi_2(Y) = i_2 \to s_2 \to i_1 \to s_1$$

Let EXΠA' be the pop-assignment obtained in this way. The new procedure eval$\Pi_1(Y)s_1$ and eval$\Pi_1(Y)s_2$ of the evaluator P' one constructs from EXΠA' are as follows : eval$\Pi_1(Y)s_1$ is the same as eval$\Pi(Y)s_1$ of Fig. 6, whereas eval$\Pi_1(Y)s_2$ in given in Fig.7 . This latter procedure does not contain the computation of (i,n2) and (s,n2) because one can be sure that eval$\Pi_1(Y)s_1(n)$ has been already executed when it is called. Similarly, eval$\Pi_2(Y)s_2$ contains the computation of (i,n2) and (s,n2) whereas eval$\Pi_2(Y)s_1$ does not.

```
Proc evalΠ1(Y)s2(n) ; node n ;
begin case production applied at n of
          3 : compute (i1,n1) ;
              call evalΠ(Y)s1 (n1) ;
              compute (s2,n) ;
          5 : compute (s2,n)
              end of case
end
```

Figure 7.

It is easy to see that P' does not recompute any attribute. Observe also tat EXΠA' is <u>not</u> a top-assignment.

In summary, <u>the general principle</u> is that, in order to avoid that some block Q of attributes is recomputed, it suffices to add to the pop-assignment in use an order among the blocks of attributes that "use" the block Q . In the previous example Q = {(i,B), (s,B)} and the new dependencies are $(s_1,Y) \to (i_2,Y)$ and $(s_2,Y) \to (i_1,Y)$.

One can easily find examples in which the above transformation for producing a non recomputing ANC-evaluator P' from a recomputing one P is such that P' has size exponential in that of P. Thus in pratice, one should use the above general principle in order to find a satisfactory compromise between the amount of recomputation and the size of the evaluator.

3 - The DNC-evaluator.

In this section we want to give at least an intuition of the way incremental DNC-evaluators can be constructed for DNC-AG's. This construction is, very similar to that of ANC-evaluators.

Let us first explain what we mean with incremental attribute evaluation (for more details see [10]). In a syntax-directed editor in which the attributes are used for checking the static semantics of the developed program, every change of this program is regarded as a substitution in its parse tree t of one subtree with another tree. Let t' be the result of this operation. In general, some, (but not all) of the attributes of t' will be inconsistent in the sense that, evaluating t' a different value would be found for them.
An incremental evaluator tries to recompute in t' only the inconsistent attributes. Let u be the node of t at which the tree substitution producing t' has taken place. Intuitively, a DNC-evaluator starts its computation in u and moving up and down recursively through t', reestablishes the consistency of the attributes of t'. For doing this, a DNC-evaluator has, in addition to the recursive procedures of an ANC-evaluator, also procedures that move up the input tree : procedures whose goal is to compute blocks of i-attributes. To construct such procedures one uses again the augmented production graphs that must exist according to Def. 3(2). DNC-evaluators are described in 2 steps. In the 1st step we describe normal (i.e., nonincremental) evaluators, called *-evaluators, that have procedures for computing blocks of i-attributes as explained above. 2ndly, we will explain how to transform *-evaluators in order to obtain DNC-ones. This 2nd step consists in adding tests for halting the computation of an *-evaluator when the attributes it would compute are already consistent. Both steps are illustrated below for the AG GEX1 and its fine pop-assignment EXΠA1.

Step 1 : the construction of an *-evaluator for GEX1.

The procedures for computing blocks of s-attributes of the *-evaluator are as those of the ANC-evaluators. We will, therefore, concentrate on those for computing blocks of i-attributes. Fig.8 shows the procedure evalΠ(X)i_2 that is constructed using the subgraphs of D(1)[Π(Z),Π(X),Π1(Y)] and D(2)[Π(X),Π(X),Π(X)] that are also given in Fig.8. These subgraphs consist of all the nodes of the corresponding augmented production graphs that are connected to (i_2,1) or (i_2,2) by a path not using edges that come from Π(X_1) and Π(X_2), respectively, see Def.2.
Observe that in the procedure of Fig.8 the knowledge of the attribute dependencies contained in the to-partitions of EXΠA1 has been used for avoiding to recompute attributes. Namely, in point(+) of evalΠ(X)i_2, cf. Fig.8, one does not need to perform the recursive call "evalΠ(X)i_1(n')" because Π(X) states that the call "evalΠ(X)i_1(n)" has been already performed whenever "evalΠ(X)i_2(n)" is executed.

```
Proc eval π(X)i2 (n) ; node n ;
 begin <<let n' be the father of n>>

   case production applied at n' of

        1 : compute (i2,n)

        2 : case n = n'j of

       (+)   j=1 : compute (i2,n)
             j=2 : compute (s1,n') ;
                   call evalπ(X)i2 (n') ;
                   compute (i2,n)
             end of case
   end of case
end
```

```
prod 1 :  (i2,1)
            ↑
          (s1,1)

prod 2 :
          ┌ (i2,1)
    j=1  <     ↑
          └ (i1,0)

          ┌ (i2,0)
          │
    j=2  <     ↑
          │ (s1,0)
          │
          │    ↑
          └ (s1,2)
```

Figure 8.

Let us now see how the main program of the *-evaluator looks like. Let t' be an input tree of GEX1 and u a node of t' labeled (for simplicity) by X. Consider the program of Fig.9. After the execution of this program all attributes of u will be evaluated together with other attributes of t' (viz. at least the attributes necessary for the evaluation of the attributes of u). For evaluating all attributes of t', the *-evaluator needs to move up to the root of t' performing, at each node encountered, only some of the recursive calls of the program of Fig. 9 (for nodes labeled by X). What calls must be done can be calculated using, once more, the augmented production graphs. Unfortunately, the method for doing this is quite technical and it could not be included in the present paper.

```
        begin
                for j : = 1 to 3 do
                        call evalπ(X)ij(u)
                        call evalπ(X)sj(u)
                                od
        end
```

Figure 9.

<u>Step 2</u> : obtaining DNC-evaluators from *-evaluators.

There are 2 types of useless computations of an *-evaluator in an incremental environment (i.e., in t' all attributes have a value that may be inconsistent) :

(a) if it performs a recursive call already executed,

(b) if it performs a recursive call, for instance $\mathrm{eval}\Pi(X)i_2(n)$, with $n \neq u$ (u is the node in which the tree substitution has taken place), when the attributes of n, needed for the call, are consistent ((s_1,n) in the example, because $s_1 \to i_2$ is in $\Pi(X)$). Observe that (s_1,n) is guaranteed to be consistent when $\mathrm{eval}\Pi(X)i_2(n)$ is executed.

Inefficiency (a) can be easily eliminated by marking every block of attributes for which a recursive call has been executed. For eliminating inefficiency (b) one needs to add to each block of attributes of t' a flag stating whether some of its attributes have changed its value or not : in our example, if (s_1,n) is not modified one can avoid to do the call $\mathrm{eval}\Pi(X)i_2(n)$, (if $n \neq u$). Note that these flags could be used also for reducing inefficiency (a), but that this use would not, in general, guarantee the complete elimination of this inefficiency : a block of attributes may be already checked without that any of its attributes has changed value.

4 - Conclusions.

The most interesting aspects of the present work are, in our opinion, the following two.

First, it gives a unifying view of ANC- and DNC-AG's and their evaluators allowing, in this way, to better understand the relation among different results concerning these classes.

Secondly, it stresses the role that the knowledge of the dependencies of an AG plays for the amount of recomputation of the corresponding evaluator. In this sense several interesting questions should be studied.

For instance, it would be interesting to device methods for transforming a given good (or fine) pop-assignment into a better one, i.e.,producting a more efficient evaluator. Such process could consist in :

(i) individuating where the recomputation may take place,

(ii) merging 2 blocks of a po-partition of a nonterminal,

(iii) adding new dependencies to a po-partition of a nonterminal, (as was done for $\Pi(Y)$ at the end of section 2).

REFERENCES

[1] K. Barbar ; Etude comparative de différentes classes de grammaires d'attributs ordonnées ; thèse de 3ème cycle Université de Bordeaux I (1982).

[2] B. Courcelle ; Attribute Grammars : definitions, analysis of dependencies, proof methods ; in Methods and Tools for compiler Construction (B. Lorho ed.), INRIA-CEC course, Cambridge University Press, pp.81-102 (1984)

[3] B. Courcelle et P. Franchi-Zannettacci ; Attribute grammars and recursive program schemes (I and II) ; Theoretical Computer Science 17, pp. 169-191 and 235-257 (1982).

[4] E. Gombas et M. Bartha ; A multi-visit characterization of absolutely noncircular attribute grammars ; Acta Cybernetica 7, pp. 19-31 (1985).

[5] M. Jourdan ; Les grammaires attribuées : implantation, applications; optimisations ; Thèse DDI, Université Paris VII, May 1984.

[6] M. Jourdan; Strongly noncircular attribute grammars and their recursive evaluation ; ACM SIGPLAN 84 Symp. on Compiler Const., Montreal, SIGPLAN Notices 19, pp 81-93 (june 1984).

[7] K. Kennedy et S.K. Warren ; Automatic generation of efficient evaluators for attribute grammars ; 3rd POPL, Atlanta, pp. 32-49 (January 1976).

[8] D.E. Knuth ; Semantics of context-free languages ; Math. Systems Theory 2, pp. 127-145 (1968). Correction : Math Systems Theory 5, pp. 95-96 (1971).

[9] H. Riis Nielson ; Computation sequences :a way to characterize subclasses of attribute grammars ; Acta Informatica 19 pp. 255-268 (1983).

[10] T. Reps, T. Teitelbaum et A. Demers ; Incremental context dependent analysis for language-based editors ; ACM TOPLAS 5, pp. 449-477 (1983).

TRANSFORMATION STRATEGIES FOR DERIVING ON LINE PROGRAMS

Alberto Pettorossi
IASI CNR
Viale Manzoni 30
00185 Roma Italy

BSTRACT

e consider a class of programs whose output is a sequence of "elemen-ary actions" or "moves". We provide some transformation strategies for eriving efficient iterative programs which exhibit a "on-line behaviour", .e. producing the output moves, one at the time, according to a given equence ordering. Our methods also give an answer to a long standing hallenge [Hay77] for deriving a very fast on-line program for the Towers f Hanoi (and similarly defined) problems.

NTRODUCTION

When one applies the transformation technique for deriving correct nd efficient programs, various strategies may be used for driving the app-ication of the basic transformation rules (see for instance [Bir84, et84]). Those strategies play a crucial role in assuring that the erived programs are equivalent to their corresponding initial versions nd they are indeed more efficient.

The application of the basic transformation rules, in fact, does not uarantee that equivalence is preserved or efficiency is improved (see Kot78]).

Unfortunately there is not a universal strategy which will always ssure the derivation of programs with better performances. The situ-tion in this area is like the one in the field of automated deduction. n fact, in a sufficiently powerful logical calculus it is impossible to efine a universal strategy which makes an automatic system to prove ny given theorem. Nevertheless one can define various strategies which re successful when the theorems to be proved are of a given specified orm.

In this paper we will consider a particular class of programs and we will provide a powerful strategy for improving them. In particular that strategy will achieve a "on-line behaviour" (which will be defined later on).

We will be concerned with recursive equation programs as the ones in [BuD77] and we will restrict our attention to the ones which produce strings as output, i.e. whose output belongs to a monoid.

That class of programs, which will be called strings producing programs, often occurs in practical situations. It includes all procedures which generate sequences of instructions or plans for solving problems, and they are encountered in the area of Artificial Intelligence, Robotic Game Playing, Compiler Construction, etc.

We think that for those programs a nice feature to possess is what we call the on-line behaviour, i.e. the output string is produced incrementally, one element at the time, in a specified sequential order (which is usual the order in which the consumer reads that string). Here are some examples of string producing programs.

The Towers of Hanoi program.

f: number × $peg^3 \rightarrow$ Moves*

f(0,A,B,C) = skip

f(n+1,A,B,C) = f(n,A,C,B):AB:f(n,C,B,A)

The output belongs to the monoid Moves*, freely generated by the set Moves={AB,BC,CA,BA,CB,AC,skip} with the concatenation operation : and the identity element skip.

f(n,A,B,C) computes the sequence of moves for moving n disks (of different size) which are stacked as a tower on a peg A (with smaller disks on top of larger ones), from that peg to peg B using peg C as an auxiliary peg. Only one disk at the time can be moved and smaller disks can be placed on top of larger disks only.

The Hilbert's curves of order n.

Hilbert: number → HMoves*

Hilbert(n) = A(n)   A(0)=B(0)=C(0)=D(0)=skip
A(n+1)=D(n):W:A(n):S:A(n):E:B(n)
B(n+1)=C(n):N:B(n):E:B(n):S:A(n)
C(n+1)=B(n):E:C(n):N:C(n):W:D(n)
D(n+1)=A(n):S:D(n):W:D(n):N:C(n)

where N,S,E,w denote the unit moves towards North, South, East, West respectively. skip is the empty sequence of moves. HMoves={N,S,E,W,skip}. For instance, Hilbert(2) is:

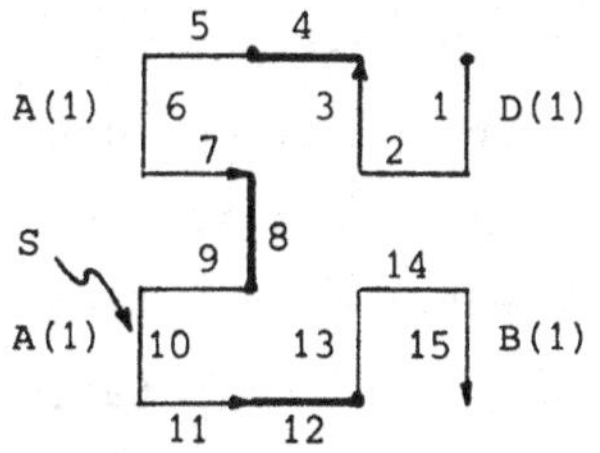

<u>N chinese rings</u> (It is a generalization of a puzzle considered in Artificial Intelligence investigations [ErG82]).
The objective of the game is to remove all rings, numbered from 1 to N, from a stick. Ring N can be removed from the stick or put back on the stick at any time. Any other ring k may be removed from the stick or put back on the stick only if rink k+1 is on the stick and rings k+2,k+3,... N are not on the stick. A solution is given by the following program.
remove: $ring^2 \rightarrow$ RMoves* with RMoves={1,...,N,$\underline{1}$,...,$\underline{N}$}, where k denotes the removal of ring k, and $\underline{k}$ denotes the putting back of the same ring for k=1,...,N. remove(k,N) denotes the sequence of moves for removing from the stick the rings k,k+1,...,N.
put: $ring^2 \rightarrow$ RMoves* The sequence of moves for putting back on the stick the rings k,k+1,...,N is computed by put(k,N).

remove(N,N)=N;
remove(N-1,N)=N-1:N;
remove(k,N)=remove(k+2,N):k:put(k+2,N):remove(k+1,N) <u>if</u> k ⩽ N-2

put(N,N)=$\underline{N}$;
put(N-1,N)=$\underline{N}$:$\underline{N-1}$;
put(k,N)=put(k+1,N):remove(k+2,N):$\underline{k}$:put(k+2,N) <u>if</u> k ⩽ N-2

ON-LINE COMPUTATIONS

In order to motivate our definition of on-line computations, let us start off by recalling the Hayes' challenge concerning the Towers of Hanoi problem [Hay77]. He provided an iterative solution to the problem and he asked to derive it by transformation. Since then, various authors studied that problem (and many variants of it) [Er83,Wal83] and they presented recursive and iterative solutions. We gave a simple transformational derivation of iterative solutions in [Pet85]. The solutions we obtained were not exactly the ones which Hayes asked for, but they have the same efficiency, because they have the same time×space complexity which is exponential).

Therefore we felt that the challenge of Hayes had a satisfactory answer. However, we also felt that Hayes' program (as well as the programs by [BuL80,Er83] for the variant problems) could claim a little advantage over ours. Hayes' program, in fact, gives the sequence of moves in exponential time, but it takes only constant time and constant space to compute the k+1st move after the computation of the k-th one. On the contrary our iterative program is less efficient when one wants to compute the k+1st move after the k-th one. The request for that last improvement motivated our investigations. The results we achieved were generalized to the class of string producing programs.

We say that a computation which produces a string of ℓ elements as output, is on-line if the production of one output element takes constant time and constant space after the production of the precedent element of the string (after a precomputation phase with time complexity at most proportional to ℓ).

One could also allow for logarithmic time and logarithmic space w.r.t. the length ℓ of the output string after a precomputation phase proportional to $\ell \cdot \log \ell$, if the behaviour of the consumer of the output string is not affected by that degradation of performances. Such computations will be called pseudo on-line.

According to our definition, Hayes' program in [Hay77] determines on-line computations, while ours in [Pet85] does not.

The on-line property may often be required because the consumer of the output string (e.g. an executing agent) is asked to work at a convenient and uniform speed. Obviously in obtaining the on-line behaviour for the derived program, we want also to keep the same time×space global performances.

THE GENERAL PROBLEM AT SCHEMA LEVEL

Now we study the problem of transforming recursive programs into on-line iterative versions "at schema level". Let us consider the following recursive schema S, which is general enough to include the above given examples and many others:

$$S: \quad z(n) = \begin{cases} \langle a_1(z_1,\dots,z_r),\dots,a_r(z_1,\dots,z_r)\rangle \ \underline{\text{where}}\ \langle z_1,\dots,z_r\rangle = z(b(n)) & \underline{\text{if}}\ p(n) \\ \langle e_1(n),\dots,e_r(n)\rangle & \underline{\text{otherwise}} \end{cases}$$

We assume that the function $z(n)$ is an r-tuple of functions. The case when $z(n)$ is a single value is obtained for $r=1$.

Having in S only one non recursive clause is not a significant restriction. Everything in what follows can be easily extended to the general case. Since the above schema should produce an output string we assume that:

$\forall i \ 1 \leq i \leq r \ \forall n \ z_i(n) \equiv a_i(z_1,\ldots,z_r) \in M^*$ where M^* is a monoid over a given set M. We also have:

i) $\forall n \ \exists s \geq 0$ s.t. $p(b^s(n))$=false, i.e. for any n the computation of $z(n)$ terminates;

ii) $\forall i \ 1 \leq i \leq r \ e_i(n)$ is an element of M , and $a_i(z_1,\ldots,z_r)$ is the concatenation of its arguments and (possibly) some constants in M, i.e. $a_i(z_1,\ldots,z_r) \in (M \cup \{z_1,\ldots,z_r\})^*$.

We also assume that the length of the i-th component of $z(n)$ for $1 \leq i \leq r$ satisfies a linear recurrence relation with constant coefficients. More formally:

$$\forall i \ 1 \leq i \leq r \quad L(z_i(n)) = \sum_{j=1}^{r} c_{ij} \ L(z_j(b(n))),$$

where $L: M^* \to N$ is the length function, which counts the number of elements of M in the given sequence in M^*. The c_{ij}'s are non-negative integers.

Finally we assume that the recurrence relations for the $L(z_i(n))$'s can be <u>inverted</u>, i.e. $\forall n \ \exists m \geq 0$ s.t. given the values of $L(z_i(b^k(n)))$ for $1 \leq i \leq r$ and $0 \leq k \leq m$ we can compute the values of $L(z_i(b^{k+1}(n)))$ for $1 \leq i \leq r$.

The above invertibility hypothesis allows the evaluation of $z(n)$ using a bounded number of memory cells (and avoiding the use of a stack). The bound m is indeed proportional to the amount of memory we need. (In what follows we will give a necessary and sufficient condition for the invertibility property).

For the schema S we can derive a recurrence relation on the length of the components of $z(n)$. By $L(z(n))$ we denote the r-tuple of those lengths, i.e. $\langle L(z_1(n)),\ldots,L(z_r(n))\rangle$. For simplicity we will write $Lz(n)$ instead of $L(z(n))$, and $Lz_i(n)$ instead of $L(z_i(n))$.

$Lz: N \to N^r$

$$Lz(n) = \begin{cases} \langle \sum_{j=1}^{r} c_{1j} \ Lz_j(b(n)),\ldots, \sum_{j=1}^{r} c_{rj} \ Lz_j(b(n))\rangle & \underline{\text{if}} \ p(n) \\ \langle \ell_1(n),\ldots,\ell_r(n)\rangle & \underline{\text{otherwise}} \end{cases}$$

where $\forall i,j$ $1 \leq i,j \leq r$ the value of c_{ij} depends on the function a_i.

Suppose we have to compute $z(N)$. We may proceed as follows. By hypothesis $\exists k \geq 0$ s.t. $p(b^k(N))$ is false. Let us call $\underline{k}$ the minimum value of k which satisfies that condition. Let n_0 be $b^{\underline{k}}(N)$.

We may tabulate the values of $Lz(n)$ for n_0 (s.t. $p(n_0)$=false), $b^{-1}(n_0)$, $b^{-2}(n_0),\dots,b^{-\underline{k}}(n_0)$, where by b^{-1} we denote the inverse of the function b i.e. $b^{-1} \cdot b = \lambda x.x$. For instance if b is $\lambda x.x-1$ then b^{-1} is $\lambda x.x+1$.

We have the following matrix LZ:

	n_0	$b^{-1}(n_0)$	$b^{-2}(n_0)$	...	$b^{-\underline{k}}(n_0)=N$
Lz_1					
⋮					
Lz_r					

The rightmost column of the matrix LZ gives us the length of the components of $z(N)$ we want to compute.

At this point we do not have yet the desired output, but we have the lengths of its components. Moreover we have to solve the problem related to the fact that the matrix LZ is unbounded, i.e. the value of $\underline{k}$ depends on the value N of the input. The use, in fact, of an unbounded matrix is equivalent to the use of a stack for evaluating the recursion and we want to avoid that.

In what follows, using the invertibility hypothesis, we will provide a solution to that problem by showing that only a bounded portion of the matrix LZ is needed.

However, for the time being, let us proceed by assuming that we may access to the values of the entire matrix LZ.

The <u>procedure FIND(j,n,k)</u> computes the k-th element of any component $z_j(n)$ for $1 \leq j \leq r$ of the output $z(n)$ which is a r-tuple of strings. j and n are given. For instance, FIND(j,n,1) produces the first element of $z_j(n)$ and FIND(j,n,$Lz_j(n)$) produces the last (i.e. $Lz_j(n)$-th) element of $z_j(n)$. The answer "no element" means that the output string does not have the k-th element. This is the case when $k > Lz_j(n)$.

Let LZ(j,n) be the entry of LZ at row j and column n.

FIND(j,n,k) =

<u>if</u> k>LZ(j,n) <u>then</u> "no element" <u>else</u>

<u>begin</u> Let $z_j(n) \equiv a_j(z_1,\dots,z_r)$ be the string s1 s2 ...sp of elements in $M \cup \{z_1,\dots z_r\}$ and let [ℓ1,...,ℓp] be the list of the lengths of those strings s1,...,sp. (The lengths of z_i for $1 \leq i \leq r$ as components of $z_j(n)$ are the entries of LZ in the column b(n) and row i).

<u>if</u> the k-th element of the sequence $a_j(z_1,\dots,z_r) \in M$

<u>then</u> the answer is that element of M

<u>else</u> call FIND(newj,b(n),newk) where newj and newk are obtained as result of NEW([ℓ1,...,ℓp],k,[s1,...,sp]).

<u>end</u>

NEW([ℓ1,...,ℓp],k,[s1,...,sp]) =

<u>if</u> k<ℓ1 <u>then</u> newk=k and newj=q where z_q=s1

<u>else</u> call NEW([ℓ2,...,ℓp],k-ℓ1,[s2,...,sp]).

The above procedure FIND always terminates because:

i) the procedure NEW terminates, because $k \leq \sum_{i=1}^{p} \ell i = LZ(j,n)$ and $k-\ell 1 \leq \sum_{i=2}^{p} \ell i$;

ii) when constructing [ℓ1,...,ℓp] we have: $\sum_{i=1}^{p} \ell i = LZ(j,n)$ and each ℓi is <u>either</u> a constant <u>or</u> an entry of the left adjacent column in the matrix LZ;

iii) each successive call of FIND reads a left-adjacent column of LZ. Eventually FIND reads the first column of LZ and stops.

<u>SOME EXAMPLES OF APPLICATION OF THE PROCEDURE FIND</u>

Let us compute the 10-th move of the drawing of the <u>Hilbert's curve</u> of order 2. It is the S (South) move as shown in the figure we gave in the introduction.

By application of the tupling strategy [Pet84] we derive:

```
Hilbert(n)=z1(n)           (i.e. the first projection of z(n))
z(n+1)=<z4:W:z1:S:z1:E:z2,   z3:N:z2:E:z2:S:z1,
       z2:E:z3:N:z3:W:z4,    z1:S:z4:W:z4:N:z3>
                                        where <z1,z2,z3,z4>=z(n)
z(0)=<skip,skip,skip,skip>
```

with the "eureka function" [BuD77] z(n)≡<A(n),B(n),C(n),D(n)>.

The linear recurrence relation on the lengths of the components of z(n) is easily derived from the program above:

Lz(n+1)=<2Lz1+Lz2+Lz4+3, Lz1+2Lz2+Lz3+3,
Lz2+2Lz3+Lz4+3, Lz1+Lz3+2Lz4+3>
where <Lz1,Lz2,Lz3,Lz4>=Lz(n)

Lz(0)=<0,0,0,0>

In this case $b=\lambda x.x-1$, $n_0=0$. Since we want to compute the 10-th move of Hilbert(2), k=2. The matrix LZ is:

	0	1	2
Lz1	0	3	15
Lz2	0	3	15
Lz3	0	3	15
Lz4	0	3	15

We call FIND(1,2,10) because: j=1 (Hilbert(n) is the first projection of z(n) and n=2 (we want to compute a move of Hilbert(2)).

We have: z1(2) = z4(1):W:z1(1):S:z1(1):E:z2(1) with the list of lengths [3,1,3,1,3,1,3]. Since the 10-th element is not an element in {W,S,E,N} (indeed it is the second element in z1(1)) in the symbolic expression of z1(2) given above, we make a recursive call of FIND. For that purpose we compute NEW([3,1,3,1,3,1,3], 10,[z4(1),W,z1(1),S,z1(1),E,z2(1)])

= NEW([1,3,1,3,1,3],10-3,[W,z1(1),S,z1(1),E,z2(1)])= ...

= NEW([3,1,3], 2,[z1(1),E,z2(1)]).

Therefore newk=2 and newj=1 (because q=1). We call FIND(1,1,2), i.e. we for the second element in z1(1). We derive :

z1(1) = z4(0):W:z1(0):S:z1(0):E:z2(0) with the list of lengths: [0,1,0,1,0,1,0]. Thus the 10-th move of Hilbert(2) is equal to the 2nd move of z1(1)=W:S:E (because z4(0)=z1(0)=z2(0)=skip). That move is S, as expected.

We take the second example of application of the procedure FIND from the (standard) Towers of Hanoi problem.

In [Pet85] by applying the tupling strategy we obtain:

f(n,A,B,C) = t1(n)

t(n+2) = <t1:AC:t2:AB:t3:CB:t1, t2:BA:t3:BC:t1:AC:t2,
t3:CB:t1:CA:t2:BA:t3> where <t1,t2,t3>=t(n)

t(1) = <AB,BC,CA>

t(0) = <skip,skip,skip>

with the "eureka function" t(n)≡<f(n,A,B,C), f(n,B,C,A), f(n,C,A,B)>.

As usual ti denotes the i-th component of t for i=1,2,3.

Notice that the recursive equations for t(n) do not fit the schema S because there are two non recursive cases: t(1) and t(0). However, the extension of the analysis we presented in the previous section is straightforward and it leads to the following solution.

Suppose we want to compute the 13-th move of f(5,A,B,C), i.e. the 13-th move when moving 5 disks from peg A to peg B.

The linear recurrence relation of the length of the components of t(n) is:

Lt(n+2) = <2Lt1+Lt2+Lt3+3, Lt1+2Lt2+Lt3+3,
Lt1+Lt2+2Lt3+3> <u>where</u> <Lt1,Lt2,Lt3>=Lt(n)

Lt(1) = <1,1,1>

Lt(0) = <0,0,0>.

In this case b=λx.x-2, n_0=0 if N (i.e. the number of disks) is even, and n_0=1 if N is odd. Since we want to compute f(5,A,B,C) it turns out that <u>k</u>=2 and n_0=1. The relevant base case is t(1) and the matrix LT has the leftmost column for n_0=1. (Now we call the matrix of lengths LT instead of LZ).

	1	3	5
Lt1	1	7	31
Lt2	1	7	31
Lt3	1	7	31

We call FIND(1,5,13). Since 13<LT(1,5)=31 we have:
t1(5) = t1(3):AC:t2(3):AB:t3(3):CB:t1(3) with the list of lengths: [7,1,7,1,7,1,7]. We call NEW([7,1,7,1,7,1,7],13,[t1(3),AC,t2(3),AB,t3(3), CB,t1(3)]) and we obtain NEW([7,1,7,1,7],5,[t2(3),AB,t3(3),CB,t1(3)]). Thus newk=5 and newj=2. We call FIND(2,3,5). Since 5<LT(2,3)=7, we have: t2(3) = t2(1):BA:t3(1):BC:t1(1):AC:t2(1) with the list of lengths [1,1,1,1,1,1,1]. Then we will call FIND(1,1,1) which produces the first component of t(1), i.e. AB.

<u>ANALYSIS AND IMPROVEMENTS OF THE TRANSFORMATION METHOD</u>

The transformation method we propose consists of two phases:

i) derivation of a recursive structure (maybe by the application of the tupling strategy) such that the lengths of the components of the output string can be arranged in a matrix (like LZ or LT above);

ii) computation of the requested k-th element of the output string by using a "decomposition" of k according to the numbers stored in the matrix of lengths and taking into consideration the recursive program structure. This second phase is time-proportional to the length of that decomposition (see the way the procedure FIND works).

Now we would like to show that there is no need for storing the entire matrix of lengths LZ. It is enough to keep the values of a fixed number of its columns only. That fact makes it possible to produces an iterative algorithm which does not simulate the use of a stack, when translating recursion into iteration.

The crucial observation is the invertibility hypothesis of the recurrence relation on the length of the output string. It is enough to store m adjacent columns of the matrix LZ to recover all of them. (The parameter m was introduced in the definition of the invertibility property).

For instance, in the case of the Towers of Hanoi the recurrence relations we have given, can be inverted as follows. For simplicity let us write ℓ' instead of Lt(n+2) and ℓ instead of Lt(n). As usual the suffixes 1,2, and 3 denote the components of a tuple.

We have: $\ell1'=2\ell1+\ell2+\ell3+3$, $\ell2'=2\ell2+\ell3+\ell1+3$, $\ell3'=2\ell3+\ell1+\ell2+3$.

We can easily derive the values of ℓ from those of ℓ' and we get:

$\ell1=(3\ell1'-\ell2'-\ell3'-3)/4$; $\ell2=(3\ell2'-\ell3'-\ell1'-3)/4$; $\ell3=(3\ell3'-\ell1'-\ell2'-3)/4$.

This result allows us to derive from a column of LT its left-adjacent column.

If the invertibility hypothesis holds, the amount of memory needed for the procedure FIND to work is constant and it does not depend on the input parameter N.

A necessary and sufficient condition for the invertibility of a given recurrence relation is stated as follows.

Let us consider without loss of generality the case of 2 equations of order 2.

$$\begin{cases} x(n) = a_1x(n-2) + a_2x(n-1) + a_3y(n-2) + a_4y(n-1) + a_0 \\ y(n) = b_1x(n-2) + b_2x(n-1) + b_3y(n-2) + b_4y(n-1) + b_0 \end{cases}$$

We get:

$$\begin{cases} a_1x(n-2) + a_3y(n-2) = c_a & \underline{\text{where}} \quad c_a=x(n)-a_0-a_2x(n-1)-a_4y(n-1) \\ b_1x(n-2) + b_3y(n-2) = c_b & \underline{\text{where}} \quad c_b=y(n)-b_0-b_2x(n-1)-b_4y(n-1) \end{cases}$$

This system of equations has a unique solution iff $d \equiv \begin{vmatrix} a_1 & a_3 \\ b_1 & b_3 \end{vmatrix} \neq 0$ (by the Rouché-Capelli Theorem).

Therefore, given $x(n), y(n), x(n-1), y(n-1)$ and $a_0, \ldots, b_0, \ldots$ we can univocally derive $x(n-2)$ and $y(n-2)$ iff $d \neq 0$.

Notice also that the procedure FIND and NEW are tail-recursive and they can easily be transformed into iterative programs without using a stack.

We close this section by giving a detailed analysis of the behaviour of the method we have proposed for deriving on-line iterative programs.

Our method requires the initialization of the matrix LZ, or at least of a fixed number of its columns, if the invertibility hypothesis holds.

The time complexity of this initialization phase is logarithmic with the value of the input parameter. (Recall that linear recurrence relaions with constant coefficients can be computed in logarithmic time.)

The space complexity is constant, if the invertibility hypothesis holds, because the tupling strategy requires a bounded number of components only.

After the initialization phase, the computation of a particular element of the output string is, in the algorithms we derive, <u>independent</u> from any other ("past" or "future") element of that string.

Our transformation method derives, in the given hypotheses, algorithms which produce the elements of the output string in a completely independent way, while for the on-line behaviour we need only to obtain the elements of the output string independently from the "future", i.e. from those which follow, if the chosen order is the canonical one: 1st,2nd,...

Unfortunately we pay that extra capability we achieve, because the time required for computing the k+1st element of the output string, after the computation of the k-th element, is not constant.

Therefore we do not manage, in general, to achieve the on-line behaviour, but a <u>pseudo on-line</u> behaviour only.

The complexity analysis of the procedure FIND shows that for lengths of the output strings which satisfy linear recurrence relations, if we want to compute the k-th element of the output string, the number of recursive calls of FIND depends on a "decomposition" of k (related to the syntactic structure of the derived program), which is logarithmic.

The evaluation of each recursive call of FIND requires a constant amount of space, i.e. the two lists which are arguments of FIND and NEW, because it is proportional to the length of the derived program.

Since usually arithmetic operations, which indeed take logarithmic time w.r.t. the value of their arguments denoted by their binary expansions, are considered to take constant time, we can assume that the evaluation of FIND takes constant time.

Therefore we can conclude that in particular for the Towers of Hanoi problem, we achieved the on-line behaviour, at least in the same sense it is achieved by Hayes' algorithm.

In the following section we will show how one can improve the performances of the procedure FIND. The purpose of that investigation is to give an even better answer to Hayes' challenge.

AN ANSWER TO HAYES' CHALLENGE

A different application of the tupling strategy (w.r.t. the one given above) gives us the following program for the Towers of Hanoi problem:

```
f(n,A,B,C) = t1(n,A,B,C)
t(n+1,a,b,c) = <t1:ab:t2, t3:bc:t1, t2:ca:t3> where <t1,t2,t3>=t(n,a,c,b
t(0,a,b,c) = <skip,skip,skip>.
```

Given a sequence s of elements in Moves={AB,BC,CA,BA,CB,AC,skip}, s.t. $s \equiv m_1 \ldots m_p$ let us denote by $\underline{s}$ the sequence $\underline{m}_1 \ldots \underline{m}_p$ where for any i each move $\underline{m}_i$ is obtained from m_i by interchanging B and C. For instance, $\underline{AB}$ is AC and $\underline{BC}$ is CB. Obviously $\underline{\underline{s}}=s$, because $\underline{\underline{m}}=m$ for any m ε Moves.

Therefore we may derive the following version of the program:

$f(n,A,B,C) = t1(n)$

$t(n+1) = \langle \underline{t1}:AB:\underline{t2},\ \underline{t3}:BC:\underline{t1},\ \underline{t2}:CA:\underline{t3} \rangle$ <u>where</u> $\langle t1,t2,t3 \rangle = t(n)$

$t(0) = \langle skip,skip,skip \rangle$.

The matrix LT of the lengths of the components of the moves is:

	0	1	2	3	4	
Lt1=Lt2=Lt3	0	1	3	7	15	...

where the h column of LT has 2^h-1. (To solve the problem for h disks we need 2^h-1 moves).

The structure of the matrix LT allows us to interpret the computation of the procedure FIND as a walk through a finite automaton. The automaton is:

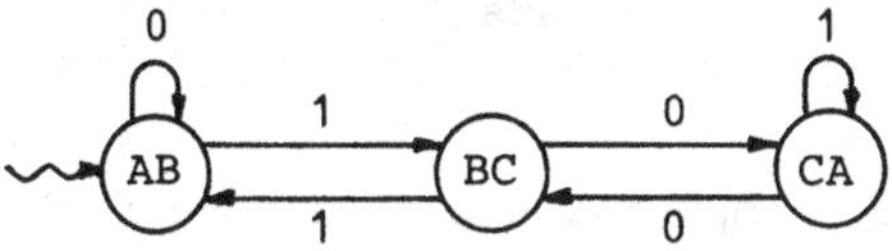

Let us explain this fact through an example. Suppose we want to compute the k-th move of f(n,A,B,C).

The structure of the automaton is related to the one of the program equation for t(n+1). t1 corresponds to the state AB, t2 to BC, and t3 to CA. For the component t1:AB:t2 we have that the state AB has a 1(0)-transition to the state BC(AB), because if $k>2^n-1$ $(k<2^n-1)$, i.e. the most significant bit of k is 1(0), then the move belongs to t2 (t1). Analogously one may derive the transitions for the states BC and CA looking at the other components t3:BC:t1 and t2:CA:t3.

The walk through the automaton is driven by the digits of $k=\sum_{i=1}^{n} k_i\cdot 2^{i-1}$ starting from k_n. If $k_n=1(0)$ we follow the 1(0)-transition. When we reach a new state, we proceed by using the following less significant digit of k. The walk ends after reaching a state for which the digits of k yet to be read denote a power of 2. In that halting state we compute the k-th move: if during the walk we followed an even number of transitions, it is the name of the halting state, otherwise it is that name, where B and C are interchanged.

Here is the informal proof of correctness of the above interpretation. The length of f(h,A,B,C) is 2^h-1. The k-th move in the sequence f(h,A,B,C)≡ f(h-1,A,C,B):AB:f(h-1,C,B,A) is computed as follows. It is AB if $k=2^{h-1}$. It is the k-th move of f(h-1,A,C,B) if $k<2^{h-1}$. It is the $k-2^{h-1}$th move of f(h-1,C,B,A) if $k>2^{h-1}$. It means that: if k is a power of 2 we know the k-th move; otherwise we have to discard the initial digit of k, whereby obtaining k', and we have to find the k'-th move of f(h-1,A,C,B) (or f(h-1,C,B,A)) if the most significant digit of k is 0 (or 1). Notice that after transforming the problem of finding the k-th move in the one of finding the k'-th move, we need to interchange B and C: in fact in the equation for t(n+1) the various components of t(n) are all underlined.

To store the automaton described above requires constant space. It can be represented by the following transition matrix A:

A:

i \ j	0	1
1	1	2
2	3	1
3	2	3

A[i,j] =1,2,3 stands for AB,BC,CA respectively.
i =1,2,3 stands for AB,BC,CA respectively.

The time requirements for computing the k-th move is logarithmic with th length of the binary expansion of k. Testing a number for being a power of 2 is assumed to take constant time.

Here is the program for computing the k-th move of f(h,A,B,C) with $1 \leq k \leq 2^h - 1$. Let $k = \sum_{i=1}^{h} k_i \cdot 2^{i-1}$ and $k^p = \sum_{i=1}^{p} k_i \cdot 2^{i-1}$.

```
p:=h;  i:=1; comment the initial state is AB, i.e. i=1;
while p≥1 do if k^p=2^(p-1) then out(h-p,i) else i:=A[i,k_p]; p:=p-1 od
comment h-p is the number of transitions followed during the walk;

out(m,i) = if even(m) then print(i) else print(i[B⇌C])
```

where i[B⇌C] denotes i with B and C interchanged. The out instruction produces an escape from the while-do loop.

Example. Suppose we want to compute the 14-th move of f(4,A,B,C). Since 14=1110 we go from state AB to state AB with 2 transitions, using the 2 initial digits '11'. In the state AB we have to use the digits '10 Since 10=2 we stop our walk, and we get: h=4; p=2; h-p=2, and 2 is even. Thus the 14-th move of f(4,A,B,C) is AB.

The 11-th move of f(4,A,B,C) is CA[B⇌C] , i.e. BA. □

Our solution computes the sequence of moves of the output string in an possible order, and it is an advantage on Hayes' program.

CONCLUSIONS

We applied a proposed transformation method to a class of programs which produce strings, i.e. elements of a monoid, as output. We presented an algorithm for obtaining equivalent iterative programs highly efficien and exhibiting the "on-line behaviour" (here defined). By that notion we mean that subsequent elements of the output strings can be computed efficiently enough (for instance, in constant time and space) after the computation of the preceding elements. Among other examples, we applie that method to the Towers of Hanoi problem, providing an answer to a long standing challenge [Hay77] of deriving by a transformational approach an efficient on-line iterative program for it.

REFERENCES

[Bir84] Bird, R.S.: "The Promotion and Accumulation Strategies in Transformational Programming" ACM Transaction on Programming Languages and Systems Vol.6 No.4 October 1984, 487-504.

[BuD77] Burstall, R.M. and Darlington, J.: "A Transformation System for Developing Recursive Programs" J.A.C.M. Vol.24, 1 (1977) 44-67.

[BuL80] Buneman, P. and Levy, L.: "The Towers of Hanoi problem" Information Processing Letters 10,(1980), 243-244.

[Er83] Er, M.C.: "An Iterative Solution of the Generalized Towers of Hanoi Problem" BIT 23 (1983), 295-302.

[ErG82] Ernst, G.W. and Goldstein, M.M.: "Mechanical Discovery of Classes of Problem-Solving Strategies" J.A.C.M. 29 (1)(1982), 1-23.

[Hay77] Hayes, P.J.: "A Note on The Towers of Hanoi Problem" Computer Journal 20, (1977), 282-302.

[Kot78] Kott, L.: "About Transformation System: A Theoretical Study" Proceed. 3ème Colloque International sur la Programmation. Dunod, Paris (1978), 232-247.

[Pet84] Pettorossi, A.: "A Powerful Strategy for Deriving Efficient Programs by Transformation" ACM Symposium on LISP and Functional Programming. Austin,Texas USA (1984), 273-281.

[Pet85] Pettorossi, A.: "Towers of Hanoi Problems: Deriving Iterative Solutions by Program Transformation" BIT Vol.25 (1985), 327-334

[Wal83] Walsh, T.R.: "Iteration Strikes Back at the Cyclic Towers of Hanoi" Information Processing Letters 16 (1983), 91-93.

PROBABILISTIC IANOV'S SCHEMES (Extended abstract)

D. Frutos Escrig
Dpto. Ecuaciones Funcionales - Fac. C. Matemáticas
Universidad Complutense - 28040 MADRID (SPAIN)

Abstract

We present probabilistic Ianov's schemes, studying their semantics and proving the equivalences between operational and denotational ones. We also study the equivalence of schemes relative to them; as usual all thes equivalence problems are decidable, and we prove it giving the appropiate decision algorithms.

Introduction

Ianov's schemes are, without any doubt, the simplest control systems. That is because when we decided to study probabilistic programs, we began by studying probabilistic Ianov's schemes. But as probabilism and nondeterminism are closely related, we have studied first nondeterministic Ianov's schemes, considering the three classical (nowadays !) ways to define a semantics of nondeterministic programs: Hoare's or angelical semantics (forget about the infinite computations) [1 , 6 , 7], Plotkin's one (consider all the computations) [9] and Smyth's one (nontermination is a disaster) [10]. In [2] and [4] you can find an exhaustive study of these schemes, whose principal results we will resume in the first section of this paper.

Our probabilistic schemes are obtained from nondeterministic ones, by labelling the ares leaving and OR-node by rational nombers, that say us what is the probability of a computation reaching that node continues going after each of these arcs. So that we have to complicate our schemes but this complication is not gratuitous, as we obtain some intuitive improvements: semantics say us what is "probable" instead of only "possible"; and also some technical ones: firstly we have a "natural" semantics, and on the other hand when we consider trace semantics, denotationa and operational ones are equal.

In the second section of this paper we will define our probabilistic schemes and their natural semantics (operational and denotational) proving their equivalence. The third one is dedicated to an study of the equivalence of schemes relative to the defined semantics, while in the fourth we will define an Smyth's - like semantics of probabilistic schemes, which induces an equivalence between schemes that is decidable too, although to prove it turn out rather difficult. Finally in the fifth section we study trace semanticns; for we need a nontrivial notion of probabilistic powerdomain, as now the base domain is not a flat one.

By lack of space some proofs have been abridged. You can find the complete ones in [2], and soon an english extended version of this paper will appear elsewhere.

1. Nondeterministic Ianov's schemes

Def. 1.1 : A nondeterministic Ianov's scheme S is given by: a set of predicate symbols P; another set of action symbols A; and a labelled (on its nodes) directed finite graph G, with labels from the set $A \cup P \cup \{STOP, OR\}$, so that if n is labelled by $a \in A$, from it leaves exactly one arc, if n is labelled by $p \in P$, there are exactly two arcs, labelled by TRUE and FALSE, going out from it, and STOP nodes are the only ones from which leaves no arc; finally G has a distinguished node r, that we will call its root, from which each node of G can be reached.

Def. 1.2 : a) An interpretation I is a 3-uple $\langle D,\varphi,\psi\rangle$, where D is a set (its domains), φ: $D \times P \to \{ TRUE, FALSE \}$ and ψ: $D \times A \to D$. We write $\varphi_p(d)$ for $\varphi(d,p)$ and $\psi_a(d)$ for $\psi(d,a)$. b) We say that I is free when $D = A^*$ and $\psi_a(d) = d\cdot a$.

Def. 1.3 : a) If we have an scheme $S = (P,A,G,r)$ and $n \in G$, we can define the subscheme $S_n = (P,A,G_n,n)$, where G_n is the complete subgraph of G whose nodes are the ones that can be reached, by a path, from n. b) We can classify the schemes paying attention to their roots: if r is labelled by $p \in P$ and n_1 and n_2 are the nodes with $(r,n_1) \in G$ labelled by TRUE, and $(r,n_2) \in G$ labelled by FALSE, we will denote it by $p(S_{n_1},S_{n_2})$; if r is labelled by a, and $(r,n) \in G$, we will denote it by $a(S_n)$;if it is labelled by OR, and $\{n_1,\dots,n_k\} = \{ n \in G \mid (r,n) \in G\}$ we will denote it by $OR(S_{n_1},\dots,S_{n_k})$; and if r is labelled by STOP, we will say that $S = STOP$.

Def. 1.4 : a) A configuration (under I) is a pair (d,S) where $d \in D$, and S is an scheme such that I can be applied to S (that is, I is defined over P' and A', and $S = (P,A,G,r)$ with $A \subseteq A'$, $P \subseteq P'$). b) A computation step of S, under I, is a pair of computations $(d_1,S_1) \overset{I}{\to} (d_2,S_2)$ verifying:

- $S_1 = STOP \Rightarrow S_2 = STOP \wedge d_2 = d_1$
- $S_1 = a(S_1') \Rightarrow S_2 = S_1' \wedge d_2 = \psi_a(d_1)$
- $S_1 = p(S_1',S_1'') \Rightarrow ((\varphi_p(d_1) = TRUE \Rightarrow S_2 = S_1') \wedge (\varphi_p(d_1) = FALSE \Rightarrow S_2 = S_1'') \wedge d_2 = d_1$
- $S_1 = OR(S_1',\dots,S_k') \Rightarrow (\exists i \in \{ 1,\dots,k \}\ S_2 = S_i') \wedge d_2 = d_1$

c) A computation c of S, under I, from $d \in D$ ($c \in C_I(S,d)$) is a sequence of computation steps c: $(d_1,S_1) \stackrel{I}{\to} (d_2,S_2) \stackrel{I}{\to} \ldots$ where $d_1 = d$ and $S_1 = S$. We say that c stops ($c\downarrow$), with result d', if there is some $i \in \mathbb{N}$ such that S_i = STOP and $d' = d_i$; we will denote this result by r(c).

d) A partial computation $c_p \in C_I^p(S,d)$ is a finite sequence of computation steps c_p: $(d_1,S_1) \stackrel{I}{\to} \ldots \stackrel{I}{\to} (d_n,S_n)$, being $S_1 = S$ and $d_1 = d$. We say that c_p stops, with result $r(c_p) = d_n$, if S_n = STOP.

Def. 1.5 : The operational semantics of S (under I) are given by:

- $S_A(S,d) = \{d' \in D \mid \exists c \in C_I(S,d),\ c\downarrow \ \wedge\ r(c) = d\}$

- $S_p(S,d) = \begin{cases} S_A(S,d) \cup \{\bot\} & \text{if } \exists c \in C_I(S,d),\ c\uparrow \text{ (that is } \neg c\downarrow) \\ S_A(S,d) & \text{otherwise} \end{cases}$

- $S_{SM}(S,d) = \begin{cases} \bot & \text{if } \exists c \in C_I(S,d),\ c\uparrow \\ S_A(S,d) & \text{otherwise} \end{cases}$

To define their denotational counterparts we need the corresponding powerdomains:

Def. 1.6 : a) The angelical powerdomain $P_A(D)$ is given by the powerset 2^D, ordered by set inclussion. b) Plotkin's powerdomain $P(D)$ is $P_F(D) \cup P_\bot(D)$, where $P_F(D) = \{A \subseteq D \ / \ |A| < \infty\}$, $P_\bot(D) = \{A \cup \{\bot\} \ / \ A \in P_F(D)\}$, ordered by Egli-Milner's order relative to the flat one over $D_\bot$. c) Smyth's powerdomain $P_{SM}(D)$ is the flat domain $(P_F(D))_\bot$.

Def. 1.7 : The denotational semantics S_X^d, where $X \in \{A,P,SM\}$, of an scheme S are defined, as usual, by means of a system of equations, whose minimal solution gives us them:

- $S = \text{STOP} \Rightarrow S_X^d[\,S\,](d) = \{d\}$

- $S = a(S_1) \Rightarrow S_X^d[\,S\,](d) = S_X^d[\,S_1\,](\Psi_a(d))$

- $S = p(S_1,S_2) \Rightarrow S_X^d[\,S\,](d) = \text{COND}(\varphi_p, S_X^d[S_1]\ ,\ S_X^d[S_2]\ )(d)$

- $S = OR(S_1,\ldots,S_k) \Rightarrow S_X^d[\,S\,](d) = \bigcup_{i=1}^{k}{}_X\ S_X^d[\,S_i\,](d)$, where each operator $\cup_X$ is defined as follows:

$\cup_A = \cup_p = \cup$(set union);

$\cup_{SM}\ A_i = \begin{cases} \bot & \text{if } \exists i \in \{1,\ldots,k\}\quad A_i = \bot \\ \cup A_i & \text{otherwise} \end{cases}$

Theorem 1.8 : The operational and denotational semantics, defined in each of the three ways, are equivalent.

To study the equivalence of schemes, we need a normal form:

Def. 1.9 :We say that S_1 and S_2 are X-equivalent, for $X \in \{A,P,SM\}$, $(S_1 \underset{X}{\sim} S_2)$, if for any interpretation of both, and for all $d \in D$, we have $S_X(S_1,d) = S_X(S_2,d)$.

Prop. 1.10 : $S_1 \underset{X}{\sim} S_2$ iff for any free interpretation $S_X(S_1,\varepsilon) = S_X(S_2,\varepsilon)$

Notation : We will say that c begins executing $\alpha \in A^*$ ($\alpha \in Pre(c)$) , when α is a prefix of the chain of actions executed along c ; if c stops, we will call Act(c) to the chain of actions executed by it ; and we say that c loops (c↻) when it does not stop, but it only executes a finite chain of actions, that will be denoted by Lact(c) .

Def. 1.11 : We say that S_1 and S_2 are strongly equivalent ($S_1 \simeq S_2$) if for any interpretation I of both, and for all $d \in D$, we have

i) $\forall \alpha \in A^* \; (\exists c \in C_I(S_1,d) \;\; \alpha \in Pre(c)) \Leftrightarrow (\exists c' \in C_I(S_2,d) \;\; \alpha \in Pre(c'))$

ii) $\forall \alpha \in A^* \; (\exists c \in C_I(S_1,d) \;\; c\downarrow ,\ Act(c)=\alpha) \Leftrightarrow (\exists c' \in C_I(S_2,d) \;\; c'\downarrow, Act(c')=\alpha)$

iii) $\forall \alpha \in A^* \; (\exists c \in C_I(S_1,d) \;\; c$↻ $, Lact(c)=\alpha) \Leftrightarrow$
$(\exists c' \in C_I(S_2,d) \;\; c'$↻ $, Lact(c')=\alpha)$

Remark : As usual, it is sufficient to check these conditions for any free interpretation, and $d = \varepsilon$.

Prop. 1.12 : $\forall X \in \{A,P,SM\}$, $S_1 \simeq S_2 \Rightarrow S_1 \underset{X}{\sim} S_2$.

We will not define here normal form schemes (see [2]). You only need to know that is very easy to associate to any one of them, say S , a regular language L(S) that characterizes the family of sets $\{ Res(c) / c \in C_I(S,d) ,\ c\downarrow \}$, for any interpretation of S , I , and all $d \in D$. And, fortunately , we have the following theorem :

Theorem 1.13 : For each Ianov's scheme S , we can construct $\overline{S}$ in normal form, and stongly equivalent to S .

Theorem 1.14 : $S_1 \underset{A}{\sim} S_2 \Leftrightarrow L(\overline{S_1}) = L(\overline{S_2})$.

Corollary 1.15 : Angelical equivalence of Ianov's schemes is a decidable property.

To decide Smyth's equivalence is more complicated, since all the infinite computations must be identified. But we can detect the nodes of normal form schemes that witness the existence of such computations, so that from $\overline{S}$ we can construct S^{SM} , verifying

Theorem 1.16 : a) $S \underset{SM}{\sim} S^{SM}$; b) $S_1 \underset{SM}{\sim} S_2 \Leftrightarrow S_1^{SM} \underset{A}{\sim} S_2^{SM}$.

Corollary 1.17 : Smyth's equivalence between Ianov's schemes is decidable.

Finally, for Plotkin's equivalence we have

Theorem 1.18 : $S_1 \underset{P}{\sim} S_2 \Leftrightarrow (S_1 \underset{A}{\sim} S_2 \wedge S_1 \underset{SM}{\sim} S_2)$.

Corollary 1.19 : Plotkin's equivalence is decidable.

Our last results concern trace semantics of Ianov's schemes.

Def. 1.20 : We will call (free) trace semantics of a Ianov's scheme S , under a free interpretation I , to the function $S_{SEQ}(S,\cdot)$ that associates to each $\alpha \in A^*$, the set of sequences of actions, eventually ended by

STOP, executed by the elements of $C_I(S,\varepsilon)$.

We have $S_{SEQ}(S,\varepsilon) \subseteq S(A) = A^* \cup A^*\cdot\{STOP\} \cup A^\infty$; so that to define a denotational counterpart of S_{SEQ}, we have to consider S(A) ordered by the prefix ordering, and then $P(S(A))$, where P denotes Plotkin's power-domain. Remind (see [9])that the elements of $P(S(A))$ are the convex and closed clausures of finitely generable subsets of S(A); we will denote by $\overline{X}$ this clausure of X.

Once more the denotational semantics is given by a system of equations

Def. 1.21 : Trace denotational semantics of Ianov's schemes, $S^d_{SEQ}[S]$, is defined by:

- $S = STOP \Rightarrow S^d_{SEQ}[S](\alpha) = \{STOP\}$
- $S = p(S_1,S_2) \Rightarrow S^d_{SEQ}[S](\alpha) = COND(\varphi_p, S^d_{SEQ}[S_1], S^d_{SEQ}[S_2])(\alpha)$
- $S = a(S_1) \Rightarrow S^d_{SEQ}[S](\alpha) = a\cdot S^d_{SEQ}[S_1](\alpha\cdot a)$
- $S = OR(S_1,\ldots,S_k) \Rightarrow S^d_{SEQ}[S](\alpha) = \overline{\bigcup_{i=1}^{k} S^d_{SEQ}[S_i](\alpha)}$

And we have

Theorem 1.22 : $\forall S \quad \forall\alpha \in A^* \quad \overline{S_{SEQ}(S,\alpha)} = S^d_{SEQ}[S](\alpha)$

Now we have two (different) new equivalence relations $\underset{SEQ}{\sim}$(operational) and $\underset{SEQ}{\overset{d}{\sim}}$ (denotational) between Ianov's schemes. We have also studied them

Prop. 1.23 : By means of its normal form, we can associate to S two more regular languages $L^f(S)$ and $L^1(S)$ that characterize its partial and looping computations; and then we have $S_1 \underset{SEQ}{\sim} S_2 \Longleftrightarrow (L(S_1) = L(S_2) \wedge L^f(S_1) = L^f(S_2) \wedge L^1(S_1) = L^1(S_2))$.

Corollary 1.24 : a) Strong equivalence and SEQ-equivalence are the same.
b) Both are decidable.

Prop. 1.25 : From $L^f(S)$ and $L^1(S)$ we can construct a new regular language $L^c(S)$ that reflects the convex clausure of S_{SEQ} to get S^d_{SEQ}, so that we have $S_1 \underset{SEQ}{\overset{d}{\sim}} S_2 \Longleftrightarrow L(S_1) = L(S_2) \wedge L^f(S_1)=L^f(S_2) \wedge L^c(S_1) = L^c(S_2)$

Corollary 1.26 : dSEQ-equivalence between Ianov's schemes is decidable.

2. Probabilistic Ianov's schemes: definitions and semantics.

Def. 2.1 : A probabilistic Ianov's scheme is a nondeterministic one, with the arcs leaving OR-nodes labelled by probabilities (positive rational numbers), so that the sum of all over the arcs that leave one of them is 1.

Intuitively when a computation reaches a choice node, it consults the value of a discrete random variable, that chooses between the arcs leaving the node, according to their probabilities.

ef. 2.2 : a) We will denote by S^- the nondeterministic scheme obtained rom a probabilistic one S, by forgetting the probabilities in it. b) We all precomputations of S, to the partial computations of S^-. Each one f them, c, has a probability prob(c) that is defined as the product of he probabilities of the arcs leaving OR-nodes choosed along c.

ef. 2.3 : a) We say that a precomputation $c:(S_1,d_1) \overset{I}{\underset{*}{\to}} (S_n,d_n)$ stops, f S_n = STOP but $S_{n-1} \neq$ STOP; then we say that d_n is its result res(c) = d). We denote the set of precomputations of S, under I, from d resp. that stop) $Pc_I(S,d)$ (resp. $Pcf_I(S,d)$). b) Given an interpretation , we define the operational semantics of S, under I, as the function $_{PROB}(S,\cdot)$ that associates to each input $d \in D$, the probability distri- ution asigning to each possible result $d' \in D$, the sum of the probabi- ities of the elements of $Pcf_I(S,d)$ with result d'; and associating to $\perp$ he complement to 1.

To define a denotational semantics we need a probabilistic powerdomain.

ef. 2.4 : $P_{PROB}(D_\perp)$ is the domain whose elements are the discrete pro- ability distributions over $D_\perp$, ordered by $p \sqsubseteq p' \iff \forall d \in D \;\; p(d) \leq p'(d)$.

rop. 2.5 : $P_{PROB}(D_\perp)$ is a cpo.

roof : It is very easy to check that if $\{p_i \mid i \in I\}$ is a directed subset f $P_{PROB}(D_\perp)$, its lub is p defined by $p(d) = \sup_{i\in I} p_i(d) \;\forall d \in D$; $(\perp) = \inf_{i\in I} p_i(\perp)$. (In general $p(x) = \lim_{i\in I} p_i(x) \;\;\forall x \in D_\perp$).

As usual, to define a denotational semantics we need an structural- ike classification of probabilistic schemes. It is the given in def.1.3), except that when r is labelled by OR, and $\{(r,n_1),\ldots,(r,n_k)\}$ is he set of arcs leaving r, and each one of them is labelled by q_i, we enote S by OR $(q_1:S_{n_1},\ldots,q_k:S_{n_k})$.

On the other hand $(q_1:x_1,\ldots,q_k:x_k)$ will denote the finite distribution concentrated over $\{x_1,\ldots,x_k\}$, such that $p(x_i) = q_i \;\forall i \; 1 \leq i \leq k$.

ef.2.6 : The denotational semantics of a probabilistic Ianov's scheme S under I) is the function $S^d_{PROB}[S]: D \to P_{PROB}(D_\perp)$, defined by

- S = STOP $\quad S^d_{PROB}[S](d) = (1:d)$
- $S = a(S_1) \quad S^d_{PROB}[S](d) = S^d_{PROB}[S_1](\psi_a(d))$
- $S = p(S_1,S_2) \quad S^d_{PROB}[S](d) = \text{COND}\,(\varphi_p,S^d_{PROB}[S_1],S^d_{PROB}[S_2])(d)$
- $S = OR(q_1:S_1,\ldots,q_k:S_k) \quad S^d_{PROB}[S](d) = \sum_{i=1}^{k} q_i\cdot S^d_{PROB}[S_i](d)$

This is a correct definition, as all the operators in it are conti- uous. To check it for $\sum_{i=1}^{k} q_i\cdot p_i$ you only have to observe that sums and imits conmute.

Theorem 2.7 : $\forall S \ \forall d \in D \ \ S^d_{PROB}[S](d) = S_{PROB}(S,d)$

Proof: First of all we will see that the operational semantics is a solution of the system defining the denotational one. We have to check the four clauses in def. 2.6, but here we will only see the last, as the others are trivial.

So, let $S = OR(q_1:S_1,\ldots,q_k:S_k)$; then, if $c \in Pcf_I(S,d)$ it begins by the step $(S,d) \overset{I}{\underset{q_i}{\to}} (S_i,d)$, for some $i \in \{1,\ldots,k\}$, and the rest of c will be an element of $Pcf_I(S_i,d)$. So that each result of same $c_i \in Pcf_I(S,d)$ will be the result of some $c \in Pcf_I(S,d)$ with $prob(c) = q_i \cdot prob(c_i)$, and conversely, so that $S_{PROB}(S,d) = \sum_{i=1}^{k} q_i \cdot S_{PROB}(S_i,d)$. Then, as the denotational semantics is defined as the least solution of the considered system, we have $S^d_{PROB}[S](d) \sqsubseteq S_{PROB}(S,d)$. To prove the converse unequality, we consider the approximations $S^n_{PROB}(S,d)$, $n \in \mathbb{N}$, of the operational semantics, defined as it, but allowing only precomputations of length no greater than n. Then we can prove $S^n_{PROB}(S,d) \sqsubseteq S^d_{PROB}[S](d) \ \ \forall n \in \mathbb{N}$, by a easy induction over n.

3. Equivalence of probabilistic Ianov's schemes.

As in the nondeterministic case, we will manage regular languages; although, of course, now we need probabilistic languages. They will be a version of the those defined by Paz [8].

Def. 3.1 a)A probabilistic finite automaton is a tuple $A = (S,I,s_o,F,A)$, where S is a finite set (of states), I is an alphabet, $s_o \in S$ is the initial state, $F \subseteq S$ the set of acceptor states, and A is the transition function, associating to any $x \in I$ an square matrix of rational non-negative number, whose dimension is the cardinal of S, so that the sum of the elements in each row is 1. b) We will call language accepted by A, L(A), to the set of pairs $(q:\alpha)$, where $\alpha \in A^*$, and q is the sum of the probabilities of the computations of A, that accept α. c) We will call probabilistic regular languages to the accepted ones by probabilistic finite automata.

Prop. 3.2 : The equivalence of probabilistic finite automata is decidabl

Proof: See [8] and [2].

We will also have a notion of strong equivalence.

Def. 3.3 : a) $Pca_I(S,d)$ is the set of precomputations whose last step executes an action (the root of S_{k-1} is labelled by some $a \in A$), more the only precomputation of length 0. b) We say that $c \in Pc_I(S,d)$ loops, if there is not any extension of it executing more actions, nor any pref of it with that property; we will denote by $Pl_I(S,d)$ the set of all of them. c) For $c \in Pc_I(S,d)$, Act(c) will denote the sequence of actions executed along it.

Def. 3.4 : We will say that S_1 and S_2 are strongly equivalent ($S_1 \sim S_2$) iff under any interpretation I, and for all $d \in D$, we have for each $\alpha \in A^*$

i) $\sum_{c\in A(S_1,I,d,\alpha)} prob(c) = \sum_{c\in A(S_2,I,d,\alpha)} prob(c)$, where

$A(S,I,d,\alpha) = \{c \in Pca_I(S,d) \mid Act(c) = \alpha\}$

ii) $\sum_{c\in F(S_1,I,d,\alpha)} prob(c) = \sum_{c\in F(S_2,I,d,\alpha)} prob(c)$, where

$F(S,I,d,\alpha) = \{c \in Pcf_I(S,d) \mid Act(c) = \alpha\}$

iii) $\sum_{c\in L(S_1,I,d,\alpha)} prob(c) = \sum_{c\in L(S_2,I,d,\alpha)} prob(c)$, where

$L(S,I,d,\alpha) = \{c \in Pl_I(S,d) \mid Act(c) = \alpha\}$

Remark: It is easy to check that i) $\wedge$ ii) $\rightarrow$ iii); nevertheless we have preserved the three conditions to keep the parallelism with def.1.10. Note that there the three conditions are independent one another; that is because from the probability of precomputations we can infer the probabilities of loops, but not the mere "possibility" of them. Look the following example.

Example 3.5 : In fig. 1 you can see on scheme S such that under any I, and for all $d \in D$, we have $Pl_I(S,d) = \emptyset$, since $\sum_{c\in F(S,I,d,\varepsilon)} prob(c) = 1$; nevertheless S^- has an infinite computation that loops.

Fig. 1 :

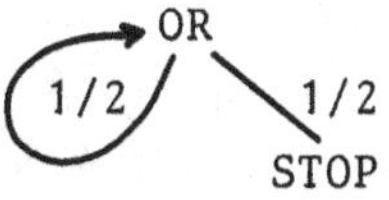

Def. 3.6 : We say that S_1 and S_2 are semantically equivalent ($S_1 \tilde{_{PROB}} S_2$) iff under any interpretation I, and for all $d \in D$, we have $S_{PROB}(S_1,d) = S_{PROB}(S_2,d)$.

Prop. 3.7 : $S_1 \simeq S_2 \rightarrow S_1 \tilde{_{PROB}} S_2$.

Now we introduce probabilistic schemes in normal form:

Def. 3.8 : We say that S is in normal form, when its graph is constituted by blocks as one in fig. 2, where $P = \{p_1,\ldots,p_n\}$,and each leaf h_{ij} represents either an STOP node, or an action node from which leaves an arc to the root of another (perhaps the same) block, or a LOOP node, that represents any scheme that neither executes actions nor stops. The root of S is the root of one of its blocks.

Theorem 3.9 : For each probabilistic Ianov's scheme S, we can construct S in normal form, and strongly equivalent to it.

Fig. 2 :

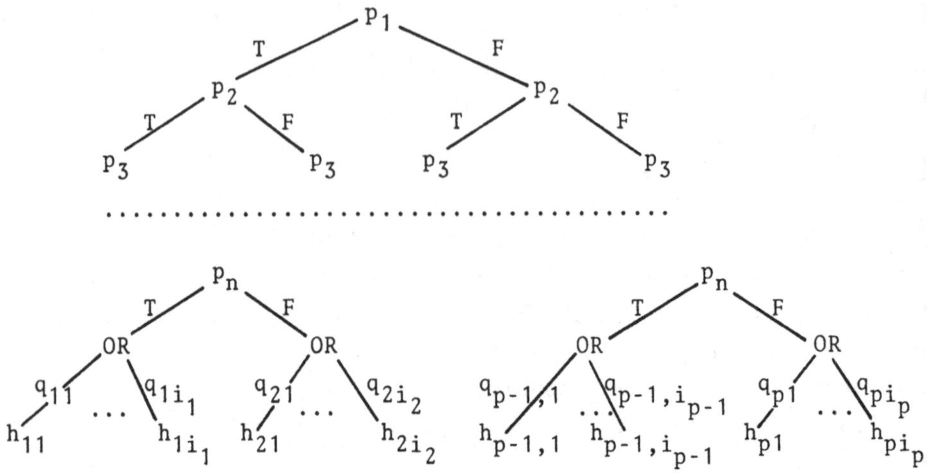

The proof of th.3.9 is based on the following lemma

Lemma 3.10 : For each probabilistic Ianov's scheme S, we can construct S strongly equivalent to it, and in the class IS^p (inductive schemes) of schemes that can be constructed by application of the following rules

- $STOP \in IS^p$
- If $S \in IS^p$, then for each $a \in A$ $a(S) \in IS^p$.
- If $S_1, S_2 \in IS^p$, then for each $p \in P$ $p(S_1,S_2) \in IS^p$.
- For all $k \in \mathbb{N}$, if $S_1,\dots,S_k \in IS^p$, $q_1,\dots,q_k \in \mathbb{Q}^+$, $\sum_{i=1}^{k} q_i = 1$, then $OR(q_1{:}S_1,\dots,q_k{:}S_k) \in IS^p$.
- And finally to allow backing arcs, when $S = (P,A.G,r) \in IS^p$, and n is one of its STOP-nodes, such that to it arrives only one arc(n',n), then $S^{-n} = (P,A,G^{-n},r) \in IS^p$, where G^{-n} is obtained from G by removing n, and adding an arc (n',r) instead of (n',n)

Proof (sketch) : The proof is by induction over the number of nodes of S that are not labelled by STOP. We replace the arcs reaching its roots, by others coming to new STOP nodes. Then we apply the induction hypothes to the subschemes that have their root below that of S, and finally we redirect the arcs arriving to the added nodes, to the root of the constructed schemes.

So that when we construct $\bar{S}$, we can suppose that $S \in IS^p$.

Construction of $\bar{S}$: It is done by structural recurrence, considering the way in that S is constructed as an element of IS^p. But a simple inductiv construction is not sufficient; we will have to define, at the same time a function f that associates STOP nodes in S to those of $\bar{S}$, so that for any STOP node in S, n, for all $\alpha \in A^*$, and under any free interpretation we have $\sum_{c \in C(n,\alpha)} prob(c) = \sum_{c \in \bar{C}(n,\alpha)} prob(c)$, where

$C(n,\alpha) = \{\, c \in Pcf_I(S,\varepsilon) \mid c$ stops over n, $Act(c) = \alpha \,\}$ and
$\bar{C}(n,\alpha) = \{\, c \in Pcf_I(\bar{S},\varepsilon) \mid c$ stops over some $n' \in f^{-1}(n)$, $Act(c) = \alpha\}$

In fig.3 you can see $\overline{STOP}$, and in fig.4 $\overline{a(S_1)}$. To define f we only have to observe that any STOP node in this scheme is in some $\bar{S}_1$, and by induction hypothesis we have already defined f over $\bar{S}_1$, obtaining STOP nodes in S_1, that are also in $a(S_1)$. Note also that the normal form is defined for a fixed P.

Fig. 3

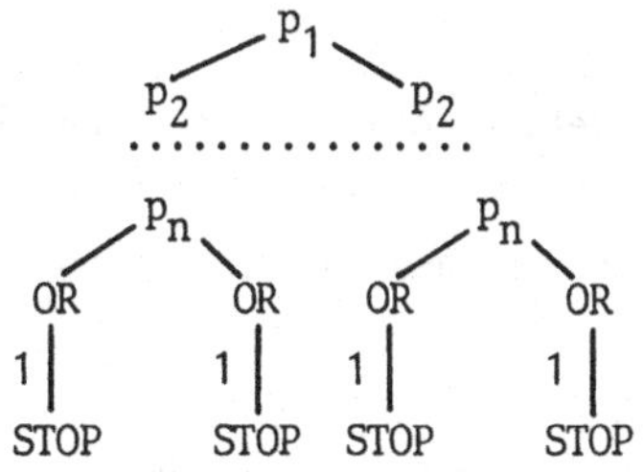

Fig. 4

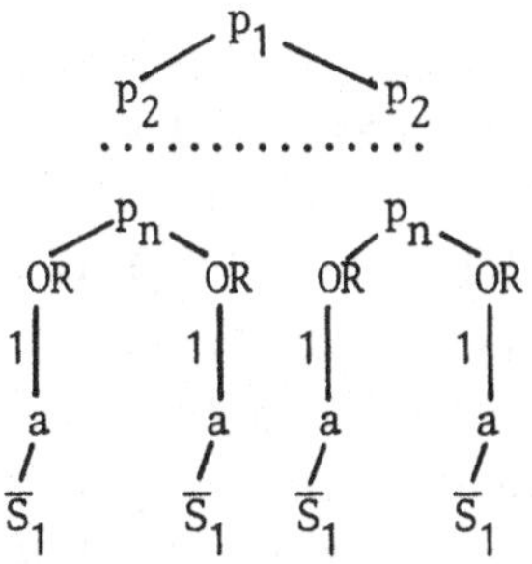

To obtain $p(S_1,S_2)$, we consider $\bar{S}_1$ and $\bar{S}_2$, setting under each OR node in the root block of it, the same subscheme that there is in the corresponding node en $\bar{S}_1$ (resp. $\bar{S}_2$) if it is in a branch including an arc labelled by TRUE (resp. FALSE) leaving a node labelled by p_i, where $p = p_i$. The normal form of $OR(q_1{:}S_1,\dots,q_k{:}S_k)$, is constructed in a very similar way. Finally, if S is obtained from S_1, removing (n',n), and adding (n',r) instead, we will construct $\bar{S}$ by introducing some modifications in $\bar{S}_1$. For we take the set N of OR-nodes in $\bar{S}_1$ from which leaves some arc, labelled by q say, to a node in $f_1^{-1}(n)$, where f_1 is the defined function for S_1. These arcs must be substituted by new ones reaching new nodes, that we have to add to $\bar{S}_1$. But we have to distinguish some cases. Let $n'' \in N$; we will consider the test (that is the tuple of boolean values over the arcs leaving the predicate nodes in the block where n'' is) that takes us to n''. Then we take as n''' the OR-node in the root block of $\bar{S}_1$, reached by following this test, and as $\{n_j \mid j \in J\}$ the set of nodes that can be reached from n''' by an arc, labelled by q_j. If for all $j \in J$, $n_j \in f_1^{-1}(n'')$ then we add only one node, labelled by LOOP, that will be accesible by an arc with probability q. Otherwise, let $t = \sum_{f_1(n_j)=n} q_j$, and we will consider the rest of n_j's. For each one of them labelled by $a_j \in A$ we will add a node so labelled, to which arrives an arc with probability $\frac{q \cdot q_j}{1-t}$, and leaves an arc to the same node that the one leaving n_j; for each labelled by STOP, but with $f_1(n_j) \neq n$, we add a new STOP node, to which we arrive with probability $\frac{q \cdot q_j}{1-t}$; and for each labelled by LOOP, we add a so labelled node accesible by an arc labelled by $\frac{q \cdot q_j}{1-t}$, too. Finally to define f we observe that $\bar{S}$ will have two kind of STOP nodes: those in $\bar{S}_1$, and the added by our construction. For the first we define f by extending f_1; and if m was added when studying n_j, we take $f(m) = f_1(n_j)$.

Proof (of th.3.9): This will only be a brief sketch. Naturally, the proof is by structural induction. All the cases are rather easy, but the last. For this we have to observe that n in S_1 must be considered as a bridge to the root of S, and so must be considered each $n'' \in N$, to obtain $\bar{S}$ from $\bar{S}_1$. But when these bridges lead us to new ones, without possible escape (that is $t = 1$) then we have fallen in an endless hole, so that we have a LOOP. Otherwise we can lose some time by crossing bridges, to finally escape, and then each exit will have probability $\frac{q \cdot q_j}{1-t}$, since $\frac{q \cdot q_j}{1-t} = \sum_{k=0}^{\infty} t^k \cdot q \cdot q_j$.

By the normal form we will associate a probabilistic automaton to each probabilistic scheme.

Def. 3.11 : Given a probabilistic Ianov's scheme, we define $A(S)$ as the probabilistic automaton whose states are the nodes of $\bar{S}$, excluding its OR-nodes; its alphabet is $P \cup \bar{P} \cup A$; s_0 is the root of $\bar{S}$; F is the set of STOP nodes; and the transition function $f: P \cup \bar{P} \cup A \to (N \times N \to \mathbb{Q})$ is defined by:

- $f(p)(n,n') = 1$ if $p \neq p_n$, $lb(n) = p$, $(n,n') \in G_{\bar{S}}$, $lb(n,n') =$ TRUE
- $f(\bar{p})(n,n') = 1$ if $p \neq p_n$, $lb(n) = p$, $(n,n') \in G_{\bar{S}}$, $lb(n,n') =$ FALSE
- $f(p_n)(n,n') = lb(n'',n')$ if $lb(n) = p_n$, $(n,n''),(n'',n') \in G_{\bar{S}}$, $lb(n,n'') =$ TRUE
- $f(p_n)(n,n') = lb(n'',n')$ if $lb(n) = p_n$, $(n,n''),(n'',n') \in G_{\bar{S}}$, $lb(n,n'') =$ FALSE
- $f(a)(n,n') = 1$ if $lb(n) = a$, $(n,n') \in G_S$
- $f(x)(n,n') = 0$ otherwise

Prop. 3.13 : a) $L(A(S))$, $L^{PROB}(S)$ from now on, accepts only pairs (α,q) where $q \in \mathbb{Q} \cap (0,1]$, and $\alpha = t^1 a^1 \ldots t^m a^m t^{m+1}$ with $a^i \in A$, $t^i = q_1^i \ldots q_n^i$, being $q_j^i \in p_j, \bar{p}_j$.
b) $(q:\alpha) \in L^{PROB}(S)$ iff under any free interpretation I such that $\varphi_{p_j}(a^1 \ldots a^l) = q_j^{l+1}$ $\forall l \in \{0,\ldots,m\}$, $j \in \{1,\ldots,n\}$, we have $\sum_{c \in F(S,I,\varepsilon,a^1 \ldots a^m)} prob(c) = q$.

Corollary 3.14 : $S_1 \simeq S_2 \Rightarrow L^{PROB}(S_1) = L^{PROB}(S_2)$

Theorem 3.15 : $S_1 \underset{PROB}{\sim} S_2 \Leftrightarrow L^{PROB}(S_1) = L^{PROB}(S_2)$

Corollary 3.16 : The semantical equivalence of probabilistic schemes is decidable.

4. Alternative treatments of nontermination in probabilistic schemes.

As our semantics of probabilistic schemes considers nontermination as a more result we could say that it is a Plotkin's-like semantics. But since the probability of nontermination can be deduced from that of stopping computations, an angelical-like semantics considering only these last computations, would be equivalent to the defined one. Nevertheless, we could consider interesting the study of an Smyth's-like semantics, that considers nontermination with positive probability as a disaster. So let us do such an study.

Def. 4.1 : a) We will take as semantical domain $P_p^{SM}(D) = P_p(D) \cup \{NT\}$, with NT as the bottom element, and the order induced by the one in $P_p(D_\perp)$, that is the trivial one, since $p_1, p_2 \in P_p(D) \subseteq P_p(D) \wedge p_1 \sqsubseteq p_2 \to p_1 = p_2$. Given a probabilistic scheme S and interpretation I, we define the Smyth's-like semantics of S, as the function $S_{PR\text{-}SM}(S,\cdot)$ given by

$$S_{PR\text{-}SM}(S,d) = \begin{cases} S_{PROB}(S,d) & \text{if } S_{PROB}(S,d)(\perp) = 0 \\ NT & \text{otherwise} \end{cases}$$

Unfortunately this semantics is not continuous relative to the natural order over the set of schemes, taking LOOP as bottom, as the following example shows.

Example 4.2 : Let S; S_n, $n \in \mathbb{N}$ as in fig.5. We have $(S_n)_{n\in\mathbb{N}} \to S$, but $S_{PR\text{-}SM}(S_n,\varepsilon) = NT$ and $S_{PR\text{-}SM}(S,\varepsilon) = \{(1/2^n: a^{n-1}) \;/\; n \in \mathbb{N}\} \neq NT$.

Fig.5 :

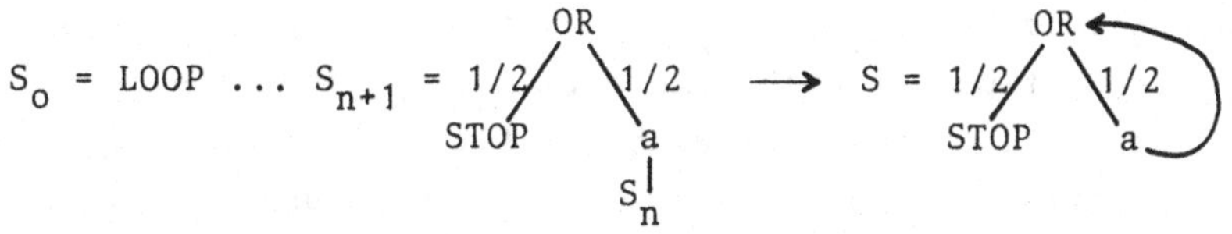

Then we can not define a denotational semantics equivalent to our Smyth's-like one. So we could think taht the equivalence induced by it is not decidable, arguing that in finite time we can only approach the semantics of S by that of some S_n. Fortunately this intuitive idea is not correct, and this new equivalence relation between schemes is decidable, too. To decide it we need a definition about set of nodes of a probabilistic scheme.

Def. 4.3 : We will call "sets of nodes with positive probability of non-termination" (s.n.p.p.n.t.) to those sets of OR-nodes $\{n_1,\dots,n_k\}$ of a normal form probabilistic scheme such that under any (free) interpretation I, there is some $i \in \{1,\dots,k\}$ such that a precomputation in $Pc_I(S_{n_i},\varepsilon)$ does not stop with some positive probability.

These sets do not depend of the probabilities over the arcs of S, but only of its nondeterministic structure.

Def. 4.4 : We will call "sets of nodes with partial computation without completion" (s.n.p.c.w.c.) to those sets of OR-nodes $\{n_1,\dots,n_k\}$ of a normal form nondeterministic scheme such that under any (free) interpretation I, there is some $i \in \{1,\dots,k\}$ and $c \in C_I^p(S_{n_i},\varepsilon)$ that can not be extended to a final computation.

Theorem 4.5 : The s.n.p.c.w.c's of a nondeterministic scheme S can be detected by an algorithm.

The proof requires some previous lemmas.

Lemma 4.6 : Given a normal form nondeterministic scheme S, and a set of roots of some blocks in S $\{r_1,\dots,r_k\}$, if there is a free interpretation I such that for all $i \in \{1,\dots,k\}$, any $c_i \in C_I^p(S_{r_i},\varepsilon)$ can be extended to some $c_i' \in C_I(S_{r_i},\varepsilon)$ that stops, then for each $i \in \{1,\dots,k\}$ we can find I_i with the same property than I, and such that there is some

$c \in C_{I_i}(S_{r_i},\varepsilon)$ that stops, and whose length, measured by the number of blocks that it traverses, is less or equal than some constant M, that only depends of the number of blocks B that constitute S.

Corollary 4.7 : Under the same hypothesis than in lemma 4.6, we will have some free interpretation I' with the same property than I, and verifying that for each $i \in \{1,\dots,k\}$, there is $c_i \in C_I(S_{r_i},\varepsilon)$ that stops and has length no greater than $k\cdot M$.

Lemma 4.8 : If $N = \{n_1,\dots,n_k\}$ is not a s.n.p.c.w.c. of S, we can find an interpretation I'' with some periodicity properties (to be explained in the proof) such that for each $i \in \{1,\dots,k\}$, any $c \in C^p_I(S_{n_i},\varepsilon)$ admits an extension to a final computation, whose length is not greater than that of c, more $2\cdot B\cdot M$, with B and M as in lemma 4.6.

Proof (sketch) : First of all we will say that I'' is periodic in the sense that we have a finite tree, whose arcs are labelled by actions, so that from each internal node leaves an arc for each action in A; and whose leaves have an "arrow" that signs to one of its ancestors; finall the internal nodes are labelled by boolean values. Then if $\alpha \in A$, $\varphi(\alpha)$ is defined by taking $\alpha = \alpha_1\cdot\alpha_2\cdot\dots\alpha_l\cdot\alpha_{l+1}$, so that i) There is a path from the root to a leaf h_1, whose arcs are labelled by the elements of α_i; ii) n_i is the internal node signed from h_i; iii) $\forall j \in \{2,\dots,l\}$ ther is a path from n_{j-1} to a leaf h_j, whose arcs are labelled by α_j; iv) There is a path from n_l to a node n, whose arcs are labelled by α_{l+1} Finally we take $\varphi(\alpha) = \text{label}(n)$. To get such a tree we construct a rela ted one in the following way.

- In its first level we have a node labelled with R_a, that is the set of roots of blocks in S, that can be reached from some $n_i \in$ by an arc labelled by a; for each $a \in A$.
- If R is the label over a node of the tree, we have (by applying corollary 4.7) an interpretation I_R such that $\forall r \in R \;\; \exists c \in C_{I_R}(S_r,$ $c\!\downarrow$, $\text{length}(c) \leq B\cdot M$. Then we take, for each $\alpha \in A^*$, R_α = {accessible blocks of S, by $c_\alpha \in C^p_{I_R}(S_r,\varepsilon)$, $\text{Act}(c_\alpha) = \alpha$, $\text{length}(c_\alpha) = B\cdot M$}, and when $R_\alpha \neq \emptyset$ we add a son to the consider node, labelled by R_α.
- Each branch of the tree will end when we add a leaf labelled as another node in the branch; that leaf will sign to this node.

Now we can give the algorithm to decide if N is a s.n.p.c.w.c.

Algorithm : We will consider the restrictions of free interpretations to sequences of length less or equal than $B\cdot M$, and we use then to try to construct a tree as the one in the proof of lemma 4.8. This is done by a backtracking procedure. If we get to construct it, then N is not a s.n.p.c.w.c., and conversely.

Proof (of th. 4.5) : Our algorithm is effective as we have finitely many restrictions of interpretations, as the considered in it. If the search fails, then lemma 4.8 say us that N is a s.n.p.c.w.c. Otherwise we have found an interpretation I'' such that each $c \in C_I^p(S_{n_i},\varepsilon)$ admits a final extension, so that N is not a s.n.p.c.w.c.

Now we will relate s.n.p.p.n.t's and s.n.p.c.w.c's.

Prop. 4.9 : The s.n.p.p.n.t's of a (n.f.) probabilistic scheme S, are exactly the s.n.p.c.w.c's of S^-.

Proof (sketch) : a) If N is a s.n.p.c.w.c. of S^-, then under any I we have $c \in C_I^p(S_n^-,\varepsilon)$, that can not be extended to a final computation, and then $c \in Pc_I(S_n,\varepsilon)$ and prob(c) > 0. b) If N is not a s.n.p.c.w.c. of S^-, we will have I'' as in lemma 4.8; so that if ε is the less probability over an arc in S, and we extend $c \in Pc_{I''}(S_n,\varepsilon)$ by more $2\cdot B\cdot M$ steps, we have a probability greater than ε^{2BM}, to obtain a final precomputation, so that $c \in Pc_{I''}(S_n,\varepsilon)$ will have length greater than $k\cdot B\cdot M$, with probability less than $(1-\varepsilon^{2BM})^k$, and so an infinite computation has probability $(1-\varepsilon^{2BM})^\infty = 0$, to get executed.

By knowing the s.n.p.p.n.t's of S, we can associate a probabilistic automata to it.

Lemma 4.10 : Given a nondeterministic scheme S, we can construct a finite automaton $A^d(S)$ that say us, given the input $t^1a^1\ldots t^ka^kt^{k+1}$ (as in prop. 3.12) which is the set of OR-nodes that can be reached, under I as in prop. 3.12 by some $c \in C_I^p(S,\varepsilon)$ that executes $a^1\ldots a^k$, and finally passes the test t^{k+1} .

Def. 4.11 : We can associate to any (n.f.) probabilistic scheme S, a probabilistic automaton $A^{SM}(S)$, that simulates A(S) and $A^d(S^-)$ in parallel, so that when the input has a prefix $t^1a^1\ldots t^{k+1}$, such that $A^d(S^-)$ gives for it a s.n.p.p.n.t. of S, then $A^{SM}(S)$ rejects the input; otherwise $A^{SM}(S)$ behaves as A(S). We denote by $L^{PR-SM}(S)$ the language accepted by $A^{SM}(S)$.

Theorem 4.12 : $S_1 \underset{PR}{\overset{SM}{\approx}} S_2 \Leftrightarrow L^{PR-SM}(S_1) = L^{PR-SM}(S_2)$

Proof (sketch): $\Rightarrow$ If $(q_1{:}\alpha) \in L^{PR-SM}(S_1)$, $(q_2{:}\alpha) \in L^{PR-SM}$, $q_1 > q_2$, we have that once we have executed α, and passed a final test, we arrive to a set of nodes N_α that is not a s.n.p.p.n.t. of S, and then we can change I, obtaining I' such that $S_{PR-SM}^{I'}(S_1,\varepsilon) \neq NT$, but maintaining I' equal to I over α and its prefixes, so that $S_{PROB}(S_1,\varepsilon)(\alpha) = q_1$ and $S_{PROB}(S_2,\varepsilon)(\alpha) = q_2$, so that $S_1 \underset{PR}{\overset{SM}{\not\approx}} S_2$.

[←] If $NT = S_{PR\text{-}SM}(S_1,\varepsilon) \neq S_{PR\text{-}SM}(S_2,\varepsilon)$, then $L^{PR\text{-}SM}(S_2)$ reflects the values of $S_{PROB}(S_2,)$ over A^* , and their sum must be 1 ; but this must not be true for $L^{PR\text{-}SM}(S_1)$, since a computation of S_1 does not stop with some positive probability.

Corollary 4.13 : SM-equivalence between probabilistic schemes is decidabl

5. Trace semantics of probabilistic Ianov's schemes

Because of lack of space, we will only outline our results about trac semantics of probabilistic Ianov's schemes.

First of all, we have to define probability distributions over a non flat domain, and for we need an adequate σ-algebra. We have considered Cantor's topology (see [9]) over D , taking as $\mathcal{B}(D)$ Borel's σ-algebra induced by it. We have done this (see [3]) as this topology has a simpl subbasis $\mathcal{T}$ such that to have a probability distribution over $\mathcal{B}(D)$, it i sufficient to have one over $\mathcal{T}$.

Then we define an operational trace semantics by a measure over $\mathcal{T}$, extending it to get one over $\mathcal{B}(D)$; while a denotational semantics defin as usual, gives us another distribution over $\mathcal{B}(D)$. And it is not very difficult to prove that both are the same. Compare this result with the obtained in th.1.22 .

Finally we study the equivalence of probabilistic Ianov's schemes, re lative to trace semantics. For we introduce a new probabilistic automato $A^f(S)$, associated to each probabilistic scheme S , which accepts a regu lar language $L^f(S)$, characterizing the set of precomputations of $\overline{S}$ unde any interpretation. So that we have $S_1 \underset{SEQ}{\overset{PR}{\sim}} S_2 \Leftrightarrow L(S_1) = L(S_2) \wedge L^f(S_1) = L^f(S_2)$; and then, trace equivalence of probabilistic Ianov's schemes is decidable.

Conclusion

We have extended many results, improving some, from nondeterministic to probabilistic Ianov's schemes. Nevertheless there is an interesting property of nondeterministic shemes, that probabilistic schemes have not sometimes. If S is a nondeterministic scheme, we can construct $\overline{\overline{S}}$, its deterministic normal form, such that under any of its choice nodes you can not find two equal action nodes. The same is not true for probabilis tic schemes (see [2]) . That is because we need sets of nodes to study probabilistic schemes, while it is sufficient to manage nodes, to study nondeterministic ones (see [4]) .

Another interesting subject is complexity of the equivalence problems that we have studied. A related paper is [5] .

Finally I want to thank K.Indermark , that directed my Ph. thesis, suggesting me the subject of this paper.

References

[1] J.ENGELFRIET "Simple program schemes and formal languages" LNCS 20 Springer Verlag (1974) .

[2] D.FRUTOS ESCRIG "Algunas cuestiones relacionadas con la semántica de construcciones probabilísticas" Tesis doctoral Fac.C.Matemáticas U.Complutense Madrid, July 1985 .

[3] D.FRUTOS ESCRIG "Some probabilistic powerdomains in SFP" STACS-86 (to appear in LNCS) .

[4] D.FRUTOS ESCRIG , K.INDERMARK "Equivalence relations of nondeterministic Ianov's schemes" Schriften zur Informatik und Angewanten Mathematik RWTH Aachen 1986 (to appear) .

[5] S.HART , M.SHARIR , A.PNUELI "Termination of probabilistic concurrent programs" ACM TOPLAS 5,3 (1983) 356-380 .

[6] Y.IANOV "The logical schemes of algorithms" Problemy Kibernet. 1 (1960) 82-140 .

[7] K.INDERMARK "On a class of schematic languages" in Formal languages and Programming (Ed.R.Aguilar) North Holland (1976) 1-13 .

[8] A.PAZ "Introduction to probabilistic automata" Academic Press (1971)

[9] G.PLOTKIN "A powerdomain construction" SIAM J.Comput. 5(3) (1976) 452-487 .

[10] M.B.SMYTH "Powerdomains" LNCS 45 Springer Verlag (1978) 537-543 .

Alternating bottom-up tree recognizers

Kai Salomaa
Department of Mathematics
University of Turku
SF-20500 Turku, Finland

1. Introduction

An alternating automaton is a natural generalization of a nondeterministic automaton. The configurations (or states) of a nondeterministic automaton can be seen as existential whereas an alternating automaton can have both existential and universal configurations.

Alternating (one-way and two-way) finite automata were discussed in [3,6] and they were shown to recognize only the regular languages. Similarly in [8] it was seen that alternating top-down tree automata recognize exactly the regular forests, (i.e. tree languages).

Here we define alternating tree recognizers that read input trees bottom-up, i.e. from the leaves to the root. Also we do not divide the state set into existential and universal states but define the recognizer in such a way that in each computation step it can both make an existential choice and do universal branching. Depending on the order in which this is done the operation mode is called either EU- or UE-alternating. It turns out that these recognizers can recognize also non-regular forests. The intuitive reason for this is that in an alternating bottom-up computation the order in which the subtrees of the input tree are read is essential.

Hence, we can define two different modes of computation: sequential and parallel. In a sequential computation the recognizer reads one (nondeterministically chosen) active node of the input tree at a time. (By an active node we mean a node that has not been read and the successors of which have all been read.) In a parallel computation the recognizer reads all presently active nodes simultaneously. In the following we concentrate on sequential alternation, the parallel computation mode is discussed briefly in the last section.

We illustrate the power of alternating bottom-up recognizers by showing that they can recognize even non-algebraic forests. Also somewhat unexpectedly it turns out that many basic questions such as the emptiness and finiteness problems are undecidable for alternating bottom-up tree recognizers.

2. Notations

When A is a set #A denotes the cardinality of A and pA the set of subsets of A.

Symbols Σ and Ω stand always for (finite) ranked alphabets. The set of m-ary ($m \geq 0$) symbols of Σ is denoted by Σ_m. Let X be a set. The set of all Σ-trees (with variables X) is denoted by F_Σ ($F_\Sigma(X)$). Subsets of F_Σ are called (Σ-)forests. The root, leaves and height of a Σ-tree t are defined in a natural fashion (cf. [4]). Differing from the usual terminology by a subtree of t we mean always a specific occurrence of a subtree in t, i.e. different subtrees of t may be identical as Σ-trees. (Formally this can be done by defining the domain of the tree t and associating with every subtree t_1 of t the address of the root of t_1.) The set of subtrees of t is denoted by SUB(t).

The function yd: $F_\Sigma \to \Sigma_0^+$ is defined inductively as follows:

i) $yd(t) = \sigma$ if $t = \sigma \in \Sigma_0$.

ii) Suppose $t = \sigma(t_1,\dots,t_m)$, $m > 0$, $\sigma \in \Sigma_m$, $t_1,\dots,t_m \in F_\Sigma$. Then

$yd(t) = yd(t_1)\cdots yd(t_m)$.

The family of regular forests is denoted by REG.

3. Alternating bottom-up tree recognizer

In this section we define the alternating recognizer under consideration. The sequential operation mode is defined in section 4 and parallel computation is considered in section 6.

Definition 3.1. An alternating bottom-up tree recognizer is a four-tuple $\underline{A} = (A,\Sigma,g,A')$, where

i) A is a finite nonempty set of states.

ii) Σ is a ranked alphabet.

iii) g is a function that associates with every element $\sigma \in \Sigma_m$ ($m \geq 0$) a mapping σ_g: $A^m \to p(pA)$. (If $\sigma \in \Sigma_0$, σ_g is interpreted to be an element of $p(pA)$.)

iv) $A' \subseteq A$ is the set of accepting final states.

The family of alternating tree recognizers is denoted by AR.

In definitions 3.2. and 3.3. the recognizer $\underline{A} = (A,\Sigma,g,A')$ is as above.

Definition 3.2. By $\underline{A}$-configurations we mean elements of $F_\Sigma(A)$. The set of active subtrees of a $\underline{A}$-configuration K is

$$\mathrm{act}(K) = \{r \in SUB(K) \mid r \text{ is of the form } \sigma(a_1,\ldots,a_m),\ m\geq 0,\ \sigma\in\Sigma_m,\ a_1,\ldots,a_m\in A\}.$$

When $f = \sigma(a_1,\ldots,a_m) \in \mathrm{act}(K)$ we write

$$f_g = \sigma_g(a_1,\ldots,a_m) \quad (\in p(pA)).$$

Definition 3.3. A configuration tree of the recognizer $\underline{A}$ is a finite directed tree whose nodes are labeled by $\underline{A}$-configurations. The set of all configuration trees of $\underline{A}$ is denoted by $CT(\underline{A})$.

4. Sequential alternation

We define two different sequential operation modes (EU- and UE-alternating) for the recognizer of the last section. In the following definitions $\underline{A}$ is as in definition 3.1.

Definition 4.1. The sequential transition relations of the recognizer $\underline{A}$ $\vdash\frac{EU}{\underline{A}}$ and $\vdash\frac{UE}{\underline{A}} \subseteq CT(\underline{A})\times CT(\underline{A})$ are defined as follows. Let $Z\in\{EU,UE\}$ and $T,T' \in CT(\underline{A})$. Then $T \vdash\frac{Z}{\underline{A}} T'$ iff the configuration tree T' is obtained from T as follows:

Choose a leaf n of T and suppose that n is labeled by a configuration $K \in F_\Sigma(A)$. Choose an active subtree $f = \sigma(a_1,\ldots,a_m) \in \mathrm{act}(K)$, $(m \geq 0)$.

(i) Case Z = EU: Choose (if possible) a nonempty set $\{c_1,\ldots,c_k\}\in f_g$, $k > 0$. Now T' is obtained from T by attaching for the node n k successors that are labeled by configurations $K_1,\ldots,K_k$. Here K_i is obtained from K by replacing f with the element c_i, $i = 1,\ldots,k$.

(ii) Case Z = UE: Let $f_g = \{D_1,\ldots,D_h\}$, $D_i \subseteq A$. Choose (if possible) from every subset D_i an element d_i, $i = 1,\ldots,h$. Now in T' the node n has h successors which are labeled by the configurations that are obtained from K by replacing f with d_i, $i = 1,\ldots,h$.

(If in (i) or (ii) the called for choices can not be made it means that the computation can not be continued from K by reading f.)

efinition 4.2. Let $K \in F_\Sigma(A)$ and $Z \in \{EU, UE\}$. The set of (sequential) -alternating K-computation trees of $\underline{A}$ is

$$COM_Z(\underline{A},K) = \{T \in CT(\underline{A}) \mid K \vdash_{\underline{A}}^{Z*} T\},$$

here $\vdash_{\underline{A}}^{Z*}$ denotes the reflexive, transitive closure of $\vdash_{\underline{A}}^{Z}$ and K s interpreted to be an element of $CT(\underline{A})$, i.e. a configuration tree with nly one node. A computation tree is accepting if all its leaves are abeled by configurations $a \in A'$. The configuration K is said to be -accepting if $COM_Z(\underline{A},K)$ contains at least one accepting computation ree. The set of Z-accepting configurations is denoted by $H_Z(\underline{A})$. The orest Z-recognized by $\underline{A}$ is

$$L_Z(\underline{A}) = H_Z(\underline{A}) \cap F_\Sigma.$$

lso we denote $\mathcal{L}_Z(AR) = \{L_Z(\underline{A}) \mid \underline{A} \in AR\}$.

Suppose that $\underline{A}$ is reading an active subtree f of a configuration . In an EU-alternating computation $\underline{A}$ first chooses existentially a set $\in f_g$ and then continues the computation in all the states of C. On the ther hand, in a UE-alternating computation $\underline{A}$ first branches universally o all the elements of f_g and thereafter makes the existential choices rom each of them.

The configurations in a t-computation tree ($t \in F_\Sigma$) stand for the ntermediate stages of the computation when $\underline{A}$ is reading the input tree . The computations of any nondeterministic bottom-up tree recognizer an be performed by both an EU- and UE-alternating AR-recognizer. In uch a computation the computation tree consists always of only one ranch.

neorem 4.3. $\mathcal{L}_{UE}(AR) \subseteq \mathcal{L}_{EU}(AR)$.

roof. Let $\underline{A} = (A,\Sigma,g,A') \in AR$. Choose $\underline{B} = (A,\Sigma,h,A')$ where h is defined s follows. Let $m \geq 0$, $\sigma \in \Sigma_m$ and $a_1,\ldots,a_m \in A$. Suppose that $_g(a_1,\ldots,a_m) = \{C_1,\ldots,C_k\}$, $k \geq 0$, $C_i \subseteq A$, $i = 1,\ldots,k$. Then

$$\sigma_h(a_1,\ldots,a_m) = \{\{c_1,\ldots,c_k\} \mid c_i \in C_i,\ i = 1,\ldots,k\}.$$

learly $L_{EU}(\underline{B}) = L_{UE}(\underline{A})$. □

The question whether $\mathcal{L}_{EU}(AR) \subseteq \mathcal{L}_{UE}(AR)$ is open but we have the ollowing weaker result. Let Σ be a ranked alphabet. Define $\Omega(\Sigma) = \Sigma \cup \overline{\Sigma}$ ere $\overline{\Sigma} = \{\overline{\sigma} \mid \sigma \in \Sigma\}$ and

$$\Omega(\Sigma)_m = \begin{cases} \Sigma_m, & \text{if } m \neq 1. \\ \Sigma_1 \cup \overline{\Sigma}, & \text{if } m = 1. \end{cases}$$

Define the tree homomorphism (cf. [4]) $h_\Sigma : F_\Sigma \to F_{\Omega(\Sigma)}$ by

$$(h_\Sigma)_m(\sigma) = \bar{\sigma}(\sigma(\xi_1,\ldots,\xi_m)),$$

$m \geq 0$, $\sigma \in \Sigma_m$.

Theorem 4.4. Let $L \subseteq F_\Sigma$. If $L \in \mathcal{L}_{EU}(AR)$ then $h_\Sigma(L) \in \mathcal{L}_{UE}(AR)$.

Proof. The proof is given in appendix 1.

Thus corresponding to every forest of $\mathcal{L}_{EU}(AR)$ there is an essentially similar forest in $\mathcal{L}_{UE}(AR)$. The forest $h_\Sigma(L)$ is obtained from L by replacing in the trees of L every node labeled by σ with two successive nodes labeled by $\bar{\sigma}$ and σ.

Theorem 4.5. REG $\subset \mathcal{L}_{EU}(AR)$.

Proof. It is evident that REG $\subseteq \mathcal{L}_{EU}(AR)$ so only the strict inclusion remains to be shown. Choose $\Sigma = \Sigma_2 \cup \Sigma_1 \cup \Sigma_0$ where $\Sigma_2 = \{\tau\}$, $\Sigma_1 = \{\sigma\}$ and $\Sigma_0 = \{\gamma\}$. Define

$$L = \{\tau(\sigma^n(\gamma),\sigma^n(\gamma)) \mid n \geq 0\}.$$

Clearly L is not regular. In appendix 2. there is given a construction of an AR-recognizer $\underline{A}$ that EU-recognizes the forest L. □

Theorem 4.6. REG $\subset \mathcal{L}_{UE}(AR)$.

Proof. Again clearly REG $\subseteq \mathcal{L}_{UE}(AR)$. Let Σ and L be as in the proof of theorem 4.5. Then $h_\Sigma(L) \in \mathcal{L}_{UE}(AR)$ by theorem 4.4. Also $h_\Sigma(L)$ is not regular because $h_\Sigma^{-1}(h_\Sigma(L)) = L$ and REG is closed with respect to inverse tree homomorphisms. □

We can show that $\mathcal{L}_{EU}(AR)$ and $\mathcal{L}_{UE}(AR)$ contain forests which are not even algebraic (cf. [1]). Let Σ be as in the proof of theorem 4.5. except that $\Sigma_1 = \{\sigma_1,\sigma_2\}$. Define the Σ-forest

$$L' = \{\tau(\sigma_2^n\sigma_1^n(\gamma),\ \sigma_2^n\sigma_1^n(\gamma)) \mid n \geq 0\}.$$

Now we can show that $L' \in \mathcal{L}_{EU}(AR)$ as follows. Define

$$L_1 = \{\tau(\omega_1(\gamma),\ \omega_2(\gamma)) \mid \omega_1,\omega_2 \in (\sigma_2)^*(\sigma_1)^*\},$$

$$L_2 = \{\tau(\omega_1(\gamma),\ \omega_2(\gamma)) \mid \omega_1,\omega_2 \in (\Sigma_1)^*,\ |\omega_1|_{\sigma_1} = |\omega_2|_{\sigma_1}\},$$

$$L_3 = \{\tau(\omega_1(\gamma),\ \omega_2(\gamma)) \mid \omega_1,\omega_2 \in (\Sigma_1)^*,\ |\omega_1|_{\sigma_1} = |\omega_2|_{\sigma_2}\},$$

$$L_4 = \{\tau(\omega_1(\gamma),\ \omega_2(\gamma)) \mid \omega_1,\omega_2 \in (\Sigma_1)^*,\ |\omega_1|_{\sigma_2} = |\omega_2|_{\sigma_2}\}\ .$$

Here $|\omega_i|_{\sigma_j}$ denotes the number of occurrences of σ_j in the string ω_i.)
he forest L_1 is regular. Each of the forests L_2, L_3 and L_4 can be EU-ecognized respectively by a recognizer $\underline{A}_i$, $i = 2,3,4$ that operates es-entially similarly as the recognizer $\underline{A}$ of the proof of theorem 4.5., $\underline{A}_i$ only passes by deterministically the unary symbols that it is not ounting). Using the above obtained AR-recognizers for the forests $_1,\ldots,L_4$ it is easy to see that also

$$L' = \bigcap_{i=1}^{4} L_i \in \mathcal{L}_{EU}(AR).$$

The recognizers of the forests $L_1,\ldots,L_4$ can always be assumed to read he leftmost leaf of the input tree first. After this observation one onstructs the AR-recognizer EU-recognizing the forest $\bigcap_{i=1}^{4} L_i$ quite traightforwardly.)

On the other hand, from the results of [1] it follows that L' is ot an algebraic forest. Also according to theorem 4.4. $h_\Sigma(L') \in \mathcal{L}_{UE}(AR)$ nd $h_\Sigma(L')$ is not algebraic because the family of algebraic forests is losed with respect to inverse linear tree homomorphisms when the ranked lphabets in question contain only symbols of rank at most two (cf.[2]).

The power of alternating bottom-up recognizers is essentially due o the fact that in different branches of the computation tree the rec-gnizer may read the subtrees of the input tree in different order. For nstance, if an AR-recognizer $\underline{A}$ is always forced to read the leftmost ctive subtree of a configuration it is easy to see that $\underline{A}$ EU- (or UE-) ecognizes a regular forest. Hence, it follows also that in the unary ase AR-recognizers recognize only the regular languages.

. Undecidability results

Given an AR-recognizer $\underline{A}$ and an input tree t the set $COM_Z(\underline{A},t)$ Z ∈ {EU,UE}) is always finite. Thus the membership problem for AR-ecognizers is trivially decidable. In the following we discuss some roblems that turn out to be undecidable.

heorem 5.1. The question whether for a given recognizer $\underline{A} \in AR$ $_{EU}(\underline{A})$ is empty is undecidable.

utline of the proof. (The full construction is too long to be given here.) The above question is reduced to the following which is known to be undecidable:

Given two λ-free context-free grammars G_1 and G_2 one has to decide whether $L(G_1) \cap L(G_2) = \emptyset$. (Here λ denotes the empty word.)

Suppose now that λ-free context-free grammars G_1 and G_2 are given. Then we can effectively construct a ranked alphabet Σ and regular Σ-forests L_1 and L_2 such that $yd(L_i) = L(G_i)$, $i = 1,2$, (cf. [4]). Define $\Omega = \Sigma \cup \{d\}$ where the rank of d is two and let

$$L = \{d(t_1,t_2) \mid t_1 \in L_1,\ t_2 \in L_2,\ yd(t_1) = yd(t_2)\}.$$

Using basically the same technique as in the proof of theorem 4.5. we can construct such an AR-recognizer $\underline{A}_1$ that in every EU-accepting $d(t_1,t_2)$-computation tree of $\underline{A}_1$, $(t_1,t_2 \in F_\Sigma)$, $\underline{A}_1$ has to read alternately the leaves of t_1 and t_2 from left to right. Then it is easy to see that $\underline{A}_1$ can check that $yd(t_1) = yd(t_2)$. Also, because $L_1,L_2 \in REG$ $\underline{A}_1$ can check that $t_i \in L_i$, $i = 1,2$. Furthermore $\underline{A}_1$ is constructed in such a way that for every $t \in L$ there exists an EU-accepting t-computation tree of $\underline{A}_1$. Thus $L_{EU}(\underline{A}_1) = L$.

Now $L_{EU}(\underline{A}_1) = \emptyset$ iff $L(G_1) \cap L(G_2) = \emptyset$ and hence the emptiness problem for AR-recognizers has to be undecidable. □

We have immediately

<u>Corollary 5.2.</u> For EU-alternating AR-recognizers the equivalence and inclusion problems are undecidable.

<u>Theorem 5.3.</u> The following questions are undecidable for $\underline{A} \in AR$:

(i) Is the forest $L_{EU}(\underline{A})$ finite ?

(ii) Is the forest $L_{EU}(\underline{A})$ regular ?

<u>Proof.</u> Suppose we are given λ-free context-free grammars G_1 and G_2. As in the proof of theorem 5.1. construct a ranked alphabet Ω and a Ω-forest $L \in \mathcal{L}_{EU}(AR)$ such that $L(G_1) \cap L(G_2) = \emptyset$ iff $L = \emptyset$. Choose a ranked alphabet Δ ($\Omega \cap \Delta = \emptyset$) and a non-regular Δ-forest M such that $M \in \mathcal{L}_{EU}(AR)$ Define the $\Omega \cup \Delta \cup \{e\}$-forest (where the rank of e is two):

$$N = \{e(t_1,t_2) \mid t_1 \in M,\ t_2 \in L\}.$$

It is easy to see that also $N \in \mathcal{L}_{EU}(AR)$ and one can effectively construct a recognizer $\underline{B} \in AR$ such that $N = L_{EU}(\underline{B})$. Now N is regular iff N is finite iff $N = \emptyset$ iff $L = \emptyset$.

Hence (i) and (ii) are undecidable. □

Using theorem 4.4. we get from the preceding results:

Corollary 5.4. The following questions are undecidable for AR-recognizers $\underline{A}$ and $\underline{B}$:

(i) $L_{UE}(\underline{A}) = \emptyset$?

(ii) $L_{UE}(\underline{A}) = L_{UE}(\underline{B})$?

(iii) $L_{UE}(\underline{A}) \subseteq L_{UE}(\underline{B})$?

(iv) Is $L_{UE}(\underline{A})$ finite ?

(v) Is $L_{UE}(\underline{A})$ regular ?

6. Parallel computation

Finally we discuss informally the parallel computation mode of AR-recognizers. Let $\underline{A} = (A,\Sigma,g,A')$ be as in definition 3.1. We only explain intuitively how the parallel transition relations $\vdash^{pEU}_{\underline{A}}$ and $\vdash^{pUE}_{\underline{A}}$ are applied to a computation tree $T \in CT(\underline{A})$, ($\vdash^{pEU}_{\underline{A}}$ and $\vdash^{pUE}_{\underline{A}}$ correspond to the sequential relations of definition 4.1.).

Choose a leaf of T and let it be labeled by $K \in F_\Sigma(A)$. Suppose that $act(K) = \{f^1,\dots,f^k\}$, $k \geq 1$.

(i) EU-case: Corresponding to every active subtree f^i, $i = 1,\dots,k$ choose (if possible) a nonempty set $C^i \in f^i_g$. Then the computation has to be continued from all the configurations that are obtained from K by replacing the subtrees f^i with each possible combination of elements from the respective sets C^i, $i = 1,\dots,k$.

(ii) UE-case: Corresponding to every possible combination of elements $C^1_j \in f^1_g,\dots,C^k_j \in f^k_g$ ($C^i_j \subseteq A$, $i = 1,\dots,k$) choose (if possible) elements $c^1_j \in C^1_j,\dots,c^k_j \in C^k_j$, $j = 1,\dots,N$. Here $N = (\#f^1_g)\cdots(\#f^k_g)$. The computation has to be continued from configurations $K_1,\dots,K_N$. The configuration K_j is obtained from K by replacing every subtree f^i with c^i_j, $i = 1,\dots,k$, $j = 1,\dots,N$.

Now one can define the forest parallelly EU- or UE-recognized by $\underline{A}$ analogously with definition 4.2. (Note that if $\Sigma = \Sigma_1 \cup \Sigma_0$ and hence every configuration contains at most one active subtree the parallel transition relations are exactly the same as the sequential ones.)

Suppose that the subtrees corresponding to two sons of a given node n of the input tree are of different height. Then in a parallel computation the recognizer has to wait at the node n for the completion of the computation of the subtree of greater height. However if the input tree is balanced (i.e. all paths from the root to one of the leaves are of equal length) this can never occur. This means that parallel computation can be performed "more naturally" on balanced input trees. In fact we have been able to show that if all input trees are supposed to be balanced then AR-recognizers parallelly EU- and UE-recognize the same family of forests. In the general case this question is open.

Also we have shown that AR-recognizers can parallelly EU- and UE-recognize some non-algebraic forests and that all the undecidability results of section 5 are valid also for parallelly operating AR-recognizers.

APPENDIX 1. In the following we present the proof of theorem 4.4. Suppose that $\underline{A} = (A,\Sigma,g,A') \in AR$, K is an $\underline{A}$-configuration, $f_1,\ldots,f_n \in act(K)$ and $c_1,\ldots,c_n \in A$ $(n \geq 1)$. Then

$$K(f_1 \leftarrow c_1,\ldots,f_n \leftarrow c_n)$$

denotes the $\underline{A}$-configuration obtained from K by replacing each active subtree f_i with the element c_i, $i = 1,\ldots,n$.

First we establish two facts.

Fact 1. Let $\underline{A} = (A,\Sigma,g,A')$ and K be an $\underline{A}$-configuration. Suppose that $f \in act(K)$ is such that

$$(*) \quad f_g = \{\{c_1\},\ldots,\{c_k\}\} \quad (k \geq 1),$$

$c_i \in A$, $i = 1,\ldots,k$. Furthermore let $f' \in act(K)$, $f' \neq f$ be such that $\emptyset \notin f'_g$ and $f'_g \neq \emptyset$. Hence we can write $f'_g = \{D_1,\ldots,D_p\}$, $\emptyset \neq D_i \subseteq A$, $i = 1,\ldots,p$, $p \geq 1$.

Let us consider the UE-alternating computation of $\underline{A}$. Suppose that in the configuration K $\underline{A}$ reads the subtree f' making the existential choices $d_i \in D_i$, $i = 1,\ldots,p$ and that in each of the thus obtained configurations $K(f' \leftarrow d_i)$ $\underline{A}$ reads next the subtree f. In this way one obtains the second generation successors

$$K_{ij} = K(f' \leftarrow d_i, f \leftarrow c_j),$$

$i = 1,\ldots,p$, $j = 1,\ldots,k$ of the configuration K.

Now we claim that the identical second generation successors K_{ij} of K can be obtained by reading first f and then f'.

Proof of fact 1. This follows immediately because f_g is defined in such a way that in a UE-alternating computation $\underline{A}$ does not make any existential choices when reading f.

Fact 2. Let $\underline{A}$, K and f be as in fact 1. Suppose that T is an arbitrary

JE-alternating K-computation tree of $\underline{A}$ and that the leaves of T are labeled by $\underline{A}$-configurations $M_1,\dots,M_r$, $r \geq 1$. Suppose that the active subtree f does not appear in any of the configurations $M_1,\dots,M_r$ (i.e. the recognizer $\underline{A}$ has read f in every branch of the computation tree T). Then there exists such a UE-alternating K-computation tree T' of $\underline{A}$ that the leaves of T' are labeled by $M_1,\dots,M_r$ and at the root of T' $\underline{A}$ reads the active subtree f of K.

<u>Proof of fact 2.</u> Use fact 1. and induction on the number of nodes of the computation tree T.

Now we proceed with the proof of theorem 4.4. Suppose that $L = L_{EU}(\underline{A})$ where $\underline{A} = (A,\Sigma,g,A') \in AR$. Choose $\underline{B} = (B,\Omega(\Sigma),d,B')$ where $B = A \cup (pA\times\Sigma)$, $B' = A'$ and d is defined as follows.

i) Let $m \geq 0$, $\sigma \in \Sigma_m$ $(\subseteq\Omega(\Sigma)_m)$ and $a_1,\dots,a_m \in A$. Then

$$\sigma_d(a_1,\dots,a_m) = \{\{(C,\sigma) \mid C \in \sigma_g(a_1,\dots,a_m)\}\}.$$

ii) If $\sigma \in \Sigma$ and $C \in pA$ then

$$\bar{\sigma}_d((C,\sigma)) = \{\{c\} \mid c \in C\}.$$

iii) If $m \geq 0$, $\tau \in \Omega(\Sigma)_m$ and $b_1,\dots,b_m \in B$ do not belong to cases (i) and (ii) define

$$\tau_d(b_1,\dots,b_m) = \emptyset.$$

In the UE-alternating computation mode $\underline{B}$ operates as follows. When reading an active subtree $f = \sigma(a_1,\dots,a_m)$ $(m \geq 0,\ \sigma \in \Sigma_m,\ a_1,\dots,a_m \in A)$ $\underline{B}$ chooses existentially a state (C,σ) corresponding to a set $C \in f_g$ and does not branch universally. When reading a unary symbol $\bar{\sigma} \in \bar{\Sigma}$ in state (C,σ) $\underline{B}$ branches universally to the elements of C. From the above observation it follows that $h_\Sigma(L) \subseteq L_{UE}(\underline{B})$.

On the other hand suppose that $t \in L_{UE}(\underline{B})$. From the definition of $\underline{B}$ one sees immediately that t is necessarily of the form $t = h_\Sigma(t_1)$, $t_1 \in F_\Sigma$. Let T be a UE-accepting $h_\Sigma(t_1)$-computation tree of $\underline{B}$. Using fact 2. and induction one sees that T can be transformed into such a UE-accepting $h_\Sigma(t_1)$-computation tree T' of $\underline{B}$ that in T' always after reading a node of the input tree labeled by $\sigma \in \Sigma$ $\underline{B}$ immediately thereafter reads the preceding node labeled by $\bar{\sigma}$. (Note that when $\sigma \in \Sigma$ and $C \in pA$ $\bar{\sigma}_d((C,\sigma))$ is of the form (*).) Hence clearly there exists also an EU-accepting t_1-computation tree of $\underline{A}$ and it follows that $t \in h_\Sigma(L)$. □

APPENDIX 2. In the following we construct a recognizer for the Σ-forest L in the proof of theorem 4.5.

Choose $\underline{A} = (A,\Sigma,g,A') \in AR$ where

$$A = \{c_1,c_2,u_1,u_2,u_3,u_4,d_1,d_2,v_1,v_2,v_3,v_4,f\}, \quad A' = \{f\}$$

and the mapping g is defined as follows:

(1) $\gamma_g = \{\{c_2\},\{d_2\}\}$
(2) $\sigma_g(c_1) = \{\{u_2\},\{c_2,u_3\}\}$
(3) $\sigma_g(d_1) = \{\{v_3\},\{d_2,v_4\}\}$
(4) $\sigma_g(c_2) = \{\{u_4\},\{c_1,u_1\}\}$
(5) $\sigma_g(d_2) = \{\{v_1\},\{d_1,v_2\}\}$
(6) $\sigma_g(u_i) = \{\{u_i\}\}$, $i = 1,\ldots,4$
(7) $\sigma_g(v_i) = \{\{v_i\}\}$, $i = 1,\ldots,4$
(8) $\sigma_g(f) = \emptyset$
(9) $\tau_g(x,y) = \{\{f\}\}$ iff $(x,y) \in \{(c_1,d_1),(c_2,d_2),(u_1,v_1),(u_2,v_2),(u_3,v_3),(u_4,v_4),(c_1,v_2),(c_2,v_4)\}$ and $\tau_g(x,y) = \emptyset$ otherwise.

(The state set of $\underline{A}$ could be made smaller but we are only aiming at a clear construction.)

In the following we show that $L_{EU}(\underline{A}) = L$. Denote $t(n,m) = \tau(\sigma^n(\gamma),\sigma^m(\gamma))$, $n,m \geq 0$. Also if n and m are known from the context we denote the configuration $\tau(\sigma^{n-i}(a_1),\sigma^{m-j}(a_2))$, $0 \leq i \leq n$, $0 \leq j \leq m$, $a_1,a_2 \in A$ shortly by $[a_1(i),a_2(j)]$.

First we show that $L \subseteq L_{EU}(\underline{A})$. Let $n \geq 0$. Then figure 1. on the next page depicts an EU-accepting $t(n,n)$-computation tree of $\underline{A}$.

Next we prove that $L_{EU}(\underline{A}) \subseteq L$. From the rules (8) and (9) one sees that $\underline{A}$ accepts only Σ-trees of the form $t(n,m)$. Suppose that there exists an EU-accepting $t(n,m)$-computation tree T of $\underline{A}$. We show that then necessarily $n = m$. The idea of the proof is that we show that the computation tree T must have one branch similar to the leftmost branch of the computation tree of figure 1. In this branch $\underline{A}$ reads symbols σ alternately from the left and right branches of the input tree $t(n,m)$.

In the following we denote by $[a_1,a_2]$ any configuration of the form $[a_1(i),a_2(j)]$, $0 \leq i \leq n$, $0 \leq j \leq m$, $a_1,a_2 \in A$, in the $t(n,m)$-computation tree T. For practical purposes we may assume that the root of T is labeled by $[c_2,d_2]$. We claim that in T there exists a branch B from the root to one of the leaves such that the nodes of B are labeled according to the following diagram:

$$\begin{array}{ccccc} & & \downarrow & & \\ f & \leftarrow & [c_2,d_2] & \rightarrow & [c_1,d_2] \\ & & \uparrow & & \downarrow \\ & & [c_2,d_1] & \leftarrow & [c_1,d_1] & \rightarrow & f \end{array}$$

Figure 1.

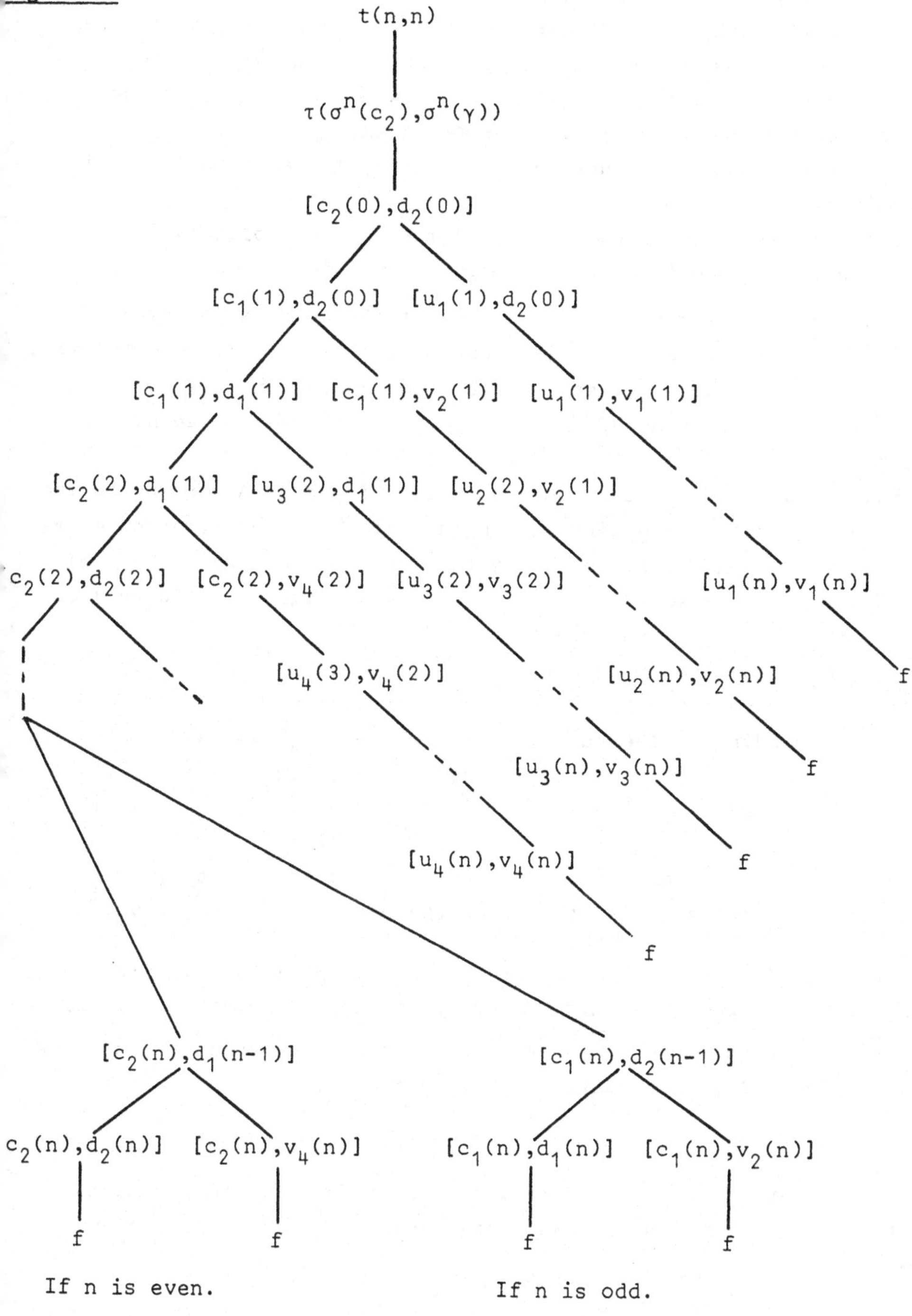
t(n,n)
τ(σ^n(c_2),σ^n(γ))
[c_2(0),d_2(0)]
[c_1(1),d_2(0)]
[u_1(1),d_2(0)]
[c_1(1),d_1(1)]
[c_1(1),v_2(1)]
[u_1(1),v_1(1)]
[c_2(2),d_1(1)]
[u_3(2),d_1(1)]
[u_2(2),v_2(1)]
[c_2(2),d_2(2)]
[c_2(2),v_4(2)]
[u_3(2),v_3(2)]
[u_1(n),v_1(n)]
[u_4(3),v_4(2)]
[u_2(n),v_2(n)]
f
[u_3(n),v_3(n)]
f
[u_4(n),v_4(n)]
f
f
[c_2(n),d_1(n-1)]
[c_1(n),d_2(n-1)]
[c_2(n),d_2(n)]
[c_2(n),v_4(n)]
[c_1(n),d_1(n)]
[c_1(n),v_2(n)]
f
f
f
f
If n is even.
If n is odd.

We verify as an example that if a configuration $K = [c_2, d_2]$ in T has not as its immediate successor the configuration f then K necessarily has a configuration $[c_1, d_2]$ as one of its immediate successors. Suppose that this is not the case. Then $\underline{A}$ would have to continue the computation from K in one of the following ways:

(i) $\underline{A}$ reads the active subtree $\sigma(c_2)$ making the existential choice $\{u_4\}$.

(ii) $\underline{A}$ reads the active subtree $\sigma(d_2)$ making the choice $\{v_1\}$.

(iii) $\underline{A}$ reads the active subtree $\sigma(d_2)$ making the choice $\{d_1, v_2\}$.

In the following we show that none of the computation steps (i) - (iii) can be made in an accepting computation tree T. We use repeatedly the observation that if at some stage a configuration $[u_l, v_k]$, $l \neq k$, $l, k \in \{1, \ldots, 4\}$, is reached then the computation tree T can not be accepting.

(i) Now K would have a successor labeled by $[u_4, d_2]$. According to the rule (6) the state u_4 remains unchanged until the root τ. At some stage of the computation the state d_2 has to be read according to the rule (5). Hence, we necessarily come either to a configuration $[u_4, v_1]$ or $[u_4, v_2]$ that are both non-accepting.

(ii) In this case the successor of K would be labeled by $[c_2, v_1]$. When continuing the computation $\underline{A}$ necessarily at some stage has to read the state c_2 according to the rule (4). The choice $\{u_4\}$ leads directly to a rejecting configuration $[u_4, v_1]$. The choice $\{c_1, u_1\}$ gives as the other successor $[c_1, v_1]$. This leads according to rule (2) necessarily to one of the rejecting configurations $[u_2, v_1]$ or $[u_3, v_1]$.

(iii) Now the other immediate successor of K is labeled by $[c_2, v_2]$. This leads according to (4) to one of the configurations $[u_4, v_2]$ or $[u_1, v_2]$.

Analogously one checks the other three cases in the previous diagram. Now in branch B the recognizer reads symbols σ alternately from the left and right branches of $t(n,m)$. Thus for the computation to terminate succesfully the equation $n = m$ must hold. □

Acknowledgement. I wish to express my gratitude to professor Magnus Steinby for constant support and encouragement at all stages of my work.

References.

[1] A.Arnold, M.Dauchet: Une théorème de duplication pour les forêts algébriques, J. Comput. Syst. Sci. 13, 223-244, (1976).

[2] A.Arnold, M.Dauchet: Forêts Algébriques et Homomorphismes Inverses, Inf. Control 37, 182-196, (1978).

[3] A.K.Chandra, D.C.Kozen, L.J.Stockmeyer: Alternation, J. ACM 28, 114-133, (1981).

[4] F.Gécseg, M.Steinby: Tree automata, Akadémiai Kiadó, Budapest, 1984.

[5] T.Kamimura: Tree Automata and Attribute Grammars, Inf. Control 57, 1-20, (1983).

[6] R.E.Ladner, R.J.Lipton, L.J.Stockmeyer: Alternating pushdown and stack automata, SIAM J. Comput. 13, 135-155, (1984).

[7] A.Salomaa: Formal Languages, Academic Press, 1973.

[8] G.Slutzki: Alternating tree automata, Proc. CAAP'83, Lect. Notes Comput. Sci. 159, 392-404, (1983).

Bottom-up Recursion in Trees

R.Casas† and J-M.Steyaert††

Abstract:

We study the average value of several parameters on trees, which are defined recursively on the structure from the leaves to the root. The emphasis is put here on the methods, both algebraic and analytic; in particular we show it is possible to study this class of problems by means of formal power series that are solutions of infinite systems of equations, and then to get asymptotics by the Darboux-Polya method. In our opinion, this approach should allow the analysis of a new class of non trivial algorithms on trees.

1.Introduction

Analyzing the average performance of programs has been for years an expanding domain of computer science; D.E.Knuth [Kn68, Kn69, Kn73] greatly influenced the domain, showing how classical methods of combinatorics and real analysis where relevent to the study of average paramaters on various data structures. Other works have then been published, which tried to develop a more systematic approach to this type of problems; in particular, under the influence of works of M.Schutzenberger, algebraic methods based on formal languages and power series proved to be an very useful tool for the combinatorial statement of problems (see also I.J.Good [Go65], e.g.). Furthermore, the use of complex analysis for deriving asymptotics from the equations obtained during the combinatorial phase was shown to be very powerful in many situations (see P.Flajolet [Fl79] and A.Odlyzko [Od82] e.g.).

In [FS83] P.Flajolet and JM.Steyaert developed a rather systematic method for analyzing algorithms which applies to a large class of recursive programs on trees; this class can be characterized by the fact that algorithms traverse input structures in a top-down way and do not modify them; in a sense, they behave merely as finite transducers on the input trees. A full development of this approach together with applications and some extensions can be found in [St84][SS84].

† Facultat de Informàtica, Universitat Politècnica de Barcelona, Barcelona 34 (Spain)

†† Centre de Mathématiques Appliquées, Ecole Polytechnique, 91128 Palaiseau (France)

However, this method does not apply to many situations where the combinatorial quantities to be evaluated are defined in a more involved manner; such examples are, for instance, the number of registers needed for algebraic expressions evaluation, the size of the representation of expressions when identical subexpressions are shared (a typical Lisp problem), the average performance of simplification algorithms for symbolic evaluation, etc...

In this paper, we want to show that a certain scheme seems to be a good candidate for evaluating parameters defined recursively bottom-up on tree structures. We strongly expect that it will be possible to extend it to a large class of tree manipulating algorithms developed in the context of symbolic computation, artificial intelligence, etc... The problems presented here should appear as a mere abstraction of more concrete situations, and the solutions we propose as a first step towards a general method.

We study here three types of parameters on binary trees; their definitions involve equality tests on subtrees, that may have been modified by the recursive bottom-up evaluation: this situation is quite common in the context of algebraic simplification and cannot be handled by the previously mentioned methods.

Section 2 is devoted to the study of a rewriting system which outputs a single copy of a subtree whenever both its left and right subtrees are identical. This kind of situation is reminiscent of an idempotent law in usual algebra; the standard *or* in boolean algebra can be taken as an example.

In section 3, we analyze a similar system which replaces such a subtree by a single extra leaf, simulating thus the substraction x-x=0, or more generally a nilpotent law.

In section 4, we show that we can analyze a class of random variables inductively defined on trees.

All the analyses show that the average size of the resulting tree or mean value of the random variable are linear in the input size. However we think that the method is more important than the results themselves; we show that such recursive definition schemes can be translated into infinite sets of equations on power series, which can be solved almost explicitly apart from a certain infinite sum: this sum in turn appears to be negligible, so that the analytic Darboux-Polya method applies.

In this extended abstract proofs will be only sketched.

2.Tree simplification (idempotent law)

In this section, we first introduce the basic definitions that are going to be used throughout the paper. These are related to the well known family ***B*** of binary trees and their enumerating power series **b(z)**. We will then turn to the study of the average size of simplified trees as a function of the original trees.

2.1 Binary trees

The family B of binary trees is classically defined by the equation

$$B = \blacklozenge + o(B, B) \qquad , \qquad (1)$$

which can be viewed either as an equation in the algebra where $\blacklozenge$, the leaf symbol, is a constant and **o** , the internal node symbol, is a binary operator which applies to an ordered pair of trees, or as a convenient shortcut for the classical definition of binary trees.

The generating power series for B , $b(z)=\sum C_n z^n$, where C_n , for n≥0, is the number of binary trees having n internal nodes, is easily seen from (1) to satisfy the equation

$$b(z) = 1 + z.b^2(z) \qquad , \qquad (2)$$

which analytically solves to

$$b(z) = (1 - (1 - 4z)^{1/2}) / 2z \qquad .$$

The Catalan numbers, C_n , are known to evaluate to

$$C_n = \frac{1}{n+1}\binom{2n}{n} = \frac{1}{\sqrt{\pi}} 4^n n^{-3/2} (1+O(1/n)) \qquad (3)$$

as n tends to infinity.

2.2 The simplification algorithm

We now consider the family D of binary trees whose leaves can be labelled by one of the two symbols a or b . The simplification procedure we study transforms these trees according to a bottom-up scheme which replaces every subtree having equal left and right subtrees (after simplification) by a single copy of one of those. More formally, in a Pascalish dialect, we state it as:

```
function simplify (tree: D ) : D ;
    if tree.degree =0   then  simplify:= tree
                        else  tl:= simplify (tree.left);
                              tr:= simplify (tree.right);
                              if   equal (tl , tr)   then  simplify := tl
                                                     else  simplify := o (tl , tr)   fi fi.
```

For instance, any tree in D , whose leaves are all labelled with a (resp. b), simplifies to the one leaf tree a (resp. b); more generally, any tree which can be obtained by substituting all the leaves of a binary tree with a unique tree t in D simplifies to a single copy of t . It should be noted that the equality test means the structural identity.

In this section we study the average size of the simplified tree w.r.t. the input tree. For this purpose we assume, as is usually the case, all the trees of a given size to be equally likely; the size of a tree t, i.e. its number of internal nodes, will be denoted by $|t|$.

Rephrasing §2.1, we get for family D the defining equation

$$D = a + b + o(D, D) \quad , \tag{4}$$

and for its generating power series $d(z) = \sum d_n z^n$, where d_n is the number of trees in D having size n, the equation

$$d(z) = 2 + z.d^2(z) \quad , \tag{5}$$

from which we derive the analytic expression

$$d(z) = (1 - (1 - 8z)^{1/2}) / 2z \quad . \tag{6}$$

Hence $d(z)$ has radius of convergence 1/8.

An easy combinatorial argument shows furthermore that

$$d_n = 2^{n+1} C_n .$$

2.3 The power series for the size of simplified trees

For any tree $t \in D$, let $simp(t)$ denote the size of the tree obtained from t after simplification; we now want to evaluate the series $S(z) = \sum_{t \in D} simp(t).z^{|t|} = \sum s_n z^n$. This series, at least apparently, does not satisfy such an easy functionnal equation as above; however, it is possible to have it defined by an infinite recursive scheme in the following way.

Let us first consider the set I of irreducible trees, that is those trees in D that cannot be simplified. For all trees $t \in I$, let $D_t(z) = \sum_{u \gg t} z^{|u|}$ be the generating power series of all trees u which simplify to t (this will be denoted by $u \gg t$).

Obviously we have

$$D_a(z) = D_b(z) = b(z) \quad , \tag{7}$$

since a tree reduces to a iff all its leaves are labelled with a, and similarly for b.

Furthermore, for any pair (u, v) of *distinct* and *irreducible* trees, the tree denoted by $u \wedge v$, whose subtrees are u and v , is still irreducible and we have

$$D_{u \wedge v}(z) = z.D_u(z).D_v(z) + z.(D_{u \wedge v}(z))^2 . \tag{8}$$

Equation (8) expresses the fact that whenever a tree simplifies to $u \wedge v$, either its left subtree simplifies to u while its right subtree simplifies to v , or both of them simplify to $u \wedge v$.

Let us first remark that

$$d(z) = \sum_{t \in I} D_t(z) \quad . \tag{9}$$

In order to evaluate S(z), we introduce the series in two variables

$$\Delta(z,y) = \Sigma_{t \in I} \ y^{|t|}.\mathbf{D}_t(z) \qquad . \qquad (10)$$

Clearly, differentiating Δ w.r.t. y and setting y = 1 in the result will give us S(z) :

$$S(z) = \Delta'_y(z,1) \qquad .$$

Equation (8) is readily rewritten as

$$y^{|u\wedge v|}.\mathbf{D}_{u\wedge v}(z) = yz.y^{|u|}.\mathbf{D}_u(z).y^{|v|}.\mathbf{D}_v(z) + z.y^{-|u\wedge v|}.(y^{|u\wedge v|}.\mathbf{D}_{u\wedge v}(z))^2 \qquad ,$$

and by summing all these equations we ultimately obtain:

$$\Delta(z,y) = 2 + yz.\Delta^2(z,y) - z.\Sigma_{t \in I} (y^{2|t|+1} - y^{|t|}).\mathbf{D}_t^2(z) \qquad . \qquad (11)$$

This last equation now yields (after differentiation w.r.t. y , setting y =1 and solving the linear equation thus obtained)

Proposition 1: *The power series* S(z) *of the sizes of simplified trees is given by*

$$S(z) = [\, \mathbf{d}(z) - 2 - z.\Sigma_{t \in I} (|t| + 1)\, \mathbf{D}_t^2(z) \,] \,.\, (1 - 2z.\mathbf{d}(z))^{-1} \,.$$

In the above equation, the denominator [1 - 2z.**d**(z)] is quite classical; it appears in a number of situations where the observed quantity is linear in the input size. The radius of convergence of S(z) being already known to be 1/8, by a direct combinatorial argument, the problem lies now in determining precisely the local behaviour of S(z) in a neighbourhood of z=1/8. The key point for this is

Proposition 2: *The series* $\delta(z) = \Sigma_{t \in I} (|t| + 1)\, \mathbf{D}_t^2(z)$ *is analytic for* $|z| \le 3/16$.

The proof of Proposition 2 involves a study of the family I of irreducible trees and of the series $\mathbf{D}_t$ as defined above; in particular, one can show that all $\mathbf{D}_t$ have radius of convergence 1/4 and that $\mathbf{D}_t(1/4) = 2$ for any $t \in I$. From an estimation of the number of irreducible trees of size n, whose generating power series **i**(z) satisfies the equation

$$\mathbf{i}(z) = 2 + z.\mathbf{i}^2(z) - z.\mathbf{i}(z^2) \qquad ,$$

it is therefore possible to conclude that $\delta(z)$ converges absolutely for $|z| > 1/8$.

We are now able to state

Theorem 1A: *The Taylor coefficients of* S(z) *have the asymptotic equivalent as* n *tends to infinity*

$$s_n = \alpha\,.8^n.n^{-1/2}\,(1+O(1/n)) \qquad ,$$

with $\alpha = [2 - \delta(1/8)/8]\,/\sqrt{\pi}$.

Theorem 1B: *The average size of the simplified tree obtained from a random tree of size* n *in D satisfies*

$$\mathbf{s}_n = s_n/\,d_n = \boldsymbol{\alpha}.n\,(1+O(1/n)) \qquad ,$$

with $\boldsymbol{\alpha} = [2 - \delta(1/8)/8]\,/2 = 0.8196...$

Part B of Theorem 1 is an immediate consequence of Part A together with Equation (3) and the value of d_n.

From Propositions 1 and 2, we have that in a neighbourhood of z = 1/8, S(z) has the expansion

$$S(z) = [2 - \delta(1/8)/8]\;(1-8z)^{-1/2} + \text{s.o.t.}$$

and satisfies the conditions of the Darboux-Polya method (see [Po37] or [He74]); whence Part A.

3.Tree reduction (nilpotent law)

The tree transformation considered in this section is a slight idealisation of the algebraic simplification **x - x = 0**. We start with binary trees whose leaves are labelled with ***a*** or ***e*** ; symbol ***e*** can be viewed as the **0** element of classical algebra, while symbol ***a*** stands for any variable or number. We do not distinguish special operators (as +, *, / ....) for the sake of simplicity, although it is not difficult to extend out treatment to any situation that could seem more realistic to the reader. The study of the average size of the reduced expression is similar to that developped in section 2; however the infinite set of equations that we establish is somehow more complex in the sense that the recurrence scheme does not start as nicely as in section 2; the analytic study is therefore less easy even if the final result looks much as the same.

3.1 The reduction algorithm

In order to have a precise definition of a reduced tree, we now give an algorithm which describes the reduction phenomenon.

```
function reduce (tree: D') : D' ;
    if tree.degree =0  then reduce:= tree
                       else tl:= reduce (tree.left);
                            tr:= reduce (tree.right);
                            if  equal (tl , tr)  then reduce := e
                                                 else reduce := o (tl , tr)  fi fi.
```

Family D' is isomorphic to family D, the only change being that symbol e replaces symbol b and plays a different role in this context. The generating function associated to D' is therefore $d(z)$ as in section 2.

From the above definition, it should be clear that any tree whose leaves are all labelled with e reduces to e, and that the only tree which reduces to a is a itself. These phenomema will be made explicit in the next paragraph.

3.2 The power series for the size of reduced trees

For any t in D', let $red\,(t)$ be the size of the tree obtained from t after reduction; in order to evaluate this average size (w.r.t. trees t of a given initial size), we have to study the series $R(z) = \sum_{t \in D'} red\,(t).z^{|t|} = \sum r_n.z^n$. Again this series cannot be simply defined; therefore we introduce the family I' of irreducible trees (in this new sense of reduction), which happens to coincide with family I of section 2 (where b 's are substituted by e 's), as can be seen by a simple argument.

For any $t \in I'$, we consider the generating power series $E_t(z)$ of all trees u in D' which reduce to t : $E_t(z) = \sum_{u \gg t} z^{|u|}$, where >> represents now the reduction operation.

Since a is the only tree which reduces to a, we have:

$$E_a(z) = 1 \qquad . \qquad (12)$$

Furthermore, any tree which reduces to e either is e itself, or is such that its left and right subtrees both reduce to the same irreducible tree t ; hence:

$$E_e(z) = 1 + \sum_{t \in I'} z.E_t^2(z) \qquad . \qquad (13)$$

Finally, for any pair (u , v) of *distinct* and *irreducible* trees, the tree denoted by $u \wedge v$, whose subtrees are u and v, is still *irreducible* and we have

$$E_{u \wedge v}(z) = z.E_u(z).E_v(z) \qquad , \qquad (14)$$

and obviously

$$d(z) = \sum_{t \in I'} E_t(z) \qquad . \qquad (15)$$

In order to obtain an expression for series R(z), we consider the series in two variables

$$H(z,y) = \sum_{t \in I'} y^{|t|}.E_t(z) \qquad . \qquad (16)$$

Clearly, with the same notations as for $\Delta(z,y)$, we have

$$R(z) = H'_y(z,1) \qquad .$$

Equations (12) (13) and (14) are easily rewritten to take into account variable y ; summing all the equations thus obtained, we get:

$$H(z,y) = 1 + E_e(z) + yz.H^2(z,y) - yz.\sum_{t \in I'} y^{2|t|}.E_t^2(z) \quad . \qquad (17)$$

We now differentiate this last equation w.r.t. y, set y =1 and solve the linear equation in R(z) thus obtained, which gives:

Proposition 3: *The power series* R(z) *of the sizes of simplified trees is given by*

$$R(z) = [\, d(z) - 2 - z.\sum_{t \in I'} (2|t| + 1)\, E_t^2(z)\,] \,.\, (1 - 2z.d(z))^{-1} \,.$$

This expression is very similar to the one obtained for S(z) in Proposition 1; the analytical study of R(z) turns to be slighty more difficult however: this is due to the global form of the system of Equations (12) (13) (14), which in a sense defines $E_e(z)$ implicitely. As in the previous section we show by direct combinatorial considerations that the radius of convergence of R(z) is 1/8; the problem is to prove that the series $\eta(z) = \sum_{t \in I'} (2|t| + 1)\, E_t^2(z)$ remains analytic for $|z| > 1/8$, which will allow us to conclude as done previously.

Letting $|t|_e$ denote the number of leaves of t labelled by e , we first remark that, for all $t \in I'$,

$$E_t(z) = z^{|t|}.(E_e(z))^{|t|_e} \qquad ; \qquad (18)$$

then

$$E_e(z) = 1 + z.\, E_e^2(z) + z.\sum_{t \in I' - e} z^{2|t|}.(E_e(z))^{2|t|_e} \qquad . \qquad (19)$$

Therefore, on the real positive axis, and as long as $z^2.\, E_e^2(z) < \rho_i$, ρ_i being the radius of convergence of the generating series i(z) of irreducible trees, we have

$$E_e(z) \leq 1 + z.\, E_e^2(z) - z + z.i(z^2.\, E_e^2(z)) \qquad ;$$

since $\rho_i > 1/8$, one can prove using the above estimations that $E_e(z)$ converges absolutely for all z, $|z| \leq \rho_\eta$, which numerically evaluates to $\rho_\eta = 0.1684....$

Hence we can state

Proposition 4: *For all* z , $|z| \leq \rho_\eta$, *the series* $\mathbf{E}_e(z)$ *and* $\eta(z)$ *are analytic.*

The analyticity of $\eta(z)$ derives immediately from that of $\mathbf{E}_e(z)$.
As we did before, we can now use the Darboux-Polya method to get asymptotic estimations for the coefficients r_n of series $\mathbf{R}(z)$, and then for the average value of the size of reduced trees.
We finally have:

Theorem 2: *The average size of reduced trees satisfies as* $n \to \infty$:

$$r_n = \beta.n\ (1+ O(1/n)) \qquad ,$$

where $\beta = 1 - \eta(1/8)\ /16 = 0.8079...$

4. Random variables on trees

It is often the case that some deterministic processes are so intricate that it is almost impossible to analyse them at first shot; it sometimes happens then that they can be approximated by random processes whose analysis is easier. In a first attempt for solving the simplification problem developped in section 2, we tried such an approximation; this lead us to the method described above. Now it appears that this approach could be useful in more complex situations, and that a general phenomenon, though not yet precisely stated, should explain many of those.

We present in this section, the original problem of analyzing the average value of a particular random variable on binary trees.

We consider the family $\boldsymbol{B}$ of pure binary trees without labels; their generating power series is $b(z)$, as previously mentioned, and the number of trees of size n is precisely the Catalan number C_n . Let $P(n)= (C_n.2^{n+1})^{-1}$, for all $n \geq 0$.

Let now $W(t)$ be a discrete random variable on $\boldsymbol{B}$ defined in the following way:

- $W(\bullet) = 0$
- $W(u \wedge v)$ = if $W(u) \neq W(v)$ then $1+ W(u) + W(v)$
  else $W(u)$ with probability $P(W(u))$ or
  $1+ W(u) + W(v)$ with probability $1- P(W(u))$.

In the sequel we will simply call W a *valuation* on trees. The problem is to determine the average value of W on all trees of size n, all trees being equally likely.

For all $t \in \boldsymbol{B}$, and $m \geq 0$, let $p_t(m)$ be the "probability" that t has valuation m; of course, when $m>|t|$, $p_t(m)=0$. These quantities are defined by the following recursive scheme, which is a direct translation of the definition scheme of W:

$$p_{\blacklozenge}(0) = 1 \tag{20}$$

$$p_{u \wedge v}(m) = \Sigma_{i+j=m-1}\, p_u(i).p_v(j) + p_u(m).p_v(m).P(m) - p_u((m-1)/2).p_v((m-1)/2).P((m-1)/2) \ .$$

It is not hard to see that $\Sigma_{m,|t|=n}\, p_t(m) = C_n$, and that the average value of valuation W we are seeking for, is precisely w_n such that

$$C_n .w_n = w_n = \Sigma_{m,|t|=n}\, m.p_t(m) \qquad . \tag{21}$$

Let us now consider the power series $F_m(z) = \Sigma_{t \in \boldsymbol{B}}\, p_t(m).z^{|t|}$, for all $m \geq 0$. They satisfy the equations

$$F_0(z) = 1+ z.P(0).F_0^2(z) \tag{22}$$

and, for $m \geq 1$,

$$F_m(z) = z.\Sigma_{i+j=m-1}\, F_i(z).\, F_j(z) \ + z.P(m).\, (F_m(z))^2 - z.P(\mu).\, (F_\mu(z))^2 \tag{23}$$

with $\mu = (m-1)/2$, this last term appearing iff m is odd.

The power series $\Phi(z,y) = \Sigma_{m\geq 0}\, y^m.\, F_m(z)$, satisfies therefore the equation

$$\Phi(z,y) = 1+ yz.\Phi^2(z,y) - z.\Sigma_{m\geq 0}\, (y^{2m+1} - y^m).P(m).(F_m(z))^2 \quad . \tag{24}$$

Since $w(z) = \Sigma\, w_n.z^n = \Phi'_y(z,1)$, we now deduce

Proposition 5: *The power series* $w(z)$ *of the average values of valuation* W *satisfies the equation*

$$w(z) = [\, b(z) - 1 - z.\Sigma_{m\geq 0}\, (m+1).P(m).(F_m(z))^2]\, .(1 - 2z.b(z))^{-1} \quad .$$

By an argument similar to the one used in section 2, we prove that the series

$$\phi(z) = \Sigma_{m\geq 0}\, (m+1).P(m).(F_m(z))^2$$

is still analytic for $|z| > 1/4$, the radius of convergence of $w(z)$, and we conclude again by a similar estimation of the average valuation.

We obtain thus

Theorem 3: *The average value of the valuation* W *on all trees of size* n *satisfies as* $n \to \infty$:

$$w_n = \gamma . n \ (1 + O(1/n))$$

with $\gamma = 1 - \phi(1/4)/4 = 0.8241...$

One should notice that γ is within less than 1% of α, as was intuitivelly expected.

As a final remark, we emphasize that the fact that $\phi(z)$ has a radius of convergence greater than 1/4 does not rely heavily on the particular shape of probability P; it appears that the behaviour of w_n shown is Theorem 3, retains a similar behaviour in a large range of variation for quantity P; whence the method is applicable to a large class of valuations W.

5. References

[Fl79] P.Flajolet : *Analyse d'Algorithmes de Manipulation d'Arbres et de Fichiers* Thèse Paris (1979), published in Cahiers du B.U.R.O. n° 34-35, Paris (1981).

[FS83] P.Flajolet & J-M.Steyaert : *A complexity calculus for recursive tree algorithms* RR n° 239, INRIA (Oct 1983) , to appear in Math.Sys.Theory.

[Go65] I.J.Good : *The Generalization of Lagrange's Expansion and the Enumeration of Trees* Proc. Camb. Phil. Soc. 117 (1965) pp 285-306.

[He74] P.Henrici : Applied and Computational Complex Analysis 2 vol, J.Wiley, New-York (1974).

[Kn68][Kn69][Kn73] D.E.Knuth : The Art of Computer Programming 3 vol, Addison Wesley, Reading (1968,1969,1973).

[Od82] A.Odlyzko : *Periodic Oscillations of Coefficients of Power Series that Satisfy Functional Equations* , Adv. in Math. 44, (1982) pp 180-205.

[Po37] G.Polya : *Kombinatorische Anzahlbestimmungen für Gruppen, Graphen und chemische Verbindungen* , Acta Mathematica, 68 (1937) pp 145-254.

[SS84] M.Soria & J-M.Steyaert : *Average efficiency of pattern-matching on Lisp-expressions* , Proceedings of 1984 CAAP (Bordeaux 1984)

[St84] J-M.Steyaert : *Structure et Complexité des Algorithmes* Thèse Paris (1984).

BASIC TREE TRANSDUCERS

Heiko Vogler*, University of Leiden
P.O.Box 9512, 2300 RA Leiden, The Netherlands

. INTRODUCTION. It is a usual matter of theoretical study to restrict the resources f objects like grammars, automata, and transducers in order to obtain a better undertanding of the way they work. The following two investigations concern the restriction "no nesting of nonterminals" in macro grammars [Fi] and context-free tree gramars [ES,Da], i.e., they concern basic macro grammars [Fi,Do,ESvL] and basic tree rammars [ES1], respectively.

a) Macro grammars are equivalent to nested stack automata [Fi,A2] and (extended) asic macro grammars are equivalent to stack-pushdown machines [ESvL], which can be onsidered as a natural restriction of nested stack automata. (b) The path languages f context-free tree languages are context-free [R1,Co,ES1] and the path languages of asic tree languages are linear [ES1].

Here we introduce basic tree transducers as a natural restriction of macro tree ransducers [E3,CF,EV1] in the sense that nesting of states is not allowed. Conceptully, basic tree transducers are basic tree grammars of which the derivations are synax-directed by an input tree (just as macro tree transducers correspond to contextree tree grammars).

This paper tries to give some insight in the influence on macro tree transducers f restricting the resource of nesting of states. Let RECOG denote the class of recogizable (or regular) tree languages. The main three results of the paper are the ollowing.

1) Basic tree transducers are characterized by "one-turn pushdown tree transducers", hich are transducers with regular tree grammars as control and one-turn pushdowns as dditional storage (cf. [E6,EV2,EV3] and Section 2 for the concept of "grammar with torage"). More precisely, the configurations of this storage are pushdowns of poiners to an input tree; once such a pushdown has decreased its length, it can never row again. In the total deterministic case, compositions of basic tree transducers re characterized by iterated one-turn pushdown tree transducers (cf. [G1,Ms,E5,DGo, V2,EV3] for the concept of iterated pushdown). Note that macro tree transducers with outside-in derivation mode) are characterized by pushdown tree transducers [EV2]; ence, the restriction to "basic" coincides with the restriction to "one-turn".

2) Path languages of images of RECOG under (compositions of) basic tree transducers re generated by (iterated) controlled linear grammars (cf. [GS2,G2/3,K1/2,DP,V1] nd Section 2 for the concept of controlled linear grammar); in particular, path

The work of the author has been supported by the Netherlands Organization for the dvancement of Pure Research (Z.W.O.).

languages of images of RECOG under basic tree transducers are linear languages.
(3) The composition hierarchy of (images of RECOG under) basic tree transducers is strict for both the nondeterministic and the total deterministic case.

Because of space restrictions, constructions and proofs are omitted. The reader is referred to [V3], where he can also find more detailed discussions of the facts presented here.

2. PRELIMINARIES. In this section we briefly recall some notation from tree language theory, the concept of controlled linear grammar [GS2,G2/3,K1/2,DP,V1], and the concepts of storage type [Sc] and grammar with storage [E6,EV2,EV3].

For a ranked set Δ, Δ_n denotes the set of symbols of Δ with rank n. The rank of a symbol is sometimes indicated as a superscript, e.g., $\delta^{(2)}$ means that δ is of rank 2. Throughout this paper, Ω denotes a countably infinite ranked set. For a ranked set Δ and an arbitrary set U, the set of (labeled) trees over Δ indexed by U, denoted by $T_\Delta(U)$, is the smallest subset of $(\Delta \cup U \cup PC)^*$, where PC denotes the set of left-parenthesis, right-parenthesis, and comma, such that (i) $U \subseteq T_\Delta(U)$ and (ii) if $\delta \in \Delta_k$ with $k \geq 0$ and $t_1, \ldots, t_k \in T_\Delta(U)$, then $\delta(t_1, \ldots, t_k) \in T_\Delta(U)$. For $U = \emptyset$, the set T_Δ of usual trees over Δ is obtained. For $t \in T_\Delta$, yield(t) denotes the yield of t, and the yield of a relation $R \subseteq U \times T_\Delta$ (where U is an arbitrary set), is the relation yield(R) = $\{(u,w) \mid (u,t) \in R$ and yield(t)=w for some t$\}$, which is also abbreviated by yR. This notion is extended to classes of relations in an obvious way. Any subset of T_Δ is a tree language and the class of recognizable tree languages [R1,Th,E1] is denoted by RECOG.

Let Δ be a ranked alphabet, i.e., a finite ranked set. The path alphabet (see, e.g., [R1,Co,ES1) associated with Δ, denoted $p(\Delta)$, is the finite set $\{(\delta,0) \mid \delta \in \Delta_0\} \cup \{(\delta,i) \mid \delta \in \Delta_n$ for some $n \geq 1$ and $1 \leq i \leq n\}$. For every ranked alphabet Δ, the path translation of Δ, denoted π_Δ, is the smallest subset of $T_\Delta \times p(\Delta)^*$ for which (i) and (ii) hold. (i) For $\delta \in \Delta_0$, $(\delta,(\delta,0)) \in \pi_\Delta$. (ii) For $t = \delta(t_1, \ldots, t_k)$ with $\delta \in \Delta_k$, $k \geq 1$, and $t_1, \ldots, t_k \in T_\Delta$, if $(t_i, p) \in \pi_\Delta$ for some $i \in [k]$, then $(t,(\delta,i)p) \in \pi_\Delta$. Let π denote the class of all path translations π_Δ for a ranked alphabet Δ. Note that π(RECOG) is contained in the class of regular languages (cf. Corollary 4.12 of [Co]).

A total deterministic bottom-up finite state relabeling (for short: relabeling, cf. Definition 3.14 of [E1]) maps any input tree s to an output tree, that has the same tree-structure as s but possibly different labels. The class of translations induced by relabelings is denoted by RELAB. Note that RECOG is closed under RELAB (cf. Corollary 3.11 of [E1]),i.e.,RELAB(RECOG) $\subseteq$ RECOG.

A controlled linear grammar K is a tuple (G,H), where G is a linear (context-free) grammar, in which the rules are labeled by symbols from some set Σ, and H is a language over Σ, i.e., $H \subseteq \Sigma^*$. The language generated by K, denoted by L(K), is the set of terminal strings w such that w is derivable from the initial nonterminal A_{in} of G, and the sequence of labels of rules which are applied in this derivation, is a string in H. Let $\mathfrak{L}$ be a class of (string) languages. A controlled linear grammar (G,H) is an $\mathfrak{L}$-controlled linear grammar if H is a member of $\mathfrak{L}$. The class of languages generated by $\mathfrak{L}$-controlled linear grammars is denoted by CTRL(LIN,$\mathfrak{L}$). The mechanism of controlling a linear grammar can be iterated as follows: given a class of languages $\mathfrak{L}$, $CTRL_0(LIN,\mathfrak{L}) = \mathfrak{L}$ and for every $n\geq 0$, $CTRL_{n+1}(LIN,\mathfrak{L}) = CTRL(LIN,CTRL_n(LIN,\mathfrak{L}))$. In [K1] it was shown that $\{CTRL_n(LIN,\mathfrak{L}_{CF}) \mid n\geq 0\}$ is an infinite strict hierarchy, which is called the geometric language hierarchy. Also $\{CTRL_n(LIN,\mathfrak{L}_{REG}) \mid n\geq 0\}$ is an infinite strict hierarchy [G2].

A storage type S is a tuple (C,P,F,I,E), where C is the set of configurations, P is the set of predicates, which are mappings C→{true,false}, F is the set of instructions, which are partial functions C→C, I is the set of input elements, and E is the set of encodings, which are partial functions I→C. In the rest of this paper S denotes the storage type (C,P,F,I,E) if not specified otherwise.

EXAMPLES. - The tree storage type TR is the storage type (C,P,F,I,E), where $C = T_\Omega$, a predicate is of the form root=σ ($\sigma\in\Omega$) and it is true on a configuration t if σ is the label of the root of t, and sel_i ($i\leq 1$) is an instruction which selects the i-th direct subtree. I = C and E = {e|e is the identity on T_Σ for some finite subset Σ of Ω}.

The trivial storage type S_0 is the tuple ({c},∅,{id},{c},{id}), where c is an arbitrary object and id(c) = c.

The pushdown of S [G1,E5,E6,EV2], denoted by P(S), is the storage type (C',P',F',I',E'), where $C' = (\Gamma\times C)^+$ and Γ is a fixed infinite pushdown alphabet, $P' = \{top=\gamma \mid \gamma\in\Gamma\} \cup \{test(p) \mid p\in P\}$, $F' = \{push(\gamma,f) \mid \gamma\in\Gamma, f\in F\} \cup \{pop\} \cup \{stay(\gamma) \mid \gamma\in\Gamma\}$, I' = I, $E' = \{\lambda u\in I.(\gamma,e(u)) \mid \gamma\in\Gamma, e\in E\}$, and for every $c' = (\delta,c)\beta$ with $\delta\in\Gamma$, $c\in C$, and $\beta\in C'\cup\{\lambda\}$, $(top=\gamma)(c') = (\gamma=\delta)$, $(test(p))(c') = p(c)$, $(push(\gamma,f))(c') = (\gamma,f(c))(\delta,c)\beta$ if f is defined on c, $pop(c') = \beta$ if $\beta\neq\lambda$, $(stay(\gamma))(c') = (\gamma,c)\beta$. The operator P can be iterated in an obvious way: $P^0(S) = S$ and $P^{n+1}(S) = P(P^n(S))$. $P^n(S_0)$ is abbreviated by P^n. (The reader should not confuse the operator P with the set P of predicates of S.)

□

Now let X be a modifier taken from the set {context-free tree, basic tree, regular tree, context-free, linear, regular} which we also abbreviate by {CFT, BT, RT, CF, LIN, REG}. The reader is assumed to be familiar with the concepts of X grammar (for CFT see [ES], for BT see [ES1], for RT see [GSt], for CF, LIN, and REG see [HU]). An X grammar is specified by a tuple (N,Δ,A_{in},R), where N is the set of nonterminals, Δ is the set of terminals, A_{in} is the initial nonterminal, and R is the finite set of rules. According to the type of X, N and Δ may be ranked alphabets. $\mathfrak{L}_X$ denotes the class of languages generated by X grammars. Since a grammar with storage induces a translation from the input set of the storage type to the set of terminal trees or strings of the involved grammar, we call it an X(S)-transducer, where X is the class of used grammars and S is the storage type.

An *X(S)-transducer* M is a tuple (N,e,Δ,A_{in},R), where N, Δ, and A_{in} are the alphabets of nonterminals, terminals, and the initial nonterminal as defined for X grammars, $e \in E$ is the encoding of M, and R is the finite set of rules; each rule has the form $\Theta \to \underline{if}\ b\ \underline{then}\ \zeta$, where $\Theta \to \xi$ is a rule of a usual X grammar, b is a boolean combination of predicates in P, and ζ is obtained from ξ by replacing every occurrence of a nonterminal B by B<f> for some $f \in F$. M is *deterministic* if for every $c \in C$ and every two different rules $\Theta \to \underline{if}\ b_1\ \underline{then}\ \zeta_1$ and $\Theta \to \underline{if}\ b_2\ \underline{then}\ \zeta_2$, $(b_1(c)\underline{and}b_2(c)) = \text{false}$.

The X grammar $G(M) = (N',\Delta,A',R')$ *associated with* M is defined by $N' = \{A<c> \mid A \in N,\ c \in C\}$ (ranks carry over from N to N'), A' is any element of N' (of rank 0), and R' is obtained as follows. If $\Theta \to \underline{if}\ b\ \underline{then}\ \zeta$ is in R, then for every $c \in C$ such that b(c)=true and such that, for every instruction f occurring in ζ, f is defined on c, the rule $\Theta_c \to \zeta_c$ is in R', where Θ_c is the result of replacing in Θ the nonterminal A by A<c>, and ζ_c is obtained from ζ by replacing every B<f> by B<f(c)> (with $B \in N$, $f \in F$). Note that an associated grammar may have infinitely many nonterminals and infinitely many rules.

The derivation relation of M, denoted $=(M)\Rightarrow$, is defined by $=(M)\Rightarrow\ =\ =(G(M))\Rightarrow$; for X=CFT we only consider outside-in derivations. The *translation induced by* M, denoted by $\tau(M)$, is the set $\{(u,v) \mid u \in I,\ v \in \Phi, \text{ and } A_{in}<e(u)> =(M)\Rightarrow v\}$ where Φ is, according to the type of X, either Δ^* or T_Δ. M is *total*, if $\text{dom}(\tau(M)) = \text{dom}(e)$. The class of translations induced by (total deterministic) X(S)-transducers is denoted by X(S) ($D_tX(S)$, respectively).

FACTS. (1) By considering the trivial storage type S_0, an $X(S_0)$-transducer can be regarded again as an X grammar. Thus, $\text{range}(X(S_0)) = \mathcal{L}_X$. (2) yRT(S) = CF(S), where we postulate a symbol of rank 0, which is interpreted by yield as λ, i.e., the empty string. (3) REG(S)-transducers are very close to usual (one-way) S automata, which have a finite control, a one-way input tape, and an auxiliary storage of type S (cf. [E6]). Actually, $\text{range}(REG(S)) = \mathcal{L}(S)$, where $\mathcal{L}(S)$ denotes the class of languages accepted by one-way S automata. (4) RT(S)-transducers can be considered as top-down tree automata with an additional storage of type S. Such automata accept range(RT(S)). In this sense, the usual finite top-down tree automata (see, e.g., [TW,E1]) coincide with $RT(S_0)$-transducers and the pushdown tree automata of [Gu] coincide with RT(P)-transducers. (5) RT(TR)-transducers and CFT(TR)-transducers are equivalent formulations of top-down tree transducers [R2,Th,E1] and macro tree transducers [CF,E3,EV1], respectively (Theorem 3.19 and Corollary 3.20 of [EV2]). □

3. CHARACTERIZATION OF BASIC TREE TRANSDUCERS. A *basic tree transducer* is a BT(TR)-transducer. Recall that basic tree grammars are context-free tree grammars, in which no nesting of nonterminals may occur. Thus a rule of a BT(S)-transducer (N,e,Δ,A_{in},R) is of the form $A(y_1,\ldots,y_k) \to \underline{if}\ b\ \underline{then}\ \zeta$, where $k \geq 0$, $A^{(k)} \in N$, b is a boolean combi-

nation over P and $\zeta \in T_{\Delta}(Y_k \cup \Lambda_k)$ with $Y_k=\{y_1,\ldots,y_k\}$ and $\Lambda_k = \{B\langle f\rangle(t_1,\ldots,t_r) \mid r\geq 0, B^{(r)}\in N, f\in F,$ and $t_1,\ldots,t_r \in T_{\Delta}(Y_k)\}$. In this section we implement BT(S)-transducers (for arbitrary S) on "one-turn pushdown S transducers", i.e., $RT(P_{1t}(S))$-transducers, where $P_{1t}(S)$ is the storage type "one-turn pushdown of S" (note that $P_{1t}(S_0)$ is the storage type of the usual one-turn pushdown automata of [GS1]). Moreover, we show that a slightly extended notion of BT(S)-transducers is equivalent to the one-turn pushdown S transducers. By specifying S to TR, we obtain the desired characterization of (extended) basic tree transducers by one-turn pushdown tree transducers.

In extended BT(S)-transducers, the parameters of nonterminals may hold regular tree languages as actual values rather than only one tree. These transducers are formalized as a special case of extended CFT(S)-transducers (cf. Definition 5.22 of [EV2]); for a symbol id, which does not occur in F, let S_{id} denote the storage type $(C,P,F\cup\{id\},I,E)$ such that for every $c\in C$, $id(c)=c$; then, an extended CFT(S)-transducer (for short: $CFT_{ext}(S)$-transducer) is a $CFT(S_{id})$-transducer M, where the use of the instruction id is restricted to those subtrees of right-hand sides of rules that have the form $B\langle id\rangle(y_1,\ldots,y_k)$ for some nonterminal B of rank $k\geq 0$. Note that, if S already contains an identity instruction, then $CFT_{ext}(S) = CFT(S)$. Instead of placing language names in the parameter positions of nonterminals, the regular tree languages are represented by "extension" nonterminals.

DEFINITION 1. An <u>extended BT(S)-transducer</u> M (for short: $BT_{ext}(S)$-transducer) is a $CFT_{ext}(S)$-transducer (N,e,Δ,A_{in},R) such that there is a partition of N into the set N_{Φ} of basic nonterminals and the set N_{Ψ} of extension nonterminals, $A_{in}^{(0)}\in N_{\Phi}$, and there is a partition of R into the set R_{Φ} of basic rules and the set R_{Ψ} of extension rules; every basic rule has the form $A(y_1,\ldots,y_k) \rightarrow$ <u>if</u> b <u>then</u> ζ, where $k\geq 0$, $A^{(k)}\in N_{\Phi}$, b is a boolean combination over P, $\zeta\in T_{\Delta}(Y_k\cup\Lambda_k\cup Z_k)$, $\Lambda_k = \{B\langle f\rangle(\psi_1\langle id\rangle\tilde{y},\ldots,\psi_r\langle id\rangle\tilde{y}) \mid r\geq 0, B^{(r)}\in N_{\Phi}, f\in F,$ and $\psi_1,\ldots,\psi_r\in N_{\Psi}$ of rank $k\}$, $\tilde{y}$ abbreviates $(y_1,\ldots,y_k)$, and $Z_k = \{B\langle id\rangle\tilde{y} \mid B^{(k)}\in N_{\Phi}\}$, and every extension rule has the form $\psi(y_1,\ldots,y_k) \rightarrow \zeta$, where $k\geq 0$, $\psi^{(k)}\in N_{\Psi}, \zeta\in T_{\Delta}(Y_k\cup Z_k)$ and $Z_k = \{\Phi\langle id\rangle(y_1,\ldots,y_k) \mid \Phi^{(k)}\in N_{\Psi}\}$. □

The class of translations induced by (total deterministic) $BT_{ext}(S)$-transducers is denoted by $BT_{ext}(S)$ ($D_tBT_{ext}(S)$, respectively).

FACTS. (6) The class range$(BT_{ext}(S_0))$ is equal to the class UltBT(2) of 2-level ultra-basic tree languages [ES1]. (7) The class range $(yBT_{ext}(S_0))$ coincides with the class EB of extended basic macro anguages [Do,ESvL], i.e., range $(yBT_{ext}(S_0)) = EB$.□

Clearly, $BT(S) \subseteq BT_{ext}(S) \subseteq CFT_{ext}(S)$. In the total deterministic case the extension does not increase the power of BT(S)-transducers, i.e., $D_tBT_{ext}(S) = D_tBT(S)$ (Lemma 4.4 of [V3]).

For the implementation of $BT_{ext}(S)$-transducers the following pushdown device is used [GS1,V1]: the <u>one-turn pushdown of S</u>, denoted by $P_{1t}(S)$, is equal to P(S) except that a push instruction is not defined after a pop instruction has occurred (i.e., after the turn). The information whether the turn lies in the future or in the past, can be indicated in the configurations of $P_{1t}(S)$ by a "turn-bit", e.g., by defining $C' = (\Gamma\times C)^{+}\times\{0,1\}$. Then the push-instructions are defined if the turn-bit is 0 and undefined if it is 1; the pop-instruction changes the turn-bit into 1. The formal definition is left to the reader.

The operator P_{1t} can be iterated $P^0_{1t}(S) = S$ and $P^{n+1}_{1t}(S) = P_{1t}(P^n_{1t}(S))$. For every $n\geq 0$, $P^n_{1t}(S_0)$ is abbreviated by P^n_{1t}. Note that $\mathcal{L}_{LIN} = \mathcal{L}(P_{1t})$ [GS1] and for every $n\geq 0$, $CTRL_n(LIN,\mathcal{L}_{REG}) = \mathcal{L}(P^n_{1t})$ [V1].

In the implementation of a $BT_{ext}(S)$-transducer M on an $RT(P_{1t}(S))$-transducer M', M' performs a symbolic expansion of the basic nonterminals of M by writing the adjacent lists of extension nonterminals on the pushdown. If a symbolic expansion is completed and a value of a parameter is required, then this value can be substituted by popping the pushdown and by symbolically expanding the appropriate extension nonterminals. Note that the latter expansions do not involve any additional push-instructions. Hence, the pushdowns are one-turn (see Lemma 5.19 of [EV2] and Lemma 5.11 of [EV2] for similar constructions). This construction preserves totality and determinism. Thus, $BT_{ext}(S) \subseteq RT(P_{1t}(S))$ and $D_tBT(S) \subseteq D_tRT(P_{1t}(S))$.

Now let us turn to the simulation of an $RT(P_{1t}(S))$-transducer M by a $BT_{ext}(S)$-transducer M'. Assume w.l.o.g. that the boolean combinations that occur in the rules of M, are mutually exclusive, and assume that M has the nonterminals $A_1,\dots,A_r$. Consider a nonterminal A and a configuration $((\gamma,c)\beta,0)$ of $P_{1t}(S)$ with the pushdown symbol γ, the configuration c of S, the pushdown β, and the turn-bit 0. Then the construct $A<((\gamma,c)\beta,0)>$ occurring in a sentential form of M is represented in M' by the tree $[A,\gamma]<c>(t_1,\dots,t_r)$, where $[A,\gamma]$ is a nonterminal and t_j represents the construct $A_j<(\beta,1)>$. Similarly, $A<((\gamma,c)\beta,1)>$ is represented by $[A,\gamma,b]<c>(t_1,\dots,t_r)$, where b is that boolean combination over predicates of S that occurs in M and is true on c (note that this b is uniquely determined). Totality and determinism are preserved in this construction.

The reason of having for every nonterminal A_j of M a representation of $(\beta,1)$ is the following. If a push instruction is applied to the pushdown β, then M does noet know in advance with which nonterminal it will return to β. Hence, for every such return nonterminal, $(\beta,1)$ has to be represented. The technique of coding finite information by preparing sufficiently many representations of an object in the parameter positions of nonterminals, was first used in Theorem 7 of [R2] to prove that every creative dendrolanguage can be generated by a one-state creative dendrogrammar; cf. also Theorem 1 of [Gu], Construction 1 of [DGu], Lemma 5.4 of [EV2], and Lemma 5.10 of [EV3] for further applications of this technique.

In the simulation of $RT(P_{1t}(S))$-transducers, the extension of basic tree transducers is essential. Instructions of the form stay(γ) are modelled by constructs like $B<id>(y_1,\dots,y_k)$ in basic rules (where B is a basic nonterminal) and constructs like $\psi<id>(y_1,\dots,y_k)$ in extension rules (where ψ is an extension nonterminal), depending on whether the stay(γ) instruction occurs before or after the turn of a pushdown.

This construction, which preserves totality and determinism, yields the inclusions $RT(P_{1t}(S)) \subseteq BT_{ext}(S)$ and $D_tRT(P_{1t}(S)) \subseteq D_tBT(S)$. Thus we obtain the characterization of $BT_{ext}(S)$-transducers in terms of $RT(P_{1t}(S))$-transducers.

THEOREM 1. $BT_{ext}(S) = RT(P_{1t}(S))$ and $D_tBT(S) = D_tRT(P_{1t}(S))$.

FACTS. (8) In particular, $BT_{ext}(S_0) = RT(P_{1t})$. Since $\mathrm{range}(BT_{ext}(S_0)) = UltBT(2)$ (Fact (6)), 2-level ultra-basic tree languages are recognized by top-down tree automata with a one-turn pushdown as storage. Note that (OI) context-free tree languages are recognized by top-down tree automata with a pushdown as storage, i.e., $CFT = \mathrm{range}(RT(P))$ (see [Gu], where these automata are called restricted pushdown tree automata). (9) Since $\mathrm{range}(yBT_{ext}(S_0)) = EB$ (Fact (7)), Theorem 1 implies that $EB = \mathrm{range}(yRT(P_{1t}))$, and since $yRT(S) = CF(S)$ (Fact (2)), $EB = \mathrm{range}(CF(P_{1t}))$. Recalling that range $(CF(P))$ is the class of indexed languages (cf. [E6]), it is easy to see that the class $\mathrm{range}(CF(P_{1t}))$ is generated by the so-called restricted indexed grammars [A1]. Hence, we reobtain the characterization of EB by restricted indexed grammars (cf. [Fl,ESvL]).

(10) Since, for every S which contains an identity, $CF(S) = REG(P(S))$ (Theorem 5.1 of [E6]) and, in particular, S_0 contains an identity, EB is also equal to $\mathrm{range}(REG(P(P_{1t}))) = \mathcal{L}(P(P_{1t}))$, i.e., extended basic macro languages are characterized by one-way $P(P_{1t})$-automata. In [ESvL] EB is characterized by stack-pushdown machines (for short: s-pd machines). Actually, it is not so difficult to see that s-pd machines are equivalent to $P(P_{1t})$-automata; this is a special case of the equivalence of nested stack automata and P^2-automata (Theorem 7.4 of [EV2]).

(11) Obviously, Theorem 1 generalizes the equivalence of linear grammars and one-turn pushdown automata [GS1] to the tree case: consider a $BT_{ext}(S_0)$-transducer M, in which the alphabets of nonterminals and terminals are monadic; then, by glueing every extension nonterminal to the corresponding basic nonterminal, the rules of M turn into rules of a linear grammar. □

Considering the tree storage type, we obtain the characterization of extended basic tree transducers in terms of one-turn pushdown tree transducers. In the total deterministic case, the composition of basic tree transducers is characterized by iterated one-turn pushdown tree transducers, i.e., $D_tBT(TR)^n = D_tRT(P^n_{1t}(TR))$. This follows by straightforward induction on n using the decomposition $D_tBT(S) = D_tRT(S) \circ D_tBT(TR)$, which is a special case of Lemma 8.5 and Lemma 8.9 of [EV2]. For two classes of translations T_1 and T_2, $T_1 \circ T_2 = \{R_1 \circ R_2 \mid R_i \in T_i\}$, and for a class of translations T and $n \geq 0$, $T^n = \{R_1 \circ \ldots \circ R_n \mid R_i \in T\}$, where $R_1 \circ R_2 = \{(x,z) \mid (x,y) \in R_1$ and $(y,z) \in R_2$ for some $y\}$.

THEOREM 2. $BT_{ext}(TR) = RT(P_{1t}(TR))$ and for every $n \geq 1$, $D_tBT(TR)^n = D_tRT(P^n_{1t}(TR))$.

FACTS. (12) Note that $CFT_{ext}(TR) = RT(P(TR))$ (Theorem 8.3 of [EV2]). Moreover, for every $n\geq 1$, $D_tCFT(TR)^n = D_tRT(P^n(TR))$ (Theorem 8.12 of [EV2]). (13) In the total deterministic case high level tree transducers [EV3] are characterized by iterated pushdown tree transducers, i.e., $D_t n\text{-}T(TR) = D_tRT(P^n(TR))$, where n-T denotes the class of n-level tree grammars [Da,EV3]. Note that 0-T = RT an 1-T = CFT. □

4. PATH LANGUAGES AND COMPOSITION HIERARCHY. In this section we study the tree-to-path translations of compositions of basic tree transducers, i.e., $BT(TR)^n \circ \pi$ with $n\geq 0$, and show inductively that $\pi(BT(TR)^n(RECOG)) \subseteq CTRL_n(LIN,\mathcal{L}_{REG})$. This forms the basis for the proof of the strictness of the composition hierarchy of basic tree transducers.

The first step in the study of $BT(TR)^n \circ \pi$ is the treatment of the class $BT(S) \circ \pi$ for an arbitrary storage type S. Clearly, one might expect that this class is included in LIN(S); just take the paths of the right-hand sides of the involved BT(S)-transducer as the right-hand sides for rules of the LIN(S)-transducer. However, the inlcusion does not necessarily hold. What is yet missing in this ad hoc construction is a facility to check whether there is a successful derivation of the BT(S)-transducer starting from those nonterminals, that are cut off from the right-hand side by considering one particular path. This facility is formalized by means of the concept of look-ahead on storage types [E6,EV4]. The storage type S with look-ahead, denoted S_{LA}, is the tuple (C,P',F,I,E), where $P' = P \cup \{\langle A,H\rangle \mid$ H is a CF(S)-transducer and A is one of its nonterminals} and for every $c\in C$, $\langle A,H\rangle(c)$ = true iff there is a $w\in\Delta^*$ such that $A\langle c\rangle \Rightarrow_{(H)}^* w$, where Δ is the terminal alphabet of H. The appropriate refinement of the ad hoc construction above proves the following two inclusions.

LEMMA 1. $BT(S) \circ \pi \subseteq LIN(S_{LA})$ and $RT(S) \circ \pi \subseteq REG(S_{LA})$.

Note that, apart from look-ahead, the involved construction is very similar to the one used in [ES1] to prove that path languages of context-free tree languages are context-free languages. The reason why the extension with look-ahead (or at least something similar) is not needed there, is the fact that a context-free tree grammar can be equivalently transformed such that from every nonterminal a terminal tree is derivable. In general, this transformation (see Theorem 3.1.5 of [Fi]) is not possible for X(S)-transducers.

FACTS. (14) $RT(TR_{LA})$-transducers and $CFT(TR_{LA})$-transducers are equivalent to top-down tree transducers with regular look-ahead [E2] and to macro tree transducers with regular look-ahead [EV1], respectively.
(15) The second inclusion of Lemma 1 is also useful to study the classes $(n+1)\text{-}T(TR) \circ \pi$ and $CFT(TR) \circ \pi$. By the characterization of n-T(TR) and CFT(TR) in terms of RT(S')-transducers for appropriate S' (Theorem 6.15 of [EV3] and Theorem 5.14 of [EV2], respectively) this inclusion show that $(n+1)\text{-}T(TR) \circ \pi \subseteq$ yield (n-T(TR)) and $\pi(CFT(TR)(RECOG)) \subseteq CF(TR)(RECOG)$ (Theorem 5.3 of [V3]). The second inclusion was

was claimed in [E3]. □

Every LIN(TR_{LA})-transducer M can be decomposed into a relabeling and a LIN(TR)-transducer. To understand this note that every look-ahead test on TR specifies a recognizable tree language in the form of the domain of a CF(TR)-transducer. Now the relabeling adds to every node of an input tree t of M some extra information; at the root of a tree s (which is a subtree of t), this information describes whether the direct subtrees of s are contained in those recognizable tree languages that are specified by the look-ahead predicates occurring in M. Hence, LIN(TR_{LA}) $\subseteq$ RELAB $\circ$ LIN(TR) (cf. also Theorem 2.6 of [E2]).

LIN(TR)-transducers have a very special feature. In every computation of a LIN(TR)-transducer M, only one path of the given input tree is considered. Hence, we can also "feed" M just with the paths of an input tree rather than with the tree itself. This allows us to describe LIN(TR) in terms of LIN(one-way)-transducers, where one-way [E6,V2] is the storage type (C,P,F,I,E) with $C = \Xi^*$ for some fixed infinite set Ξ, $P = \{sym{=}a \mid a \in \Xi\} \cup \{empty\}$, F = {read}, for every $bw \in \Xi^*$ with $b \in \Xi$ and $w \in \Xi^*$, (sym=a)(bw) = true iff (b=a) and (sym=a)(λ) = false, empty(bw) = false and empty(λ) = true, read(bw) = w and read(λ) is undefined, $I = \Xi^*$, and $E = \{\lambda w \in \Sigma^*.w \mid \Sigma \text{ is a finite subset of } \Xi\}$. It is easy to see that LIN(TR) $\subseteq \pi \circ$ LIN(one-way) and hence, BT(TR) $\circ\, \pi \subseteq$ RELAB $\circ\, \pi \circ$ LIN(one-way) (using Lemma 1 and the decomposition LIN(TR_{LA}) $\subseteq$ RELAB $\circ$ LIN(TR)).

Actually, this connection between basic tree transducers and LIN(one-way)-transducers can be generalized to a connection between the compositions of both classes of translations. In the inductive proof of this generalization, the closure of BT(TR)-transducers under right-composition with relabelings is used, i.e., BT(TR) $\circ$ RELAB $\subseteq$ BT(TR). This can be shown by a direct construction, which is very similar to the one used in [Fi] to prove the closure of IO-macro languages under intersection with regular languages.

LEMMA 2. For every $n \geq 1$,
$BT(TR)^n \circ \pi \subseteq RELAB \circ \pi \circ LIN(one\text{-}way)^n$.

Controlled linear grammars (viewed as "grammar-directed translators" [G2]) and LIN(one-way)-transducers are very closely related and there are only two small differences between the two concepts. First, the former device has to consume the whole control string, whereas the latter device can ignore suffixes of the input, and second, the former device cannot detect the end of the control string whereas the latter can. The first difference can be overcome by forcing every LIN(one-way)-transducer to read its whole input string (using an additional nonterminal). The second difference is overbridged by requiring that the class $\mathcal{L}$ of languages, of which the images are studied, is closed under right-marking and finite substitution. These closure properties are guaranteed if $\mathcal{L}$ is a full semi-AFL [Gi]. Then, for every $n \geq 0$ and every full semi-AFL $\mathcal{L}$, $CTRL_n(LIN,\mathcal{L}) = LIN(one\text{-}way)^n(\mathcal{L})$. Since RECOG is closed under relabeling and since $\pi(\mathrm{RECOG}) \subseteq \mathcal{L}_{REG}$, it follows from this characterization and Lemma 2 that the path languages of images of RECOG under compositions of basic tree transducers can be generated by iterated controlled linear grammars starting from $\mathcal{L}_{REG}$. In particular, since $\mathrm{CTRL}(\mathrm{LIN},\mathcal{L}_{REG}) = \mathcal{L}_{LIN}$ [G2],the path languages of images of RECOG under basic tree transducers are linear.

THEOREM 3. For every $n \geq 0$, $\pi(\mathrm{BT}(\mathrm{TR})^n(\mathrm{RECOG})) \subseteq \mathrm{CTRL}_n(\mathrm{LIN},\mathcal{L}_{REG})$ and $\pi(\mathrm{BT}(\mathrm{TR})(\mathrm{RECOG})) \subseteq \mathcal{L}_{LIN}$.

This theorem shows the connection between basic tree transducers and controlled linear grammars as predicted in [V1].

By means of Theorem 3 and the well-known fact that $\{\mathrm{CTRL}_n(\mathrm{LIN},\mathcal{L}_{REG})\ n \geq 0\}$ is a strict hierarchy [G2], we can prove the strictness of the composition hierarchy of basic tree transducers. For every $n \geq 0$, we define the language $\tilde{L}_n = \{(\#,1)(a,1)^k(d_1,1)(a,1)^k(d_2,1)\ \ldots\ (a,1)^k(d_{\exp(n+1)},0)\ \mid\ k \geq 0\}$, which is very close to the language L_n used in [K1] to prove that $\mathrm{CTRL}_n(\mathrm{LIN},\mathcal{L}_{LIN}) - \mathrm{CTRL}_{n-1}(\mathrm{LIN},\mathcal{L}_{CF}) \neq \emptyset$. Note that this implies that $L_n \in \mathrm{CTRL}_{n+1}(\mathrm{LIN},\mathcal{L}_{REG}) - \mathrm{CTRL}_n(\mathrm{LIN},\mathcal{L}_{REG})$. Since for every $n \geq 0$, $\mathrm{CTRL}_n(\mathrm{LIN},\mathcal{L}_{REG})$ is a full semi-AFL [G2] and since L_n is the homomorphic image of $\tilde{L}_n$, it follows immediately that $\tilde{L}_n \notin \mathrm{CTRL}_n(\mathrm{LIN},\mathcal{L}_{REG})$. However, $\tilde{L}_n$ can be realized as the path language of the image of a recognizable tree language L under τ for some $\tau \in D_t\mathrm{BT}(\mathrm{TR})^{n+1}$, i.e., $\tilde{L}_n = \pi_\Delta(\tau(L))$ for some Δ. The desired basic tree transducers $M_1,\ldots,M_{n+1}$ are just the "monadic tree versions" of the controlled linear grammars, which Khabbaz used in [K1], to generate L_n. Hence, for every $n \geq 0$, $\tau(L) \in D_t\mathrm{BT}(\mathrm{TR})^{n+1}(\mathrm{RECOG})$ and $\tau(L) \notin \mathrm{BT}(\mathrm{TR})^n(\mathrm{RECOG})$ (because otherwise, by Theorem 3, $\pi_\Delta(\tau(L)) = \tilde{L}_n \in \mathrm{CTRL}_n(\mathrm{LIN},\mathcal{L}_{REG})$ and hence also $L_n \in \mathrm{CTRL}_n(\mathrm{LIN},\mathcal{L}_{REG})$, which is a contradiction). Now the properness of the composition hierarchy of basic tree transducers (for both the nondeterministic and the total deterministic case) is an immediate consequence.

THEOREM 4. For every $n \geq 0$,

(i) $\mathrm{BT}(\mathrm{TR})^n(\mathrm{RECOG}) \subsetneq \mathrm{BT}(\mathrm{TR})^{n+1}(\mathrm{RECOG})$ and
$D_t\mathrm{BT}(\mathrm{TR})^n(\mathrm{RECOG}) \subsetneq D_t\mathrm{BT}(\mathrm{TR})^{n+1}(\mathrm{RECOG})$,

(ii) $\mathrm{BT}(\mathrm{TR})^n \subsetneq \mathrm{BT}(\mathrm{TR})^{n+1}$
$D_t\mathrm{BT}(\mathrm{TR})^n \subsetneq D_t\mathrm{BT}(\mathrm{TR})^{n+1}$. □

5. CONCLUSION. We have investigated the class BT(TR) of translations induced by basic tree transducers. These devices form a natural restriction of macro tree transducers. The characterization of extended basic tree transducers by one-turn pushdown tree transducers ($\mathrm{BT}_{ext}(\mathrm{TR}) = \mathrm{RT}(P_{1t}(\mathrm{TR}))$ and $D_t\mathrm{BT}(\mathrm{TR})^n = D_t\mathrm{RT}(P^n_{1t}(\mathrm{TR}))$), the study of tree-to-path translations of basic tree transducers ($\pi(\mathrm{BT}(\mathrm{TR})^n(\mathrm{RECOG})) \subseteq \mathrm{CTRL}_n(\mathrm{LIN},\mathcal{L}_{REG})$), and the characterization in [V1] of controlled linear grammars by

one-turn pushdown automata ($CTRL_n(LIN,\pounds_{REG}) = \pounds(P^n_{1t})$) link the following three formalisms

- basic tree transducers
- one-turn pushdown tree transducers / automata
- controlled linear grammars

The next diagram shows the connections between the classes of images of RECOG under compositions of top-down tree transducers, of basic tree transducers, and of macro tree transducers. Solid ascending lines denote strict inclusion.

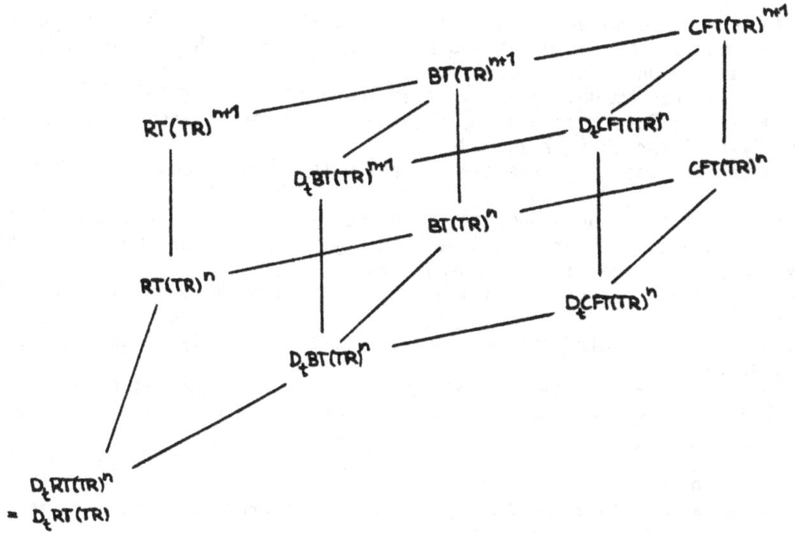

The correctness of the diagram (with respect to strictness of inclusions) follows immediately from (a) $RT(TR)^n \subsetneq BT(TR)^n$ (by Theorem 5.17(ii) of [V3]), (b) $BT(TR)^n \subsetneq CFT(TR)^n$ (by Theorem 5.15 (ii) of [V3]), (c) the strictness of the composition hierarchy of top-down tree transducers (by Theorem 3.14 of [E4]), (d) $BT(TR)^n \subsetneq BT(TR)^{n+1}$ and $D_tBT(TR)^n \subsetneq D_tBT(TR)^{n+1}$ (by Theorem 4), (e) the strictness of the composition hierarchy of nondeterministic and total deterministic macro tree transducers (by Theorem 4.16 of [EV1]), and (f) the fact that, for $X \in \{RT(TR)^n, BT(TR)^n, CFT(TR)^n\}$, D_tX is a class of mappings whereas X contains partial functions.
To show the correctness of the diagram also with repsect to incomparability of non-connected classes, it remains to prove that for every $n \geq 1$, (g) $RT(TR)^{n+1} - CFT(TR)^n \neq \emptyset$ and (h) $D_tBT(TR)^{n+1} - CFT(TR)^n \neq \emptyset$. The other "incomparability results" are proved in [V3]. The proofs of (g) and (h) may be the subject of further research.

Acknowledgment

The author wishes to thank Joost Engelfriet for many helpful discussions.

REFERENCES

[A1] A.V. Aho; Indexed grammars, an extension of context-free grammars; JACM 15 (1968), 647-671.

[A2] A.V. Aho; Nested stack automata; JACM 16 (1969), 383-406.

[Co] B. Courcelle; A representation of trees by languages I and II; TCS 6 (1978), 255-279 and TCS 7 (1978), 25-55.

[CF] B. Courcelle, P. Franchi-Zannettacci; Attribute grammars and recursive program schemes I, II; TCS 17 (1982), 163-191 and TCS 17 (1982), 235-257.

[Da] W. Damm; The IO- and OI-hierarchies; TCS 20 (1982), 95-206.

[DGo] W. Damm, A. Goerdt; An automata-theoretic characterization of the OI-hierarchy Proc. 9th ICALP, 1982, Aarhus, pp. 141-153.

[DGu] W. Damm, I. Guessarian; Combining T and level n; Proc. of the 9th Mathematical Foundations of Computer Sciences 1981, LNCS 118, p. 262-270, Springer-Verlag.

[Do] P.J. Downey; Formal languages and recursion schemes; Ph.D. Thesis, Rep. TR-16-74, Harvard University, Cambridge, Mass., 1974.

[DP] J. Duske, R. Parchmann; Linear indexed languages; TCS 32 (1984), 47-60.

[E1] J. Engelfriet; Bottom-up and top-down tree transformations - a comparison; Math. Syst. Theory 9 (1975), 198-231.

[E2] J. Engelfriet; Top-down tree transducers with regular look-ahead; Math. Syst. Theory 10 (1977), 289-303.

[E3] J. Engelfriet; Some open questions and recent results on tree transducers and tree languages; in: "Formal language theory; perspectives and open problems", (R.V. Book, Ed.), New York, Academic Press, 1980.

[E4] J. Engelfriet; Three hierarchies of transducers; Math. Syst. Theory 15(1982), 95-125.

[E5] J. Engelfriet; Iterated pushdown automata and complexity classes; Proc. 15th STOC, April 1983, Boston, pp. 365-373.

[E6] J. Engelfriet; Context-free grammars with storage; Technical Report, Institute of Applied Mathematics and Computer Science, University of Leiden, The Netherlands.

[ES] J. Engelfriet, E.M. Schmidt; IO and OI; JCSS 15 (1977), 328-353 and JCSS 16 (1978), 67-99.

[ESl] J. Engelfriet, G. Slutzki; Bounded nesting in macro grammars; Inf. and Contr. 42 (1979), 157-193.

[ESvL] J. Engelfriet, E.M. Schmidt, J. van Leeuwen; Stack machines and classes of nor nested macro grammars; JACM 27 (1980), 96-117.

[EV1] J. Engelfriet, H. Vogler; Macro tree transducers; JCSS 31 (1985), 71-146.

[EV2] J. Engelfriet, H. Vogler; Pushdown machines for the macro tree transducer; Report Nr. 84-13, Institute of Applied Mathematics and Computer Science, University of Leiden, to appear in TCS; see also: Regular characterizations of the macro tree transducer; in: "Ninth Colloquium on Trees in Algebra and Programming", March 1984, Bordeaux, France (ed. B. Courcelle), Cambridge Univers Press, 103-117.

[EV3] J. Engelfriet, H. Vogler; High level tree transducers and iterated pushdown machines; Report Nr. 85-12, Institute of Applied Mathematics and Computer Science, University of Leiden, The Netherlands, May 1985,; see also: Characterization of high level tree transducers; Proc. of ICALP 1985, Nafplion,Greece.

[EV4] J. Engelfriet, H. Vogler; Look-ahead on pushdowns; Report Nr. 85-14, Institut of Applied Mathematics and Computer Science, University of Leiden, The Netherlands, July 1985.

[Fi] M.J. Fischer; Grammars with macro-like productions; Ph.D. Thesis, Harvard University, USA, 1968.

[Fl] G. File; The characterization of some language families by classes of indexed grammars; M.Sc.Th., Dept. of Comp. Sci., Pennsylvania University Park, PA., 1

[Gi] S. Ginsburg; "Algebraic and Automata-Theoretic Properties of Formal Languages

North-Holland, Amsterdam, 1975.
[G1] S.A. Greibach; Full AFLs and nested iterated substitution; Inf. and Contr. 16 (1970), 7-35.
[G2] S.A. Greibach; Control sets on context-free grammar forms; JCSS 15 (1977), 35-98.
[G3] S.A. Greibach; One way finite visit automata; TCS 6 (1978), 175-221.
[GS1] S. Ginsburg, E.H. Spanier; Finite-turn pushdown automata; SIAM Control 3 (1966), 429-453.
[GS2] S. Ginsburg, E.H. Spanier; Control sets on grammars; Math. Syst. Theory 2 (1968), 159-177.
[GSt] F. Gecseg, M. Steinby. "Tree automata", Akademiai Kiado, Budapest, 1984.
[Gu] I. Guessarian; Pushdown tree automata; Math. Syst. Theory 16 (1983), 237-263.
[HU] J.E.Hopcroft, J.D. Ullman; "Formal languages and their relation to automata"; Addison-Wesley, Reading, Mass. 1979.
[K1] N.A. Khabbaz; A geometric hierarchy of languages; JCSS 8 (1974), 142-157.
[K2] N.A. Khabbaz; Control sets on linear grammars; Inf. and Contr. 25 (1974), 206-221.
[Ms] A.N. Maslov; Multi-level stack automata; Probl. of Inf. Transm. 12 (1976), 38-43.
[R1] W.C. Rounds; Tree-oriented proofs of some theorems on context-free and indexed languages; 2nd Symposium on Theory of Computing (1970), 109-116.
[R2] W.C. Rounds; Mappings and grammars on trees; Math. Syst. Theory 4 (1970), 257-287.
[Sc] D. Scott; Some definitional suggestions for automata theory; JCSS 1 (1967), 187-212.
[Th] J.W. Thatcher; Generalized^2 sequential machine maps; JCSS 4 (1970), 339-367.
[TW] J.W. Thatcher, J.B. Wright; Generalized finite automata theory with an application to a decision-problem of second-order logic; Math. Syst. Theory 2 (1968), 58-81.
[V1] H. Vogler; Iterated linear control and iterated one-turn pushdowns; Report Nr. 85-04, Institute of Applied Mathematics and Computer Science, University of Leiden, The Netherlands, March 1985; see also: Proc. of FCT 1985, Cottbus, GDR.
[V2] H. Vogler; The OI-hierarchy is closed under control; Report Nr. 85-20, Institute of Applied Mathematics and Computer Science, University of Leiden, The Netherlands, August 1985.
[V3] H. Vogler; Basic tree transducers; Report Nr.85-21, Institute of Applied Mathematics and Computer Science, University of Leiden, The Netherlands, October 1985.

TRIE PARTITIONING PROCESS: LIMITING DISTRIBUTIONS

Philippe JACQUET

Mireille REGNIER

INRIA
Rocquencourt
78153-Le Chesnay (France)

ABSTRACT

This paper is devoted to the well-known trie structure. We consider two basic parameters: depth of the leaves and height when the trie is formed with n items. We prove the convergence of their distributions and of their moments of any order when $n \to \infty$ to a limit distribution. We exhibit the limits: a periodic distribution or a normal distribution. The results are given for uniform or biased data distributions for Bernoulli and Poisson models. Our reasoning is based on generating and characteristic functions. We make an extensive use of analytic functions and asymptotic methods.

This paper provides a uniform framework to establish limit distributions for trie parameters. Tries are a tree structure which appears in quite a large number of applications in computer science. A trie may be used as an index to access data on secondary memory. This is Dynamic Hashing Algorithms [13,2] , or k-d-tries and Grid-File Algorithms in the multidimensional case [15] . Tries also are an underlying structure in problems as diverse as: communication protocols, polynomial factorization, radix exchange sort, simulation algorithms, Huffman's algorithm...[3, 9, 12] . Usual tree parameters, e.g. **size, external path length, height** have in these applications a simple interpretation as execution cost as it will be detailed below in Section 1.

We shall prove the convergence of the distribution of these basic parameters to simple distributions. We assume that data have either a *uniform* or a *biased* distribution. We consider successively the Bernoulli model and its approximation, the Poisson model. Our basic tool is generating or characteristic functions. We consider first the Poisson approximation. We derive a non linear functional equation for characteristic functions. Studying such equations, using notably their *Mellin transforms*, provides their asymptotic properties. Then, by the Levy's *Continuity Theorem* we establish the convergence to either a periodic distribution or a normal distribution. Finally, applying general theorems, we prove equivalent results for the Bernoulli model. In passing, we get the asymptotics for the moments of any order, and in particular, the mean and the variance, extending some previous results, [5, 1, 6, 10, 17] .

The plan of the paper is the following. Section 1 sets the problem. We first define tries and our probabilistic models. Then, we define the parameters to e studied and give their algorithmic interpretation. Finally, we introduce ome notations and recall a basic theorem. In Section 2, we deal with the *lepth of the leaves* and the *internal path length*. Section 3 is devoted to the *eight* of the tries. Section 4 is a conclusion.

. DEFINITIONS.

.1. The Trie Structure:

A *trie* can be viewed as a tree associated to a *recursive partitioning pro-:ess*, that occurs for data that form a subset S of 0–1 strings from $\{0,1\}^*$. Vhenever $|S|\le 1$, the trie T is reduced to a single leaf, possibly containing the lata ($|S|=1$) or empty ($|S|=0$). If $|S|>1$, S can be naturally divided into two ubsets consisting of the strings beginning with a 0 and strings beginning with 1. "Forgetting" this first bit, we get two subsets of $\{0,1\}^*$, S_l and S_r, and we uild T as a root with the two subtries recursively associated to S_l and S_r. 'his recursive process is to be continued until we individuate subsets of car-linality at most 1. For instance, if S is $\{00100,01000,01111,11000,11101\}$, we et the trie:

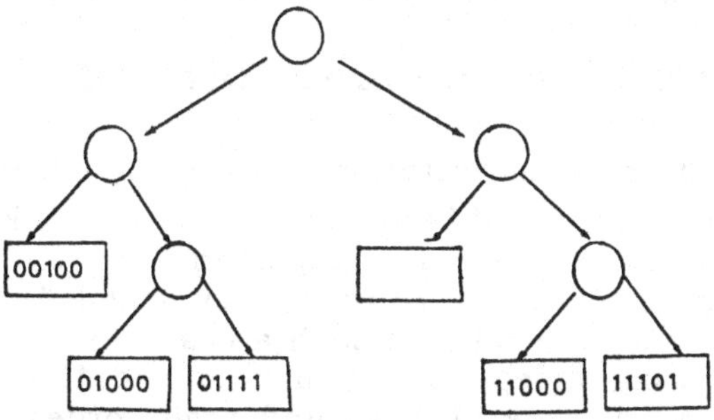

Ve can individuate as well subsets of cardinality at most b by changing the alting condition ($|S|\le 1$ replaced by $S\le b$) of the recursive splitting process. 'or instance,for $b=2$ and S as above, we get:

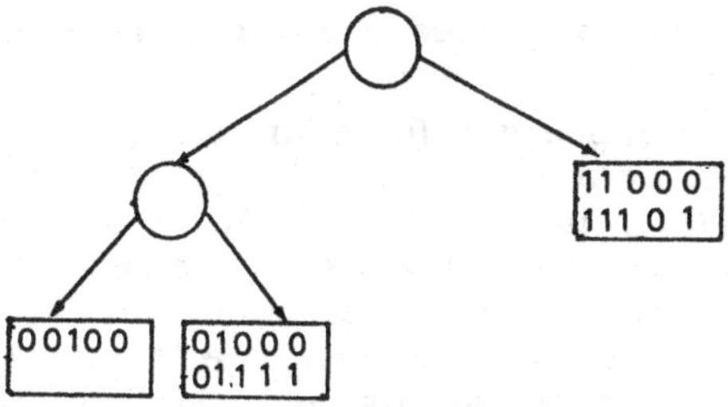

Every recursive call of the partitioning process creates an internal node nd one more level in the tree. Notice that a partition at level i involves the -th bit of the strings to be separated.

This remark enables us to define our probabilistic models for further nalysis. In the *Bernoulli* model, the number of elements in a trie is a fixed arameter n. The data distribution is fully defined as soon as the bit distribu-ion in the sequence is. We shall assume that the bit sequences are indepen-lent sequences of Bernoulli trials with the same parameters (n,p). More pre-isely, if we note $P(b_i=0)$ (respectively $P(b_i=1)$) the probability for the i-th it of a string to be equal to 0 (resp. 1), one of the two following hypotheses olds:

$$P(b_i=0) = P(b_i=1) = \frac{1}{2} \qquad \text{(H1)}$$

$$P(b_i=0) = p\,,\ P(b_i=1) = q\,,\ p+q=1,\ p \neq q\ . \qquad \text{(H2)}$$

In the first case, the distribution is **uniform**, in the second one it is **biased.**

A good approximation is provided by the Poisson model. We keep supposing either $(H1)$ or $(H2)$ holds. The number of elements in the trie is not fixed but follows a Poisson law with parameter n. That is to say, the probability $P(p)$ that a trie be built from p elements is:

$$P(p) = e^{-n} . \frac{n^p}{p!}$$

One can find more details in [4, 11].

Three parameters, namely the *size* and the *height* of tries and the *depth* of the leaves, reveal meaningful in analysis of algorithms.

Definition 1: *The* **depth of a leaf** *is the number of internal nodes on the path from the root to this leaf.*

It is closely related to the so called *internal path length.*

Definition 2: *The* **internal path length** *of a tree is the sum of the depths of all its leaves.*

Definition 3: *The* **height** *of a tree is the maximal depth of its leaves.*

We give now their algorithmic interpretations for three applications.

The **communication protocol** of Capetanakis-Tsybakov-Mikhailov assumes that several transmitters are sharing a unique channel to send information. If at some instant n users, $n>1$, try to transmit, they "collide". The packets are destroyed and the transmission fails. In such a case, the algorithm dictates users should start tossing independently a coin and thus be associated different strings 0−1. A trie may be built and prefix order determines the order in which users are allowed to transmit. The algorithm works whether the dices are biased or not. There are slightly modified algorithms which are more efficient when $p \neq q$. The size of the trie corresponds to the *length of a session* that separates n users. It is also the time the channel is closed to others transmitters. The depth of the leaves is a parameter intervening in the analysis of the *waiting time* of customers.

The trie structure also appears in **file addressing on secondary memory,** for example in databases. Data are characterized by one or several *keys*. These keys -or their binary representation- can be assumed to be infinite sequences from $\{0,1\}^*$. In this case, b stands for the capacity of the pages in secondary memory and a trie is used as an index: in the leaves one finds pointers to the data. Then, retrieval amounts to a traversal of the trie. This structure supports dynamic modifications, as insertions and deletions only modify locally the partition and the trie. When a single key is used, we get exactly the partition process described above (see *Dynamic* and *Extendible Hashing*). When several keys are used, shuffling their bits reduces to that 1-dimensional scheme (see *Grid-File Algorithms*). In 1-dimensional and multidimensional cases, we can assume biased or uniform distributions. A nice generalization of the biased case is provided when keys are strings of characters [14]. Data may be characterized by a set of keywords whose concatenation makes a key. Then, using some binary representation, one can build a trie as an index. Notice that the probabilities $P(a,b)$ for finding a b after an a in a word are not equal and they induce biased probabilities that depend on the level. We have then a *markovian process.* The results we will establish when p and q are independent of the level apply here as well. The size of the trie describes the *space* occupied in secondary memory. The height provides an indication on the space necessary to store the index. The depth of insertion allows for *cost evaluation* of some operations performed in central memory, as the traversal of the index.

For **radix exchange sort**, the depth of insertion stands for the number of bits comparisons to be done during an insertion.

We shall study the limiting distributions of two parameters: the depth of the leaves and the height of the tries. The first parameter is of particular interest, as the nature of the limiting distribution depends on the type of the data distribution .More precisely, in the uniform case, the distribution converges to a periodic distribution, while in the biased case, it converges to the normal distribution.

I.2. Basic Tools:

We begin this section by introducing our notations for the generating and characteristic functions.

Definition: *Let m be some discrete parameter. Let P_n^k the probability that the parameter m be k when n records are in the structure. We note :*

$$\begin{cases} P_n(u) = \sum_{k \geq 0} P_n^k u^k \\ P(z,u) = \sum_{n \geq 0} P_n(u) \frac{z^n}{n!} e^{-z} . \end{cases}$$

$P_n(u)$ is the ordinary generating function. $P(z,u)$ is the Poisson generating function of the parameter m.

Remark *A priori*, these expressions are well defined for $|u|<1$ and any z in C. When we set $u=e^{it}, t \in R, P_n(e^{it})$ is the **characteristic function** of the distribution under the Bernoulli model with first parameter n. When z is real, $P(z,e^{it})$ is the characteristic function of the parameter m when the number of records in the trie follows a Poisson law with mean z.

To prove the convergence of the distributions and of their moments, we make use of the two following theorems [4 , vol2, XV.3] .

Theorem 1 (Continuity Theorem): *In order that a sequence $\{F_n\}$ of probability distributions converge properly to a probability distribution F it is necessary and sufficient that the sequence $\{\varphi_n\}$ of their characteristic functions converges pointwise to a limit φ, and that φ is continuous at the origin.*

Theorem 2: *If a probability distribution has a characteristic function $\varphi(t)$ analytic in a neighbourhood of 0, then it is fully determined by its derivatives in 0 or, equivalently, by its moments.*

Thus, to get the conditions of Theorem 2, we shall consider t as a complex variable. We shall also consider z as a complex variable. Then, via a Tauberian theorem, we will be able to derive Bernoulli results from Poisson results.

II DEPTH OF INSERTION

We focuse now on the *depth of insertion*, defined below, which is related to the external path length. We use generating functions, first studying Poisson distributions, and then translating the results to Bernoulli distribution. This case is of interest as it introduces a new way to prove the convergence of Poisson and Bernoulli distributions with a systematical use of the Mellin Integral Transform.

II.1, RECURSION AND FUNCTIONAL EQUATION

Our aim in this section is to derive a functional equation satisfied by the Poisson generating function $P(z,u)$. P_n^k is here the conditional probability that a record be inserted at depth k in a trie with $n+1$ records.

Remark: We find that : $(n+1)P_n'(1)$ is the average path length, when we count the number of *internal* nodes from the root to every leaf in a trie with $n+1$ records.

We can extend the set of definition of $P(z,u)$ with respect to u and state the following proposition.

Proposition II.1: *$P(z,u)$ is defined and analytical with respect to z and u in the domain of* C×C *such that :*

$$|u| < \frac{1}{p^{b+1}+q^{b+1}}.$$

It satisfies the recurrence equation :

$$P(z,u) = u\,[pP(pz,u)+qP(qz,u)] + (1-u)e_{b-1}(z)e^{-z}\,. \tag{E1}$$

where $e_b(z)$, b being integer, is the function

$$1 + z + \frac{z^2}{2} + \cdots + \frac{z^b}{b!}\,.$$

Proof: We will first prove the functional equation.

Let K_n^k be the number of records at depth k in a trie with n records. This is a random variable of tries with n records. Let u be a complex number, *a priori* $|u| < 1$, and $K_n(u)$ be the random polynomial related to tries with n records defined by:

$$K_n(u) = \sum_k K_n^k u^k$$

When $n \le b$ we have of course $K_n(u) = n$. If $n > b$ the partitioning process is called and a root is created, thus:

$$K_n(u) = u.(K_{n_1}(u) + K_{n_2}(u))\,,$$

where n_1 and n_2 are respectively the numbers of records placed in the left sub-tree and in the right sub-tree.

Let $E[\mathbf{K}_n(u)]$ be the expectation of $\mathbf{K}_n(u)$. This is an analytical function *a priori* defined for $|u| < 1$. We get from the above additive recursion:

$$\begin{cases} E[\mathbf{K}_n(u)] = u \sum_{n_1+n_2=n} \binom{n}{n_1,n_2} p^{n_1} q^{n_2} (E[\mathbf{K}_{n_1}(u)] + E[\mathbf{K}_{n_2}(u)]) \ , n > b, \\ E[\mathbf{K}_n(u)] = n \ , \ n \le b \ . \end{cases} \qquad \text{(E2)}$$

observing that

$$zP(z,u) = \sum_n E[\mathbf{K}_{n+1}(u)] \frac{z^{n+1}}{(n+1)!} e^{-z} \ ,$$

$(E1)$ follows when we translate $(E2)$ in terms of generating functions. Directly using $(E2)$, we also prove by induction: $|E[\mathbf{K}_n(u)]| < \alpha^{n-1}$, where $\alpha > 1$ satisfies: $(p+\frac{q}{\alpha})^b + (q+\frac{p}{\alpha})^b < \frac{1}{m}$ with $m = \sup_D \frac{|u|}{|1-u(p^{b+1}+q^{b+1})|}$. ∎

We notice that $(E1)$ is an *additive* equation. This will make it possible to derive directly an analytic expansion of $P(z,e^{it})$ in a more convenient way than in the size of the tree ([11]).

II.2. ASYMPTOTIC ANALYSIS, USE OF THE MELLIN TRANSFORM

To expand $P(z,u)$, we use Mellin transform techniques [8, 17]. We note $P_u^*(s)$ the Mellin transform of $P(z,u)-1$ with respect to the variable z, i.e.:

$$\int_0^\infty (P(z,u)-1) z^{s-1} dz = P_u^*(s) \ .$$

Using the elementaries properties of Mellin transform, mentionned in [8], equation $(E1)$ translates into a functional equation for $P^*(s,u)$:

$$P_u^*(s) = u \ [p \ p^{-s} P_u^*(s) + q \ q^{-s} P_u^*(s)] + (1-u) \frac{\Gamma(s+b)}{s(b-1)!} \ ,$$

which is easy to solve:

$$P_u^*(s) = -\frac{(1-u)\Gamma(s+b)}{s(b-1)!(1-u(p^{1-s}+q^{1-s}))} \ .$$

This equation holds in the domain where the right hand side function is analytical, and the integral above is absolutely convergent. This leads us to the study of the roots of $(E4)$:

$$1-u(p^{1-s}+q^{1-s}) = 0 \qquad \text{(E4)}$$

If $u \neq 0$, we can set $u = e^{it}$ and study the roots of

$$e^{-it} = p^{1-s} + q^{1-s} \ .$$

We just mention the following theorems which are proved in detail in [11]:

Theorem II.2.:

1) *The set of roots of* $(E4)$ *is a countable set* $\{s_k(t); k \in Z\}$. *Moreover, for every compact set for u and for every strip for s* : $\{s : a < \mathrm{Re}(s) < b\}$, *there exists a constant* $\delta(p,q)$ *such that:*

$$\forall (k,k'): \ |s_k(t) - s_{k'}(t)| \ge \delta(p,q) \ .$$

(2) *There exist a neighbourhood $V(0)$ and an analytical function: $t \dashrightarrow s_0(t)$ such that $s_0(t)$ be a root of (E4) and $s_0(t) \underset{t\to 0}{\dashrightarrow} 0$. Moreover:*

$$s_0(t) = -\frac{it}{h} + \frac{t^2}{h^3}(h_2-h^2) + O(t^3)$$

with:

$$h = -(p\log p + q\log q)\,,\ h_2 = p\log^2 p + q\log^2 q\ .$$

(3) *When $p=q=\frac{1}{2}$, we have: $\{s_k(t);k\in Z\}=\{s_0(t)+\frac{2il\pi}{\ln 2};k\in Z\}$. When $p\neq q$, we have, for any t and k : $Re(s_k(t)) \geq s_0(i\,\mathrm{Im}(t))$.*

(4) *As $|u|<\alpha = \frac{1}{p^{b+1}+q^{b+1}}$, all roots are contained in a half plan $\{s:\sigma_1(\alpha)<\mathrm{Re}(s)\}$. If $\beta<|u|<\alpha$, there exists a strip $\{s:\sigma_1(\alpha)<\mathrm{Re}(s)<\sigma_2(\beta)\}$ containing all the poles of (E4).*

Theorem II.3.: *There exist a complex compact neighbourhood of 0 $V(0)\subset \mathbf{C}$ and a cone $C_{0,\vartheta}$ such that, for t in $V(0)$ and z in $C_{0,\vartheta}$:*

$$P(z,e^{it}) = [z^{-s_0(t)}r_0(t) + \sum_{k\neq 0} z^{-s_k(t)}r_k(t) + O(\frac{1}{z})],$$

where $r_k(t)$ is the residue in the simple pole $s_k(t)$ of $P^\bullet_u(s)$, i.e.: $-\frac{(1-e^{it})}{s_k(t)}\frac{\Gamma(s_k(t)+b)}{(b-1)!}\cdot\frac{1}{h}$. Then we can bound:

$$P(z,e^{it}) = z^{-s_0(t)}(1+O(t.|z|^{-At^2})).$$

or

$$P(z,e^{it}) = e^{-\frac{it}{h}\log z - \frac{t^2}{2}\frac{(h_2-h^2)}{h^3}\log z + O(t^3.\log z)} \times (1+O(t.|z|^{-At^2})).$$

Propf: (Sketch) The first expression is directly derived from the Mellin inversion formula and residue theorem applied to the sequence of poles in theorem II 2 (1) (see [8]). Then applying II.2.(3) and (4) the second expression follows. And we reach the last expression with the expansion of $s_0(t)$ in II.2.(2).

And we can derive directly:

Theorem II.4.: *Under the Poisson model of parameter z the mean $X(z)$ and the variance $v(z)$ of the depth of insertion satisfy asymptotically, for $z\in R^+$:*

$$X(z) = \frac{1}{h}\log z + O(1),$$

$$v(z) = \frac{h_2-h^2}{h^3}\log z + O(1)\,,\ p\neq q\ .$$

$$v(z) = O(1)\,,\ p=q=\frac{1}{2}\ .$$

We can now state our results on limiting distributions.

First case: Biased case ($p \neq q$).

Theorem II.5.: Convergence to the Normal Law when $p \neq q$

The Poisson distribution of the depth of insertion, once centered and normalized, converges to the normal distribution, when $p \neq q$. Moreover the moments of any order of centered and normalized distribution converge to the corresponding moments of the normal distribution.

Proof: This an extension of the classical Central Limit Theorem, we introduced in[11]. The characteristic function of the centered and normalized distribution of depth of insertion is:

$$e^{-iX(z)\frac{t}{\sqrt{v(z)}}} P(z, e^{\frac{it}{\sqrt{v(z)}}}).$$

According to theorem II.3 and 4 this expression evaluates to:

$$e^{-itO(\frac{1}{\sqrt{\log z}}) - \frac{t^2}{2}(1 + O(\frac{1}{\sqrt{\log z}})) + O(\frac{t^3}{\sqrt{\log z}})} (1 + O(\frac{t}{\sqrt{\log z}})).$$

which uniformly converges to

$$e^{-\frac{t^2}{2}}$$

with t confined in any compact neighbourhood of 0. We recognize in this last expression the characteristic function of the gaussian distribution. Then, applying theorem I.1. and 2., the proof is completed.

Second case: Uniform case ($p = q = \frac{1}{2}$).

Theorem II.6.: Periodic Behaviour of the Distribution when $p = q = \frac{1}{2}$

When $z \to \infty$, the generating function satisfies asymptotically:

$$P(z, e^{it}) = G_z(t) + O(\frac{1}{z})$$

where:

$$G_z(t) = \frac{z^{\frac{-it}{\ln 2}}}{\ln 2}\left[\frac{\Gamma(b - \frac{it}{\ln 2})}{(b-1)!} + \sum_{k \neq 0} \frac{\Gamma(b - \frac{it - 2ik\pi}{\ln 2})}{\Gamma(b)} z^{\frac{2ik\pi}{\ln 2}}\right]$$

and $O(.)$ is uniform w.r.t. u in any closed set included in $B(0, \frac{1}{p^{b+1} + q^{b+1}})$.
The moments of any order converge to the corresponding moments of $G_z(t)$

Proof: As above, we use Mellin Transform as in [8] and Theorem II.2., (4).

Corollary: When $z \in R^+$ the generating function is the generating function of the process under Poisson hypothesis of parameter z. When $z \to \infty$, the centered distribution is asymptotically arbitrarily close to a centered distribution which depends only of fractional part of $\ln z / \ln 2$.

II.3. EQUIVALENCE POISSON-BERNOULLI : USE OF RICE 'S INTEGRALS

We proceed now with the Bernoulli case, establishing:

Theorem II.7.: *The Poisson and Bernoulli distribution converge. More precisely, for u in a compact neighbourhood of* 0 *and* 1 *such that* $|u| < (p^{b+1}+q^{b+1})^{-1}$, *we have the uniform convergence with* $u = e^{it}$

$$P_n(u) = P(n,u) + O(n^{-s_0(i\mathrm{Im}(t))-1})$$

when $n \to \infty$. *Consequently, when* $p \neq q$, *under either Poisson or Bernoulli hypotheses, the distribution of the depth of insertion, centered and normalized, converges to the normal law.*

Proof: It uses Rice 's integral and Newton series (see [11,8,16 ch. 8] . For any fixed $u = e^{it}$, the following formula holds for $\mathrm{Im}(t)>0$:

$$a_n = P_n(u) = \frac{n!}{2i\pi}\int_{A_1} \frac{P_u^*(s)}{\Gamma(s+n+1)} ds$$

where A_1 is some contour around $0,-1,\cdots,-n,\cdots$ which does not contain any pole of P_u^*. As described in [11], we deform this contour in a new rectangular one, with boundaries going to infinity, and applying the residue theorem one more time we get the evaluation analogous to theorem II 3:

$$P_n(e^{it}) = P(n,e^{it}) + O(\frac{1}{n} n^{-s_0(i\mathrm{Im}(t))}) .$$

The first assertion of theorem being proved, it is now easy to translate the results about the Poisson distribution to the Bernoulli distribution. Let X_n and v_n be respectively the mean and variance of the Bernoulli distribution, we have analogously to theorem II 4:

$$\begin{cases} X_n = \dfrac{\log n}{h} + O(1) = \dfrac{h_n}{h} + O(1) \\ v_n = \dfrac{h_2 - h^2}{h^3} \log n + O(1) \end{cases}$$

and Theorems II 5 and 6, especially the normal convergence when $p \neq q$:

$$P_n(e^{\frac{it}{\sigma(n)}})\, e^{-\frac{itX(n)}{\sigma(n)}} = (1+O(\frac{t}{\sqrt{\log n}}))e^{-\frac{t^2}{2}} .$$

This complete the proof of this last theorem.

III HEIGHT OF TRIES AND NUMBER OF COLLISIONS

This section is devoted to the analysis of the distribution of ***height of the tries*** with n keys. This provides an application of the tauberian Theorem established in [11] .

The derivation is based on an approximation of the Poisson generating function $P(z,u)$ by an other -simpler- generating function $F(z,u) = \sum_k F^k(z) u^k$. First, we state general properties of P and F. Then, we approximate their difference. Finally, we establish asymptotic results on F and further on P. And we finally treat the Bernoulli case.

III.1. PROPERTIES OF GENERATING FUNCTIONS:

We note here:

$$f_b(z) = e_b(z)e^{-z}$$
$$g_b(z) = -\log(f_b(z)) .$$

Proposition III.1: *The probability generating function $P(z,u)$ satisfies:*

$$P(z,u) = (1-u)\sum_{k\geq 0} H^k(z)u^k \tag{4.1}$$
$$= (u-1)\sum_{k\geq 0}(1-H^k(z))u^k+1$$

where $H^k(z)$ is the generating function for the "cumulated probabilities":

$$H^k(z) = \sum_{n\in N} \text{Proba}(height\leq k / n\ keys)\frac{z^n}{n!}e^{-z}$$
$$= \prod_{k1+k2=k} f_b(zp^{k_1}q^{k_2})^{\binom{k}{k_1,k_2}}$$
$$H^k(z) = \exp\left(-\sum_{k1+k2=k}\binom{k}{k_1,k_2}g_b(zp^{k_1}q^{k_2})\right). \tag{4.2}$$

It is defined and analytical for $z\in C$ and complex u such that:

$$|u|<\frac{1}{p^{b+1}+q^{b+1}}.$$

Proof: Let H_n^k be the probability that a trie with n keys has a height less than or equal to k. As both subtrees, if they exist, have then a height less than or equal to $k-1$, we get, considering the bit distribution:

$$\begin{cases} H_n^k = \sum_{n_1+n_2=n}\binom{n}{n_1,n_2}p^{n_1}q^{n_2}H_{n_1}^{k-1}H_{n_2}^{k-1} \ ,k\geq 1 \\ H_n^0 = \chi_{n\leq b} \end{cases}$$

Thus, $H^k(z)=\sum_n H_n^k\frac{z^n}{n!}e^{-z}$ satisfies the functional equations:

$$\begin{cases} H^k(z) = H^{k-1}(pz).H^{k-1}(qz) \\ H^0(z) = f_b(z) = \exp(-g_b(z)) . \end{cases}$$

From this, we get the explicit solution (4.2).

Remark: When $p=q=\frac{1}{2}$, this reduces to: $H^k(z)=f_b(z2^{-k})^{2^k}$. This case was extensively studied in [5,17].

Now, we have:

$$P_n(u) = \sum_{k\geq 1}(H_n^k-H_n^{k-1})u^k+H_n^0$$
$$= (1-u)\sum_{k\geq 0}H_n^ku^k = (u-1)\sum_{k\geq 0}(1-H_n^k)u^k+1$$

and (4.1) follows.

We claim[17] :

Lemma III.2. : *In any compact neighbourhood of* 0, *one has:* $g_b(z) = \frac{z^{b+1}}{(b+1)!} + O(z^{b+2})$.

Thus:

Corollary:

$$H^k(z) = \exp\left(-\frac{z^{b+1}(p^{b+1}+q^{b+1})^k}{(b+1)!}\right) + O(z^{b+2}(p^{b+2}+q^{b+2})^k)$$

when $k \to \infty$.

As it is analytical, and $(1-H^k(z))u^k) \sim z^{b+1}\frac{(u(p^{b+1}+q^{b+1}))^k}{(b+1)!}$, we get the domain of analyticity of $P(z,u)$.

Definition: *We note:*

$$\begin{cases} F^k(z) = \exp\left(-\frac{z^{b+1}(p^{b+1}+q^{b+1})^k}{(b+1)!}\right) \\ F(z,u) = (1-u)\sum_{k\geq 0} F^k(z)u^k = (u-1)\sum_{k\geq 0}(1-F^k(z))u^k + 1 \end{cases}$$

$F(z,u)$ is analytical in the same domain as $P(z,u)$.

III.2. SOME ASYMPTOTICS:

Let $V(1)$ be an arbitrary compact neighbourhood of 1 included in $B(0,(p^{b+1}+q^{b+1})^{-1})$. Let $C_{0,\vartheta}$ be the cone defined as the set of complex numbers z such that $\mathrm{Re}(z) \geq 0$ and $|Arg(z)| < \vartheta$ (see [11]). We deal here with the asymptotic expansion of $P(z,u)$ when u ranges in $V(1)$ and z in $C_{0,\vartheta}$. ϑ is chosen such that $|f_b(z)| \leq 1$ or $Re(g_b(z)) \geq 0$. We need asymptotics in such a cone in order to apply the semi tauberian Theorem to get $P_n(u)$ from $P(z,u)$ (see III.3. below).

We state the main Theorem:

Theorem III.3.: *There exists some constant c such that:*

$$P(z,u) = F(z,u) + O(|z|^{b+2}F(c|z|,|u(p^{b+2}+q^{b+2})|)),$$

when z ranges in $C_{0,\vartheta}$ and u in some compact neighbourhood of 1.

Proof: We only draw a scheme. We write:

$$P(z,u)-F(z,u) = H^{<1>}(z,u) + H^{<2>}(z,u)$$

with:

$$\begin{cases} H^{<1>}(z,u) = \sum_{|z|^{b+2}(p^{b+2}+q^{b+2})>1} [H^k(z)-F^k(z)]u^k, \\ H^{<2>}(z,u) = \sum_{|z|^{b+2}(p^{b+2}+q^{b+2})\leq 1} [H^k(z)-F^k(z)]u^k, \end{cases}$$

first we show that there exist $\varepsilon > 0$ such that

$$H^{<1>}(z,u) = O(e^{-|z|^{\varepsilon}}).$$

Second, we consider $H^{<2>}(z,u)$. According to Lemma III.2, one sees easily that, for k in this range, that

$$H^k(z)-F^k(z) = e^{-\frac{z^{b+1}(p^{b+1}+q^{b+1})^k}{(b+1)!}}.O(|z|^{b+2}(p^{b+2}+q^{b+2})).$$

As $Re(z^{b+1})>c|z|^{b+1}$, we get, after summation:

$$H^{<2>}(z,u) = O(|z|^{b+2}F(c|z|,|u(p^{b+2}+q^{b+2})|)).$$

■

We study now asymptotics of $F(z,u)$, when $z\to\infty$ inside $C_{0,\vartheta}$. We make use of *complex* Mellin transform:

$$F^*(s,u) = \int_0^\infty z^{s-1}(F(z,u)-1)dz ,$$

where the integration is done along any half line from 0 to $C_{0,\vartheta}$.

Proposition III.6.: *The Mellin transform* F^* *is defined and analytical for:* $-1<Re(s)<inf(0,\frac{\log|u|}{\log(p^{b+1}+q^{b+1})})$ *with:*

$$F^*(s,u) = [(b+1)!]^{\frac{s}{b+1}}\frac{(1-u)\Gamma(\frac{s}{b+1})}{1-u(p^{b+1}+q^{b+1})^{-\frac{s}{b+1}}}$$

Proof: We can develop formally:

$$F^*(s,u) = (u-1)\sum_k u^k(b+1)!^{\frac{s}{b+1}}(p^{b+1}+q^{b+1})^{-\frac{ks}{b+1}}\int_0^\infty z^{s-1}(1-e^{-z^{b+1}})dz$$

$$= \frac{(b+1)!^{\frac{s}{b+1}}}{b+1}.(u-1).\sum_k[u(p^{b+1}+q^{b+1})^{-\frac{s}{b+1}}]^k.(-\Gamma(\frac{s}{b+1})) ,$$

provided that both integral and Dirichlet geometric series are *absolutely* convergent.

Proposition III.7.: *Let* z *and* u *range respectively in* $C_{0,\vartheta}$ *and in* $V(1)$. *Then* $F(z,u)$ *satisfies asymptotically:*

$$F(z,u) = z^{-\frac{t(b+1)}{\log(p^{b+1}+q^{b+1})}}G(\log z,t) + O(z^{-m})$$

where:

$$G(x,t) = \frac{(u-1)}{\log(p^{b+1}+q^{b+1})}\sum_{k\in Z}\Gamma\left(\frac{t+2ik\pi}{\log(pqb)}\right)e^{-\frac{2ik\pi}{\log(p^{b+1}+q^{b+1})}((b+1)x-\log(b+1)!)}$$

$t = \ln u$,

$O(z^{-m})$ *is uniform in the cone. This expression is to be continued when* $u=1$ *and in the paradoxal case* $u=0$.

Proof: This is steadily proved [7]. All poles are simple when $u\neq1$. These poles are 0 (with residue 1) and $\{\frac{-\log u+2ik\pi}{\log(p^{b+1}+q^{b+1})}\}_{k\in Z}$.

Interpretation: The distribution defined by F, shifted by $\frac{(b+1)\log z}{\log(p^{b+1}+q^{b+1})}$, is asymptotically the one associated to the characteristic function $G(\log z, t)$. G is a periodic function of $\log z$ with period $\frac{\log(p^{b+1}+q^{b+1})^{-1}}{b+1}$. Thus, F has for mean:

$$-\frac{(b+1)\log z}{\log(p^{b+1}+q^{b+1})} + P(\log z) + O(z^{-m})$$

and for variance: $Q(\log z) + O(z^{-m})$, when P and Q are periodic with period $\frac{\log(p^{b+1}+q^{b+1})}{b+1}$.

Theorem III.8.: $P(z,u) = z^{-\frac{t(b+1)}{\log(p^{b+1}+q^{b+1})}}(G(\log z, t)+O(z^{-\gamma})$ *with* $\gamma = (b+2)-(b+1)\frac{\log(p^{b+2}+q^{b+2})}{\log(p^{b+1}+q^{b+1})} > 0$, *$G$ defined as above, z in $C_{0,\vartheta}$ and u in a neighbourhood of* 1 *not containing* 0.

Proof: From Theorem III.3. , one has:

$$P(z,u) = F(z,u) + O(|z|^{b+2}F(c\,|z|, |u(p^{b+2}+q^{b+2})|)) .$$

As G and $\ln u$ are bounded, we get from Proposition III.7.:

$$|z|^{b+2}F(c\,|z|, |u(p^{b+2}+q^{b+2})|) = O(|z|^{-\frac{(b+1)(\ln u+\log(p^{b+2}+q^{b+2}))}{\log(p^{b+1}+q^{b+1})}+b+2})+O(z^{b+2-m})$$

$$+\ O(|z^{-\frac{(b+1)\ln u}{\log(p^{b+1}+q^{b+1})}}|.|z|^{-\gamma}) .$$

III.3. BACK TO BERNOULLI CASE:

Theorem III.9.: *The distribution of the height of tries with n keys converge* in the sense of distributions *to the distribution with characteristic function* $F(n,u)$.

Proof: We use a general *semi tauberian* theorem developed in [11].

Semi Tauberian Theorem (adapted):

Definition

Let $\{a_n\}_{n\in N}$ be a real positive sequence such that a_n is monotonous and uniformly bounded with respect to t. Let $a(z)$ be its Poisson generating function

$$a(z) = \sum_n a_n \frac{z^n}{n!} .$$

$a(z)$ is *logarithmically* varying when for every $\alpha > 0$ there exists β such that, for n real positive and z complex:

$$|z-n| < \alpha n : \quad |a(z)-a(n)| < \beta \frac{|z-n|}{n} .$$

Theorem: *we have the convergence:*

$$a_n - a(n) \to 0$$

when $n \to \infty$. This convergence is uniform on every class whose sequences experience a same given (α,β).

This theorem is formally a *semi* tauberian theorem because the definition of the logarithmic variation implies an extension of the analysis of $a(z)$ in a non negligible part of the complex plan. Thus we understand why we analyzed $P(z,u)$ in a specific cone $C_{0,\vartheta}$.

Indeed, the sequence $\{H_n^k\}_{n\in N}$ is decreasing (a new insertion can only increase the height). Thus, one may readily claim:

$$H_n^k - \exp(-(p^{b+1}+q^{b+1})^k)\frac{n^{b+1}}{(b+1)!}) \underset{n\to\infty}{\to} 0$$

for functions $\exp(-(p^{b+1}+q^{b+1})^k)\frac{x^{b+1}}{(b+1)!})$ clearly have a logarithmic variation.

But for the analysis of the convergence *in sense of distribution*, we need introducing the following evaluation

$$\begin{cases} <\Delta_n|\varphi> = \sum_{k\in N} H_n^k \varphi(k) \\ <\Delta(z)|\varphi> = \sum_{k\in N} H^k(z)\varphi(k) \end{cases}$$

with φ being any test positive function with compact support such that $\int\varphi=1$. It is also clear that the sequence $\{<\Delta_n,\varphi>\}_{n\in N}$ is positive, decreasing and upper bounded by 1 and that $<\Delta(x),\varphi>$ is a logarithmically varying function (and (α,β) does not depend on φ). Thus:

$$<\Delta_n|\varphi> - <\Delta(n)|\varphi> \underset{n\in N}{\to} 0\,,$$

uniformly, whatever the test function φ is. And the theorem is proved.

■

The Limiting Distributions provides an accurate method for asymptotic analysis of algorithmic parameters. In particular, when dealing with the depth of insertion, we point out a solution of continuity between the biased and uniform case. This does not appear when the analysis is restricted to the average values. As a matter of fact, the mean is $O(\log n)$ in both cases. But one finds a gaussian dilatation around the mean in the biased case while in the uniform case the distribution remains centered around its mean with a $O(1)$ deviation.

There are many poles of interest in the analysis of trie parameters. We are actually working at the following topics.

* The distribution of the depth of insertion for a given key (a fixed sequence of bit) when it is inserted among n other random keys. When $p \neq q$, the gaussian behaviour mentioned above disappears and gives place to a centered distribution around a mean $O(\log n)$ determined by the sequence of bits of the given key.
* The previous results may be extended to data generated by some *markovian* process. Markovian process can be adequate model when the keys are alphabetical. The transition matrix is then formed with the transition probabilities from one letter to another. Our analysis is a particular case

of a stationary process over a binary alphabet.

In this paper the systematical use of generating functions and the reference to complex analysis prove to be a valuable and powerful tool.

References

1. Ph. Flajolet and C. Puech, "Tree Structure for Partial Match Retrieval," pp. 282-288 in *Proc. 24-th I.E.E.E. Symp. on FOCS*, (1983). To appear in JACM
2. R. Fagin, J. Nievergelt, N. Pippenger, and H.R Strong, "Extendible Hashing:A Fast Access Method for Dynamic Files," *ACM TODS* **4,3** pp. 315-344 (1979).
3. G. Fayolle, Ph. Flajolet, M. Hofri, and Ph. Jacquet, "Analysis of a Stack Algorithm for Random Multiple-Access Communication," *IEEE Trans. on Information Theory* **IT-31,2** pp. 244-254 (1985).
4. W. Feller, *An Introduction to Probability Theory and its Applications*, Wiley-third Edition-1971 (1957).
5. Ph. Flajolet, "On the Performance Evaluation of Extendible Hashing and Trie Searching," *Acta Informatica* **20** pp. 345-369 (1983).
6. Ph. Flajolet, M. Régnier, and D. Sotteau, "Algebraic Methods for Trie Statistics," *Annals of Discrete Mathematics* **25** pp. 145-188 (1985).
7. Ph. Flajolet, M. Regnier, and R. Sedgewick, *Some Uses of the Mellin Transform Techniques in the Analysis of Algorithms* , Springer NATO ASI SEr. F12, Combinatorial Algorithms on Words (1985).
8. Ph. Flajolet, M. Regnier, and R. Sedgewick, "Mellin Transform Techniques for the Analysis of Algorithms ," *Monography in preparation*, (1986).
9. Ph. Flajolet and N. Saheb, "Digital Search Trees and the Complexity of Generating an Exponentially Distributed Variate," in *Proc. Coll. on Trees in Algebra and Programming*, Lecture Notes in Computer Science, L'Aquila (1983). to appear
10. Ph. Flajolet and J.M. Steyaert, "A Branching Process Arising in Dynamic Hashing, Trie Searching and Polynomial Factorization," pp. 239-251 in *Proceedings ICALP 82* , Lecture Notes in Computer Science (1982).
11 Ph. Jacquet and M. Régnier, *Limiting Distributions for Trie Parameters*, in preparation 1985.
12. D. Knuth, *The Art of Computer Programming*, Addison-Wesley,Reading,Mass. (1973).
13. P.A. Larson, "Dynamic Hashing," *BIT* **18** pp. 184-201 (1978).
14. W. Litwin, "Trie Hashing," pp. 19-29 in *Proc. ACM-SIGMOD Conf. on MOD,.* , Ann Arbor,Mich. (1981).
15. J. Nievergelt, H. Hinterberger, and K.C. Sevcik, "The Grid-file : an Adaptable Symmetric Multi-Key File Structure," *ACM TODS* **9,1**(1984).
16. N.E. Norlund, *Vorlesungen Uber Differenzenrechnung*, Chelsea Publishing Company (1954).
17. M. Régnier, "Evaluation des performances du hachage dynamique," *These de 3-eme cycle,Universite d'Orsay*, (1983).

RANDOM WALKS, GAUSSIAN PROCESSES AND LIST STRUCTURES

G. Louchard
Laboratoire d'Informatique Théorique
Université Libre de Bruxelles, CP 212
Boulevard du Triomphe - 1050 Brussels
Belgium

Abstract

An asymptotic analysis of list structures properties leads to limiting Gaussian Markovian processes. Several costs functions are shown to have asymptotic Normal distributions.

1. Introduction

List structures are well-known objects in Computer Science, let us mention : dictionaries, priority queues, symbol tables, linear lists, stacks, etc. (see Flajolet [7] ch.IV for detailed description). We will consider here lists of length 2n, i.e. initially of size 0 and returning to size 0 at step 2n, on which some operations are performed, such as: insertions, deletions, successful queries, unsuccessful queries. Let us call N_{2n} the total number of such lists, of any type, with all possible operations. We define a probability measure on lists by assigning to each history (sequence of values of the list and operations performed) the probability $1/N_{2n}$. Some cost functions can be defined on each history: storage cost (total integrated size on [0,2n]), time cost (total cost related to the operations performed).

This last cost depends of course on the list implementation. Only a few results are available on probability distributions of these variables: the stack storage is analysed in Louchard [14], [15] and [16]: it is shown to be asymptotically ($n \to \infty$) equivalent to a Brownian Excursion Area. Flajolet, Puech and Vuillemin [9] obtain exact mean and variance of storage and time cost for dictionaries and priority queues. The purpose of the present paper is to develop general techniques to derive asymptotic distributions of list structures cost functions. We actually obtain limiting Gaussian Markovian processes for the histories and Normal variables for linear and polynomial cost functions.

The paper is organized as follows: in Section 2, we summarize basic notations. Section 3 describes some classical list structures and known results. Section 4 is devoted to the simplest list: the linear one, that we treat in some details. Priority queues and dictionaries are analysed in our detailed technical report [17] and lead to similar results. Section 5 concludes the paper.

2. Basic notations

The following notations will be used throughout the paper.

- 2n:= size of the structure
- LL:= Linear list
- PQ:= Priority queue
- D:= Dictionary
- $\sim$:= asymptotic to } for $n\to\infty$
- $\to$:= converges to } for $n\to\infty$
- $\Rightarrow$:= weak convergence of random functions in the space of all right continuous functions having left limits and endowed with the Skorohod metric (see Billingsley [2] ch.III)
- M:= mean of some random variable (R.V.)
- V:= variance of some R.V.
- μ_k:= k-th moment of some R.V.
- $Y, \tilde{Y}, Y^-, \hat{Y}$:= classical random walks
- Y^*:= weighted random walk
- $E_a(B)$:= expectation of event B for a random walk starting from a at time 0
- $E^*(B)$:= expectation of event B for a weighted random walk
- $\hat{E}(B)$:= expectation of event B for product of rectangular R.V.
- G.F.:= generating function
- $\mathcal{N}(M,V)$:= the Normal (or Gaussian) R.V.
- B.M.:= the classical Brownian Motion (see Ito and McKean [12] for a good introduction and Louchard [13] for some complexity applications)
- B.E.:= the standard Brownian Excursion (see Chung [3])
- X(.):= Markovian, Gaussian process with 0 mean
- E_{2n}:= the 2n-th Euler number (or secant number) with exponential G.F.:
  $\sum_0^\infty E_{2n} z^{2n}/(2n)! = \sec z$ (see Flajolet [7] p.142), $E_{2n} \sim 4^{n+1}(2n)!/\pi^{2n+1}$ (1)
  (see Abramowitz and Stegun [1] equ.(4.3.69) and (23.1.15))
- $n?$:= 1.3.5...(2n-1), $n? \sim \sqrt{2}\, e^{-n}(2n)^n$ (2)
- C_{2n}:= the n-th Catalan number $\equiv \binom{2n}{n}/(n+1)$ with G.F. :

  $\sum_0^\infty C_{2n} z^n = (1-\sqrt{1-4z})/2z$ (see Flajolet [6] p.135), $C_{2n} \sim 4^n/(\sqrt{\pi}\, n^{3/2})$ (3)
- $\hat{M}_{2n}$:= the number of paths (see Section 3) of length 2n with upward, downward and <u>two</u> types of level steps, with G.F.: $\sum_0^\infty \hat{M}_n z^n = (1-2z-\sqrt{1-4z})/2z^2$,
  $\hat{M}_n \sim 4^{n+1}/(\sqrt{\pi}\, n^{3/2})$ (see Appendix of [17]) (4)

  This is a generalization of classical Motzkin numbers.
- $\{x\}$:= $x - \lfloor x \rfloor$
- i.i.r.v:= independent identically distributed random variables

3. Some list structures

This section summarizes the main properties of the structures we will analyse in the sequel.

i) Following Flajolet et al. [9], we define a schema (or path) as a word $\Omega := O_1 O_2 \ldots O_{2n} \in \{I, D, Q^+, Q^-\}^*$ such that for all j, $1 \leq j \leq 2n$:

$$|O_1\ O_2 \ldots O_j|_I \geq |O_1\ O_2 \ldots O_j|_D \qquad (5)$$

A schema is to be interpreted as a sequence of 2n requests (the keys operated on not being represented) where I, D, Q^+, Q^- represent respectively an insertion, a deletion, a positive (successful) query and a negative (unsuccessful) query.

(5) means that the size of the structure is always ≥ 0 . In the case of LL and PQ, only insertions and deletions are performed.

A structured history is a sequence of the form :

$h := O_1(r_1)\ O_2(r_2) \ldots O_{2n}(r_{2n})$ where $\Omega = O_1\ O_2 \ldots O_{2n}$ is a schema, and the r_j are integers satisfying: $0 \leq r_j < pos(\alpha_{j-1}(\Omega))$

where

- $\alpha_j(\Omega) := |O_1\ O_2 \ldots O_j|_I - |O_1\ O_2 \ldots O_j|_D$ is the size (level) of the structure at step j
- pos is a possibility function (defined for each request), given in Table 1.
- r_j is the rank (or position) of the key operated upon at step j.

We shall only consider schemas and histories with initial and final level 0 (The general case can be treated with similar techniques).

The number N_{2n} of such histories of length 2n is given in Table 1 as well as S_{2n}: the number of schemas of length 2n.

	pos(k)				N_{2n}	S_{2n}
	I	D	Q^+	Q^-		
LL	k+1	k	0	0	E_{2n}	C_{2n}
PQ	k+1	1	0	0	n?	C_{2n}
D	k+1	k	k	k+1	(2n)!	$\hat{M}_{2n}$

Table 1

ii) To any history h, we will associate cost functions: C(h).

Two cost functions will be firstly considered in this paper: the storage cost functions: $\sigma(h) := \sum_{j=1}^{2n} \alpha_j(h)$ and the time cost function $\tau(h)$. This last function depends on the implementation of our lists structures. We will use two implementations only: the sorted list (SL) and the unsorted list (UL).

The time cost functions dealt with by Flajolet et al. [9] for PQ and D are summarized in Table 2 as well as the time cost we use for LL.

	$\tau(h)$	
	SL	UL
LL		$\sum_{I+D} r_j(h)$
PQ	$\sum_{I} r_j(h)$	$\sum_{D} \alpha_j(h)$
D	$\sum_{I+D+Q^+ +Q^-} r_j(h)$	

Table 2

Remark 1: Asymptotically, $\tau_{PQ,UL}(h) \sim \frac{1}{2}\sigma_{PQ}(h)$: indeed to each insertion at some level ℓ, corresponds one deletion at level $\ell-1$. ▼

Other cost functions are analysed in Section 4.6. To any cost function $C(h)$, we associate a random variable C^* defined as follows :

$$Pr[C^*=\kappa] := [\mathrm{card}\{h : C(h)=\kappa,\ h \in \tilde{N}_{2n}\}]/N_{2n} \qquad (6)$$

where, for any structure, $\tilde{N}_{2n}$ is the (finite) set of histories of length 2n.
For further use, let us call $\tilde{S}_{2n}$ the set of all schemas of length 2n.
Expectation of any event related to C^* will be denoted by E^*. Mean M^* and variance V^* of C^* for storage σ and time cost τ have been investigated in Flajolet et al.[9] for PQ and D. They use continued fractions and orthogonal polynomials. The main term in their results is given in Table 3.
The main term for LL, also given in Table 3, is deduced from our results in Section 4.

	M^*			V^*		
	σ	τ		σ	τ	
		SL	UL		SL	UL
LL	$8n^2/\pi^2$		$4n^2/\pi^2$	$n^3(128/\pi^2-12)/\pi^2$		$n^3(32/\pi^2-8/3)/\pi^2$
PQ	$2n^2/3$	$n^2/6$	$n^2/3$	$8n^3/45$	$n^3/45$	$2n^3/45$
D	$2n^2/3$	$n^2/3$		$8n^4/9$ (*)	$2n^3/45$	

Table 3

Remark 2: (*): this result seems erroneous, we obtain in [17] : $4n^3/45$. ▼

For more details on structure histories (notably in connection with generating functions, continued fractions, permutations, etc.) the reader is invited to consult Flajolet [7] ch.IV and V and [6].

4. Linear lists

In this paper, we will only deal with the simplest list: the linear one. After establishing a sequence of lemma's, we will obtain the asymptotic properties of this structure. This case will be treated in some details, preparing the way to other structures analysis (see [17] for a complete description). Cost functions are also analysed.

4.1 Large deviations

The linear list is asymptotically equivalent, by Table 1, to a classical random walk $Y(i)$, of length $2n$, from 0 to 0, constrained to remain non negative and with weight $\prod_{i=1}^{2n-1} Y(i)$ (corresponding to all histories of a given schema).

Without this weight, the random walk would be asymptotically equivalent to a B.E. (see Louchard [12]). But here, the Brownian limit is no longer applicable: after some approximations experiments, we actually found that the main weight was related to large deviations in the random walk, of order $O(n)$ (this is precisely justified in Section 4.4).

We will thus start with the probability analysis of such deviations, which is solved by the following

Lemma 1

Let $Y(i):=\sum_1^i \xi_j$ where ξ_j are independent identically distributed random variables (i.i.r.v.) with $\Pr(\xi_j=-1)=\Pr(\xi_j=+1)=\frac{1}{2}$.

Then, for $0<u<1$, $m\to\infty$,

$$E_0[\frac{Y(m)}{m}\in du]=(1-u^2)^{-m/2}\left(\frac{1+u}{1-u}\right)^{-mu/2}\sqrt{m}\,du[1+\frac{\rho(u)}{m}+O(m^{-2})]/\sqrt{2\pi(1-u^2)}$$

$$=\varphi(m,u)[1+\rho(u)/m+O(m^{-2})]du, \text{ say} \tag{7}$$

where $\rho(u):=[1-3(1-u^2)/2]/[(1-u^2)/2]$ (8)

Proof: One could use asymptotics for the classical binomial distribution but it is more elegant to proceed as in Feller [5] p.548 or Greene and Knuth [11] p.79. The detailed proof may be found in [17]. ▼

4.2 Constrained random walk

We must now take into account the constraint of non negativity for the random walk. We may proceed as Chung did for the passage from B.M. to B.E. (see Chung [3] p.167), and obtain the

Lemma 2

Let $Y(i):=\sum_{1}^{i}\xi_j$ and $\Lambda:=\min(n\geq 1:Y(n)=0)$. Define $\tilde{Y}(i):=[Y(i)|\Lambda=2n]$.

Let also $\quad Y_a^-(i):=[a+Y(i)|\Lambda>i]$ (9)

(random walk starting at a and killed at 0).

We denote by B an event belonging to the Borel field generated by Y^-, and define

$$\Pi(v,x,y)dy:=E_{nx}[B,\frac{y_{nx}^-(nv)}{n}\in dy].$$

Let $0<\varepsilon,\delta<1$. The event B for $\tilde{Y}(i)$, $i\in[n\varepsilon,n(2-\delta)]$, has the following asymptotic probability:

$$E_0\left[\frac{\tilde{Y}(n\varepsilon)}{n\varepsilon}\in du_1,\ B,\ \frac{\tilde{Y}(n(2-\delta))}{n\delta}\in du_2\right]\sim$$

$$\sqrt{8\pi}(2n)^{3/2}\phi(u_1,\varepsilon)\phi(u_2,\delta)\Pi(2-\varepsilon-\delta,\varepsilon u_1,\delta u_2)du_1\,du_2 \tag{10}$$

where $\phi(u,\varepsilon):=\log\left(\frac{1+u}{1-u}\right)\varphi(n\varepsilon,u)$ if $u<1$, $\phi(u,\varepsilon):=2^{-(n\varepsilon+1)}$ if $u=1$

Proof: the proof may be found in [17].

4.3 Weighted random walk Y*

The weight defined by (6) and Table 1 must now be taken into account.

Let us tentatively set $\sqrt{n}X_n(v)=Y^*([nv])-ny(v)$, $v\in[0,2]$ (11)

where y(.) is a (deterministic) continuous non-negative symmetric function (with $y(0)=y(2)=0$) and $X_n(.)$ a random process with asymptotic 0 mean (coefficients will be justified in the sequel). Moreover as $Y^*(1)=Y^*(2n-1)=1$ and $Y^*\leq n$, we must have $\lim ny(\frac{1}{n})=\lim ny(2-\frac{1}{n})=1$, $ny(.)\leq n$.

The constraints we put on y(.) in the sequel with be summarized by y-Conditions (C.y): $y\in C^1$, $y(0)=y(2)=0$, $y(.)\leq 1$, $y'(0)=-y'(2)=1$.

According to (6), we put on each trajectory of type (11) a total measure which is the product of probability measures as defined by (10) and the weight $W=\prod_{i=1}^{2n-1}Y^*(i)$.

We firstly must find y(.), then establish the stochastic properties of $X_n(.)$ and justify (11). Let us firstly look for an asymptotic formula for W along ny(.). We have the

Lemma 3

Let y(.) satisfy C.y.

Let $W:=\exp[\sum_{i=1}^{2n-1}\log(ny(\frac{i}{n}))]=\exp(Z)$ say and $W(j,k):=\exp[\sum_{i=j}^{k-1}\log(ny(\frac{i}{n}))]$.

Then

$$W \sim \exp[2n\ \log n + n\int_0^2 \log(y(v))dv + 2 + \int_{1/n}^{2-1/n} B_1(\{nv\}) \frac{y'(v)}{y(v)}\,dv] \tag{12}$$

$$W(j,k) \sim \exp[(k-1-j)\log n + n\int_{j/n}^{k/n} \log(y(v))dv$$

$$- \tfrac{1}{2}[\log(y(\tfrac{k}{n})) - \log(y(\tfrac{j}{n}))] + \int_{j/n}^{k/n} B_1(\{nv\})\frac{y'(v)}{y(v)}\,dv] \tag{13}$$

where B_1 is the first Bernoulli polynomial.
By convention, we say that the Convergence Condition (C.C.) is satisfied if the last terms of (12) and (13) converge.

Proof: the proof, by Euler's sommation formula, may be found in [17].

4.4 Determination of y(.)

This problem is solved in Theorem 1 at the end of this sub-section.
i) To obtain y(.), we use Laplace asymptotic method on functional space. Let ε, δ be small positive constants. Along any trajectory of type (11), for $i \in [2n\varepsilon, 2n(1-\delta)]$, it is clear that, asymptotically, Y^- (see (9)) and Y are identically distributed (the hitting probability of the 0 boundary is exponentially small).
Let us assume that y satisfies C.y (this will have to be checked later on). When we let ε, $\delta \to 0$, it follows from (10) that the total probability measure on (11) is asymptotically $(n\to\infty)$ the same measure as the limiting one (ε, $\delta\to 0$) deduced from (7) for $i \in [2n\varepsilon, 2n(1-\delta)$, multiplied by $\sqrt{8\pi}(2n)^{3/2}\ 2^{-n(\varepsilon+\delta)}/4$ (14)
We now define $x_1, x_2 \in [\varepsilon, 2-\delta]$, $x_2 = x_1 + \Delta$, (with $\Delta > 0$) $y_1 = y(x_1)$, $y_2 = y(x_2)$.
The transition probability from $[nx_1, ny_1]$ to $[nx_2, n(y_2+dy_2)]$, as given by (7), leads (if we neglect $\rho(u)$) to

$$(1-u^2)^{-n\Delta/2}\left(\frac{1+u}{1-u}\right)^{-n\Delta u/2} \sqrt{n}dy_2/\sqrt{2\pi\Delta(1-u^2)} \tag{15}$$

with $\quad u = n(y_2-y_1)/(n\Delta)$

To apply the Laplace method, we must obtain the dominant term in the log of this density, which is given by: $\frac{-n\Delta}{2}\left[\log(1-u^2) + u\ \log\left(\frac{1+u}{1-u}\right)\right]$.
Now as $\Delta\to 0$ (in some sense to be precised later on), $u \to y'(x_1)$. The dominant term in the log of the total asymptotic probability measure along ny(.) is now easily derived as

$$-\frac{n}{2}\int_{\varepsilon}^{2-\delta}[\log(1-(y'(v))^2) + y'(v)\log\left(\frac{1+y'(v)}{1-y'(v)}\right)]dv \tag{16}$$

Three problems remain to be solved:
- We must take into account, in (15), the factor

$$\sqrt{n}/\sqrt{2\pi\Delta(1-u^2)} \tag{17}$$

As we will see in Section 4.5, this is related to the asymptotic distribution of $X_n(.)$.

- The factor $2^{-n\varepsilon}$ in (14) must be considered.
  It can easily be checked that, if the function y(.) satisfies the Expansion Condition (C.E.): $y'(v)=1+Kv^j+o(v^j)$, $j\geqslant 1$, then:
  $2^{-n\varepsilon} \underset{\varepsilon\to 0}{\sim} \exp\left[-\frac{n}{2}\int_0^\varepsilon[\log(1-(y')^2)+y'\log\frac{1+y'}{1-y'}]dv\right].$
  The integral in (16) can thus be extended to [0,2].
- We must check that we can neglect the contribution from $\rho(u)$ in (7). This can be done be carefully coupling Δ with n.
  To ease the analysis, let $\varepsilon=\delta=\Delta=2/k$ say.
  By (8), we must add to (16) a term which is $\sim \frac{1}{n\Delta}\sum_{i=1}^{k-2}\rho[y'\left(\frac{2i}{k}\right)]$.
  By Euler's sommation formula, this is equivalent to

$$\frac{1}{n\Delta}\int_1^{k-1}\rho[y'\left(\frac{2x}{k}\right)]dx \qquad \text{(Term ①)}$$

$$-\frac{1}{2n\Delta}\left[\rho[y'(2-\Delta)]-\rho[y'(\Delta)]\right] \qquad \text{(Term ②)}$$

$$+\frac{1}{n\Delta}\int_1^{k-1}B_1(\{x\})\partial_x\rho[y'\left(\frac{2x}{k}\right)]dx \qquad \text{(Term ③)}$$

  Let us assume that C.E. is satisfied. Then, by (8), $\rho[y'(v)]=K/v^j+o(1/v^{j-1})$ for some K.
  Term ① becomes: $\int_\Delta^{2-\Delta}\rho[y'(v)]dv/(n\Delta^2)$ and $\to 0$ if $n\Delta^{j+1}\underset{n\to\infty}{\to}\infty$

  Term ② $\sim 1/(n\Delta^j)\to 0$ if $n\Delta^j\to\infty$

  Term ③ $\to 0$ if $1/(n\Delta^{j+1})\to 0$.

  The contribution from $\rho(.)$ can thus be neglected by letting $n\to\infty$, $\Delta\to 0$ with $n\Delta^{j+1}\to\infty$. Note that, with this last condition, Term ② $\to 0$ even if the measure in (16) is computed on $[\varepsilon,v]$, $v<2$.

Summarizing our results, we derive the

Lemma 4

The dominant term in the log of the total asymptotic probability measure along ny(.), between v_1 and v_2, is given, if C.E. is satisfied, by

$$-\frac{n}{2}\int_{v_1}^{v_2}[\log(1-y'(v)^2)+y'(v)\log\left(\frac{1+y'(v)}{1-y'(v)}\right)]dv \qquad (18)$$

ii) Turning now to the weight, we see that, if C.C. of Lemma 3 is satisfied, the dominant term in the log of the weight along ny(.), between v_1 and v_2, is given by

$$(v_2-v_1)n \log n+n\int_{v_1}^{v_2} \log(y(v))dv \tag{19}$$

Collecting the results of (18) and (19), we finally obtain the

Lemma 5

The dominant term in the log of the asymptotic total measure (L.A.T.M.) along ny(.), between v_1 and v_2, is given, if C.y, C.C. and C.E. are satisfied (with $v_1 \equiv j/n$, $v_2 \equiv k/n$), by:

$$(v_2-v_1)n \log n+n\int_{v_1}^{v_2}\left[-\frac{1}{2}\log(1-(y'(v))^2)-\frac{1}{2}y'(v)\log\left(\frac{1+y'(v)}{1-y'(v)}\right)+\log(y(v))\right]dv$$

$$=(v_2-v_1)n \log n+n\int_{v_1}^{v_2} f(y,y')dv, \text{ say.} \tag{20}$$

iii) We can now justify the first part of (11): (18) and (19) are balanced in the sense that their variable (in y(.)) part is linear in n. It is easily seen that any other function h(n)y(v), with h(n)=o(n) would lower the first term in (12) without of course increasing the probability measure (18) beyond O(1). The total measure in (20) would only be smaller.

On the other side, $Y^* \leqslant n$ and we cannot go beyond h(n)=n. The determination of y(.) is finally solved by the

Theorem 1

For LL,

$$y(v)=\sin(C_1(v-C_2))/C_1 \text{ where } C_1 \text{ and } C_2 \text{ are some constants} \tag{21}$$

The boundary conditions y(0)=y(2)=0 lead to: $C_1=\frac{\pi}{2}$, $C_2=0$.
Conditions C.y, C.C. and C.E. are obviously satisfied.

Proof: The Laplace method leads to the following question : find the function y(.) maximizing (20), this is classical variational problem.
As f in (20) does not depend on v, we know that the associated Euler-Lagrange equation possess an integral of the form $f-y'f_{y'}=L$ (L constant) which easily leads to (21). ▼

As a verification, we obtain the log of the asymptotic form of (1) from the sum of the L.A.T.M. (20) along $y(v)=\frac{2}{\pi}\sin(\frac{\pi}{2}v)$ and $\log C_{2n}$ (see (3)), we obtain after some computation $\sim 4^{2n}n^{2n}e^{-2n}/\pi^{2n}$ which corresponds to (1).

4.5 Asymptotic properties of $X_n(.)$

These properties will be completely analysed in Theorem 2 below, after establishing some convenient lemma's.

i) Let us firstly analyse the distribution of $X_n(x)$, $x \in [0,2]$. Set $\theta = X_n(x)/\sqrt{n}$. Assuming (11) to be valid, it is clear from Lemma 5 and Theorem 1 that the L.A.T.M. along any trajectory from 0 to $n[y(x)+\theta]$ must be computed on $[nz(v,y(x)+\theta,x)+\sqrt{n}\chi(v)]$ where, by Theorem 1, $z(v,u,x)=\frac{1}{C}\sin(Cv)$ with $C(u,x)$ such that

$$z(x,u,x)=\sin(C\,x)/C=u \quad \text{and} \quad \chi(.) \in C^1 \text{ with } \chi(0)=\chi(x)=0.$$

To simplify notations set $\eta(v):=\frac{\chi(v)}{\sqrt{n}}$, $y:=y(v)$, $y^*:=y(x)$, $y':=\frac{\partial y}{\partial v}$.

By symmetry, we can use $z(2-v,y^*+\theta,2-x)$ for $v \in [x,2]$ (22)
and, of course, $z(v,y^*,x)\equiv y(v)$.

The first result we need is summarized in the following

Lemma 6

The dominant term in the L.A.T.M. along $n[z(v,y^*+\theta,x)+\eta(v)], v\in[0,x]$ and along (22), $v\in[x,2]$ is given by: $2n\log n + n\int_0^2 f(y,y')dv - \frac{n\theta^2\pi^2}{4\gamma(x)\gamma(2-x)}$ with $\gamma(v):=\frac{\pi}{2}\cos(\frac{\pi}{2}v)v-\sin(\frac{\pi}{2}$

Proof: the proof, based on Euler-Lagrange equation properties, may be found in [17].▼

We can now draw two important conclusions from Lemma 6.

- The second part of (11) is now justified: θ is clearly a term of order $1/\sqrt{n}$.
- Normalizing the result of Lemma 6, in the sense of (6), the limit of $X_n(x)$ is obviously a Gaussian variable with mean 0 and variance

$$V^*=2\gamma(x)\gamma(2-x)/\pi^2 \tag{23}$$

ii) Lemma 6 is of course not sufficient to draw definitive conclusion on the process $X(.)$: we must still compute the covariance of $X(.)$ and check the tightness of the sequence $X_n(.)$ to obtain weak convergence. But, as Y is a random walk, the tightness verification is classical (see Billingsley [2] ch.III).
The covariance problem is solved by the following

Lemma 7

The covariance of $X_n(.)$ is asymptotically given by $(x_1 \leq x_2)$:

$$E^*[X_n(x_1)X_n(x_2)] \sim 2\gamma(x_1)\gamma(2-x_2)/\pi^2 = C^*_{12} \text{ say} \tag{24}$$

C^*_{12} is clearly factorized, the limiting process $X(.)$ is thus Markovian.

Proof : the proof may be found in [17].

iii) Collecting our results for (11), (21), (23), (24) we finally state the

Theorem 2

Let $\gamma(v):=\frac{\pi}{2}\cos(\frac{\pi}{2}v)v-\sin(\frac{\pi}{2}v)$. For LL, $[Y^*([nv])-n\frac{2}{\pi}\sin(\frac{\pi}{2}v)/\sqrt{n}\Rightarrow X(v)$

where $X(.)$ is a Markovian Gaussian process with mean 0 and covariance $C^*_{12}=2\gamma(x_1)\gamma(2-x_2)/\pi^2$. $X(.)$ can of course be written as $X(v)=\frac{\sqrt{2}\gamma(2-v)}{\pi}B\left(\frac{\gamma(v)}{\gamma(2-v)}\right)$ for some B.M.: $B(.)$. ▼

(17) can now be justified: it corresponds to the asymptotic ($\Delta\to0$) density factor for the Gaussian. Let $x_1,x_2\in[0,2]$. Set $\Delta:=x_2-x_1>0$.

The conditioned variance of $[X(x_2)|X(x_1)]$, is given by $V^*_2(1-C^{*2}_{12}/V^*_1V^*_2)$. By Theorem 2, this is equivalent to $2[\gamma(x_2)\gamma(2-x_2)-\gamma(2-x_2)\gamma(x_1)/\gamma(2-x_1)]/\pi^2$.

For small Δ, this gives $\Delta\sin^2(\frac{\pi}{2}x_1)+o(\Delta)$. But (17) leads to $\sqrt{n}/\sqrt{2\pi\Delta(1-(y')^2)}+o(\Delta)$.

By Theorem 1, the identification is immediate.

4.6 Asymptotic distributions of cost functions

Storage and time cost functions are analysed: their asymptotic properties are obtained in three typical theorems.

i) We firstly consider the storage cost σ^*_{LL}. Its asymptotic properties are given by the

Theorem 3

$[\sigma^*_{LL}-n^2\nu_1]/(n^3\nu_2)^{1/2}\sim\mathcal{N}(0,1)$ where $\nu_1=8/\pi^2$, $\nu_2=(128/\pi^2-12)/\pi^2$.

Proof: from (11) and Theorem 1, the asymptotic mean of σ^*_{LL} is clearly given by $n^2\nu_1$, with $\nu_1:=\int_0^2 y(v)dv$ and $y(v)=\frac{2}{\pi}\sin(\frac{\pi}{2}v)$.

From Theorem 2, the random variable $\int_0^2X(v)dv$ is a classical stochastic integral which is Gaussian, with mean 0 and variance $\nu_2:=\int_0^2du_1\int_{u_1}^2du_2 2\gamma(u_1)\gamma(2-u_2)/\pi^2$ which gives $(128/\pi^2-12)/\pi^2$. ▼

ii) The time cost τ^*_{LL} will be firstly analysed through its mean and variance. They are related to some properties of Y^*. We obtain the

Lemma 8

Let $\nu_3:=\int_0^2 y^2(v)dv=4/\pi^2$

$E^*(\tau^*_{LL})\sim\frac{1}{2}E^*(\sigma^*_{LL})\sim\frac{1}{2}n^2\nu_1=4n^2/\pi^2$

$E^*(\tau^{*2}_{LL})\sim\frac{1}{4}E^*(\sigma^{*2}_{LL})+\frac{1}{12}E^*[\sum_1^{2n-1}Y^{*2}(i)]\sim\frac{1}{4}[n^3\nu_2+n^4\nu_1^2]+\frac{1}{12}n^3\nu_3$

$V^*(\tau^*_{LL})\sim\frac{1}{4}n^3\nu_2+\frac{1}{12}n^3\nu_3=n^3[32/\pi^2-\frac{8}{3}]/\pi^2$

Proof: from (6), Tables 1 and 2, we deduce

$$E^*(\tau^*_{LL}) \sim \left[\sum_{\tilde{\gamma}\in\tilde{S}_{2n}} \sum_{i=1}^{2n-1}\left[\sum_{j=1}^{\tilde{\gamma}(i)} \prod_{k\neq i} \tilde{\gamma}(k)\right]\right]/E_{2n}$$

$$\sim \left[\sum_{\tilde{\gamma}\in\tilde{S}_{2n}} \left[\sum_{i=1}^{2n-1} \frac{\tilde{\gamma}(i)}{2} \prod_{k} \tilde{\gamma}(k)\right]\right]/E_{2n}$$

$$= \frac{1}{2} E^*(\sigma^*_{LL}) \text{ by the very definition of } \sigma^*_{LL}.$$

Similarly,

$$E^*(\tau^*_{LL})^2 \sim \left[\sum_{\tilde{\gamma}\in\tilde{S}_{2n}} \sum_{j_1=1}^{\tilde{\gamma}(1)} \sum_{j_2=1}^{\tilde{\gamma}(2)} \cdots \sum_{j_{2n-1}=1}^{\tilde{\gamma}(2n-1)} \left[\sum_{i=1}^{2n-1} j_i\right]^2\right]/E_{2n}$$

$$\sim \left[\sum_{\tilde{\gamma}\in\tilde{S}_{2n}} \left[\frac{1}{4}\left(\sum_{1}^{2n-1} \tilde{\gamma}(i)\right)^2 + \frac{1}{12}\sum_{1}^{2n-1} \tilde{\gamma}^2(i)\right] \prod_{k} \tilde{\gamma}(k)\right]/E_{2n}$$

$$\sim \frac{1}{4} E^*(\sigma^{*2}_{LL}) + \frac{1}{12} E^*\left[\sum_{1}^{2n-1} Y^{*2}(i)\right].$$

But $E^*(\sigma^{*2}_{LL}) = V^*(\sigma^*_{LL}) + (M^*(\sigma^*_{LL}))^2 \sim n^3(\frac{128}{\pi^2} - 12)/\pi^2 + (M^*(\sigma^*_{LL}))^2$ by Theorem 3,

and $E^*[\sum_{1}^{2n-1} Y^{*2}(i)] \sim nE^*\int_0^2 [ny(v)+\sqrt{n}\, X(v)]^2 dv \sim n^3\nu_3 + o(n^3)$

where $\nu_3 := \int_0^2 y^2(v)dv = 4/\pi^2$. ▼

iii) We now come to a surprising result: τ^*_{LL} is also distributed as a Gaussian variable! We obtain the

Theorem 4

$$[\tau^*_{LL} - 4n^2/\pi^2]/\left[\frac{n^3}{\pi^2}\left(\frac{32}{\pi^2} - \frac{8}{3}\right)\right]^{1/2} \sim \mathcal{N}(0,1)$$

Proof: the proof, based on variance analysis tricks and Central Limit Theorem, may be found in [17]. ▼

Actually, the proof of Lemma 8 can be extended mutatis mutandis to any polynomial cost functions. Leaving the details, we then obtain the

Theorem 5

Let ${}^s\tau_{LL}(h) := \sum_{I+D} [r_j(h)]^s$ (see Section 3 for notations)

$${}^{s}\nu_1 := \int_0^2 y^s(v)dv \qquad {}^{s}\nu_2 := V^*(\int_0^2 y^{s-1}(v)X(v)dv) \qquad {}^{s}\nu_3 := \int_0^2 y^{2s}(v)dv$$

then

$$E^*({}^{s}\nu^*_{LL})/n^{s+1} \sim {}^{s}\nu_1/(s+1)$$

$$[{}^{s}\tau^*_{LL} - E^*({}^{s}\tau^*_{LL})]/n^{(2s+1)/2} \sim \mathcal{N}\left(0, \left[\frac{s^2}{(s+1)^2}\ {}^{s}\nu_2 + \frac{s^2}{(s+1)^2(2s+1)}\ {}^{s}\nu_3\right]\right) \qquad \blacktriangledown$$

Other cost functions can be analysed by similar methods.

5. Conclusion

We have developed a general methodology to analyse asymptotic distributions of list structures histories and cost functions: the techniques are illustrated by three classical lists: the linear one, the priority queue and the dictionary (see [17] for the last two cases). Starting from a classical random walk, we put a weight on it and obtain the limiting process as a superposition of a deterministic function (of order n) and a Gaussian Markovian process (of order $\sqrt{n}$).

Cost functions are then related to stochastic integrals on these processes (which lead to Gaussian variables).

Three generalizations are possible:

- putting other boundary conditions on the paths
- defining other cost functions
- using more general random walks (several keys insertion or deletion, etc.).

Our techniques can be easily adapted to such situations.

Acknowledgments

The author is indebted to P. Flajolet for asking some questions which led to our interest into list structures analysis.

REFERENCES

[1] ABRAMOWITZ, M. and STEGUN, I.A. "Handbook of mathematical functions", 1965, Dover Publications

[2] BILLINGSLEY, P. "Convergence of probability measures", 1968, Wiley

[3] CHUNG, K.L. "Excursions in Brownian Motion", Ark. Mat. 14, 1976, pp. 155-177

[4] FELLER, W. "An introduction to probability theory and its applications", Vol.I, 1970, Wiley

[5] FELLER, W. "An introduction to probability theory and its applications", Vol.II, 1971, Wiley

[6] FLAJOLET, P. "Combinatorial aspects of continued fractions", Discr. Math. 32, 1980, pp. 125-161

[7] FLAJOLET, P. "Analyse d'algorithmes de manipulation d'arbres et de fichiers", Cahiers du BURO, 1981

[8] FLAJOLET, P. and ODLYZKO, A. "The average height of binary trees and other simple trees", J. Comp. Syst. Sc. 25, 2, 1982, pp. 171-213.

[9] FLAJOLET, P., PUECH, C. and VUILLEMIN, J. "The analysis of simple list structures", to appear in Information Sc.: and Int. J.

[10] FLAJOLET, P. and PRODINGER, H. "Register allocation for unary-binary trees", to appear in SIAM J. Comput.

[11] GREENE, D.H. and KNUTH, D.E. "Mathematics for the analysis of algorithms", 1981, Birkhaüser

[12] ITO, K. and Mc KEAN, H.P.Jr "Diffusion processes and their sample paths", Sec. Ed., 1976, Springer-Verlag

[13] LOUCHARD, G. "The Brownian Motion: a neglected tool for the complexity analysis of sorted tables manipulations", RAIRO, Inf. Th. 17, 4, 1983, pp. 365-385

[14] LOUCHARD, G. "Kac's formula, Levy's local time and Brownian excursion", J. Appl. Prob. 21, 1984, pp. 479-499

[15] LOUCHARD, G. "The Brownian excursion area: a numerical analysis", Comp. and Math. with Appl. 10, 6, 1984, pp. 413-417

[16] LOUCHARD, G. "Brownian Motion and Algorithms Complexity", to appear in B.I.T.

[17] LOUCHARD, G. "Random Walks, Gaussian Processes and List Structures", L.I.T. Technical Report N°152

RANDOM WALKS ON TREES

R. SCHOTT
C.R.I.N.
U.E.R. Sciences Mathématiques
UNIVERSITE DE NANCY I
54506 VANDOEUVRE LES NANCY CEDEX
FRANCE

Abstract.

Random walks or Brownian motions appear as a useful tool in algorithms analysis. Recently P. Flajolet ([2]) obtained a complete and detailed analysis of the two stacks problem with the help of properties of simple random walks on lattices. G. Louchard ([7], [8]) proved that the Brownian motion permits to give easily asymptotic results on the complexity of manipulation algorithms for sorted tables, dictonaries and priority queues. In [4], J. Françon and the author proved that random walks on some homogeneous trees can be analysed with simple combinatorial technics : generating functions, continued and multicontinued fractions, orthogonal polynomials, theorem of Darboux etc.. In this paper we show that random walks on more general trees can be related to random walks on $\mathbb{N}$ and that on general Cayley graphs (i.e. graphs corresponding to finitely generated groups with relations between the generators) the asymptotic behavior of the random walks can be obtained using proporties of the Brownian motions on Riemannian manifolds and a simple criteria can be given in terms of $\gamma(n)$ the number of different words was length is less than or equal to n .

Keywords

Random walk, tree, generating function, generator, asymptotic behavior, dynamic data structure, method of Darboux.

INTRODUCTION

The analysis of dynamic data structures requies often the introduction of new analytical, combinatorial or probabilistic models (see [2], [7], [8]). Simple dynamic algorithms can be the source of difficult mathematical questions : as illustration of this the reader can see the paper of A. Jonassen and D.E. Knuth ([6]). Another example was given by P. Flajolet [2].He studied the following storage allocation algorithm : assume that two stacks are to be maintained inside a shared (contiguous) memory area of a fixed size m . A trivial algorithm will let them grow from both ends of that memory area until their cumulated sizes fill the initially allocated storage (m cells) and the algorithm stops having exhausted its allocated memory. He compared this storage allocation algorithm to another option, namely allocating separate zones of size $m/2$ to each of the two stacks, this separate storage allocation method will halt as soon as any one of the two stacks reaches size $m/2$. This problem has a natural formulation in terms of random walks in a triangle in 2-dimensional space : a state is the couple (x,y) formed with the size of both stacks, the random walk has two reflecting barriers (a deletion takes no effect on an empty stack) and one absorbing barrier (the algorithm stops when the combined sizes of the stacks exhausts the available storage). The expected value of $\max(x,y)$ is a measure of the efficiency of the storage allocation. Let p be the probability that an insertion occurs in a stack, P. Flajolet proved that if $0 < p < \frac{1}{4}$, memory sharing is a real-advantage.

More general dynamic data structures like priority queues and dictionaries can be analysed in the same way : the cost function is a random variable X and the mean value of X is called the integrated cost, we have to do with a special random walk in $\mathbb{N} \times \mathbb{N}$, on each trajectory we put a total measure which is defined by some weight function (see [8]). Using the central limit theorem G. Louchard find that X is asymptotically $(n \longrightarrow +\infty)$ equivalent to a Brownian excursion area for the stack storage, some gaussian processes are abtained for priority queues and dictionaries.

A behavior of length n of the dynamic data structure can be considered as a word of the monoïd generated by the operations I (insertion), D (delection), Q^+ (successful query), Q^- (unsuccessful query). For the analysis in the mean case these operations are random variables and we have basically to do with some birth and death process. The same kind of problem appears in the analysis of operating systems (see [3]):a behavior is correct if some rules concerning partial commutativity between the operations are respected.

So the starting point of this paper was my effort to determine the behavior of random walks on partial commutative monoïds, more precisely to evaluate $p_n(x,x)$ the probability that starting from a point x we return to x in n steps.

Let A be a finite alphabet, if we consider the finitely generated discrete group associated to A some results can be given for $p_n(x,x)$ this is the object of the second part :

- first we remember briefly some facts concerning free groups, combinatorial methods developped by P. Flajolet ([1]) are sufficient here, the saddle point technic gives the behavior of $p_n(x,x)$ or $n \longrightarrow +\infty$ and we have to do with homogeneous trees.
- if there exist relations between the generators more sophisticated methods are needed, we apply results of [5] concerning random walks on Lie groups and of [10] concerning the potential theory on **Riemannian** manifold,some basic definitions of Riemannian geometry are also remembered.

Of course the general case concerning partial commutative monoïds still open, nevertheless we can make a conjecture as concluding remark.

II.- Random walks on $\mathbb{N}$ and applications

We consider random walks on $\mathbb{N}$ where 0 is reflecting and one-step-transitions can only occur to neighbooring states. The one-step transitions probabilities are :

$P(0 \rightarrow 1) = p_o = 1$

$P(i \rightarrow i+1) = p_i$, $P(i \rightarrow i-1) = q_i$ with $0 < p_i < 1$ and $p_i + q_i = 1$

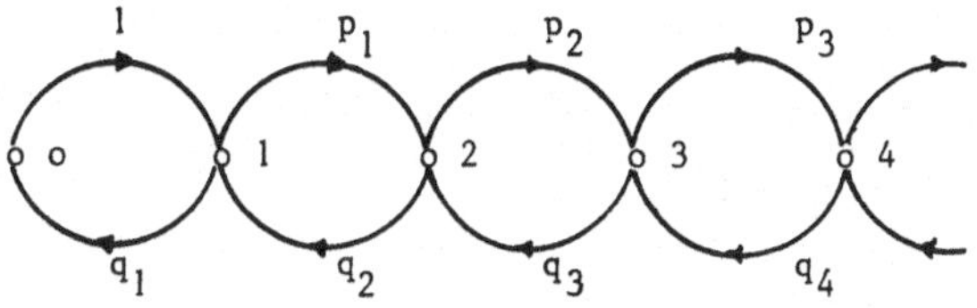

(1)

Remark II.1. (1) represents the evolution of stacks under random insertions and delections.

Let $P_{o,o}(n)$ be the probability to reach 0 in n steps if starting in 0 and $F_{o,o}(n)$ the probability to reach 0 for the first time in n steps if starting in 0 .

Consider the generating functions :

$$G_o(z) = \sum_{o}^{\infty} P_{o,o}(2n)\, z^n \quad , \quad P_{o,o}^{(o)} = 1$$

$$F_o(z) = \sum_{o}^{\infty} F_{o,o}(2n)\, z^n$$

The following relation is obvious : $G_o(z) = \dfrac{1}{1 - F_o(z)}$.

We have the following continued fractions expansions : ([11])

$$F_o(z) = \cfrac{q_1 z}{1 - \cfrac{p_1 q_2 z}{1 - \cfrac{p_2 q_3 z}{\cdots}}} \quad ; \quad G_o(z) = \cfrac{1}{1 - \cfrac{q_1 z}{1 - \cfrac{p_1 q_2 z}{\cdots}}}$$

several results can be proved for random walks on $\mathbb{N}$, in particular :

Proposition II.2.

If $\lim\limits_{n \to +\infty} (p_n \cdot q_{n+1}) = 0$ then $P_{o,o}^{(2n)} \underset{n \to +\infty}{\sim} C.R^{-n}$

where C is a positive constant and $R > 1$ is the radius of convergence of $G_o(z)$

Proof :

If $\lim\limits_{n \to +\infty} p_n q_{n+1} = 0$ then $G_o(z)$ is meromorphic and the result is a direct application of the method of Darboux.

We consider below an example in which the hypothesis of proposition II.2 is not fulfilled.

Example II.3

Let $0 < a < c$ and put $q_{2n-1} = \frac{a+n-1}{c+2n-2}$, $q_{2n} = \frac{n}{c+2n-1}$, $p_n = 1-q_n$

In this case $\lim\limits_{n \to +\infty} p_n q_{n+1} = \frac{1}{4}$

we obtain : $G_o(z) = 1 + \frac{a}{c} z + \frac{a(a+1)}{c(c+1)} z^2 + \ldots$

the hypergoemetric series of Gauss. Therefore :

$$p_{oo}^{(2n)} = \frac{a(a+1)\ldots(a+n-1)}{c(c+1)\ldots(c+n-1)} \underset{n \to +\infty}{\sim} \frac{\Gamma(c)}{\Gamma(a)} \frac{1}{n^{c-a}}$$

Different kind of behavious can be obtained.

Example II.3

If $0 < r < 1$, $q_{2n-1} = 1-r$, $q_{2n} = r$, $p_n = 1-q_n$

We find with the help of Darboux's method :

$$p_{oo}^{(2n)} \underset{n \longrightarrow +\infty}{\sim} \frac{C}{\sqrt{n}} \quad , C \text{ stricly positive constant}$$

Applications.

Consider now trees with root 0 such that all vertices at distance n from 0 have the same degree d_n $(d_n \in \mathbb{N})$

Example II.5

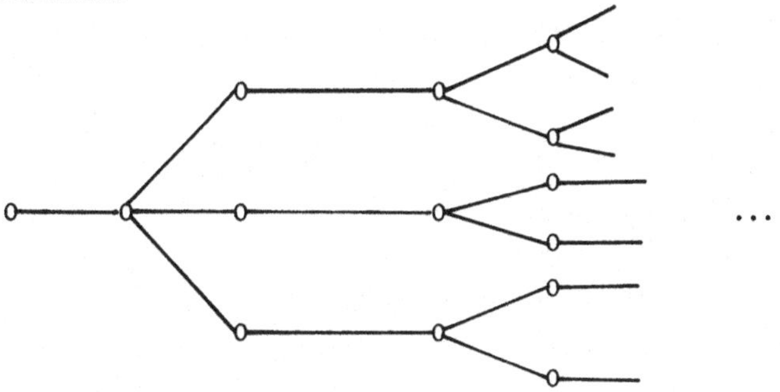

$d_o = 1 \quad d_1 = 4 \qquad d_2 = 2 \quad d_3 = 3 \ \ldots$

A random walk on such kind of trees is equivalent to a random walk on $\mathbb{N}$ defined by : $p_o = 1$, $q_n = \frac{1}{d_n}$, $p_n = 1 - \frac{1}{d_n}$

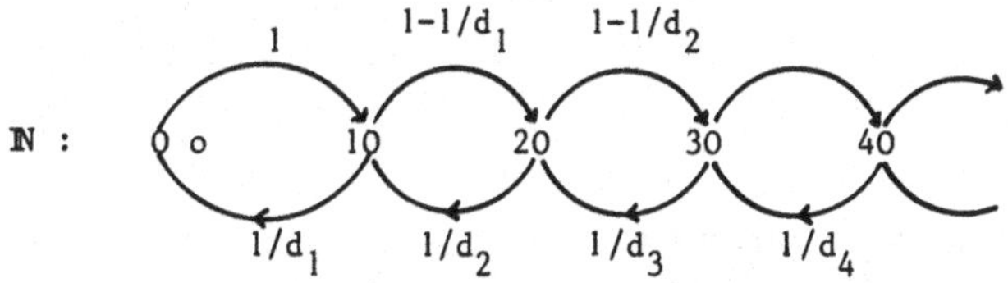

Therefore we can apply well-known results concerning random walks on $\mathbb{N}$ in order to obtain the behavior of $P_{o,o}(2n)$ for such kind of trees.

Example II.6 : if $d_1 = 3$, $d_2 = d_3 = \ldots = 2$

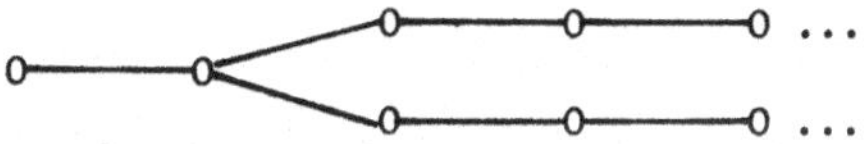

A simple calculation gives :

$$P_{o,o}(2n) \underset{n \to +\infty}{\sim} . \ c\, n^{-3/2} \quad , \quad c > 0$$

This result is not surprising and can be obtained with the methods of [1] .

Example II.7

Let $d_1 = 2$, $d_2 = d_3 = \ldots = r > 2$.

Proposition II.8

The generating function $G_o(z)$ is given by :

$$G_o(z) = \frac{4r-3rz+6z-8-z\sqrt{r^2-4(r-1)z}}{10z+4r-6rz-8+2rz^2-2z^2}$$

The radius of convergence R is : $R = 8/9$ if $r = 3$

$$R = \frac{(r-1)}{2r-4} \quad \text{if} \quad r \geq 4$$

and $P_{o,o}^{(2n)} \underset{n \to +\infty}{\sim} c\,(\frac{8}{9})^n .\, n^{-3/2}$ for $r = 3$, $c > o$ (constant)

$$\sim c\,(\frac{3}{4})^n .\, n^{-1/2} \quad \text{for} \quad r = 4$$

$$\sim c \left(\frac{r-1}{2r-1}\right)^n \quad \text{for} \quad r > 4$$

Proof :

Some calculations proves that the generating function $G_o(z)$ has the following continued fraction expansion

$$G_o(z) = \cfrac{1}{1 - \cfrac{z/2}{1 - \cfrac{z/2r}{1 - \cfrac{(\frac{r-1}{r^2})z}{\cdots}}}}$$

and the right member of this equality is exactly equal to the announced result (see [11]).

The behavior of $P_{o,o}^{(2n)}$ is obtained with Darboux's method.

Remark II.9

From the examples one can see that random walks on different types of non homogeneous trees can be treated as random walks on $\mathbb{N}$. Many variations of this kind are possible.

III.- Random walks on general Cayley graphs.

Consider an alphabet $A = \{a_1, a_2, \ldots, a_k\}$, which generates a group G , P a probability measure supported by $\{a_1, a_2, \ldots, a_k, \bar{a}_1, \bar{a}_2, \ldots, \bar{a}_k\}$ (i.e. $\sum_{i=1}^{k} P(a_i) + P(\bar{a}_i) = 1$ where $\bar{a}_i$ is the symmetric element of a_i .

Let e be the neutral element of G .

To G we can associate the following graph (sometimes called Cayley-graph in the literature).

ex. : k = 2

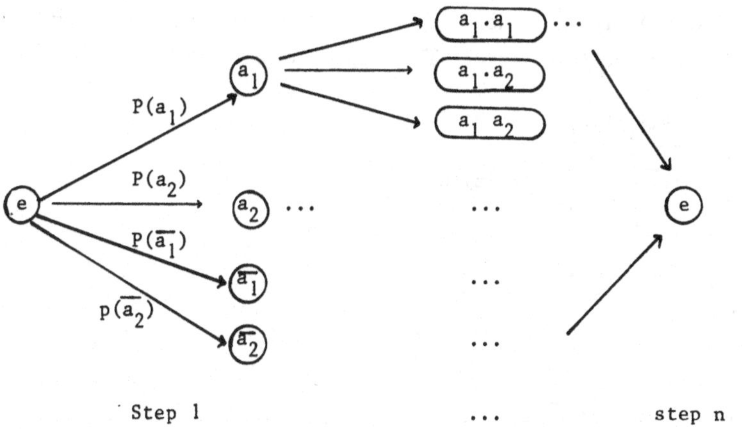

The problem is now to find the behavior of $P_{e,e}^{(2n)}$ as $n \to +\infty$.

The quantity $U = \sum_{n=o}^{\infty} P_{e,e}^{(2n)}$ is called the potential of the random walk.

If U is finite, e (and any other $g \in G$) appears only a finite number of times.

If U is infinite we say that the random walk is recurrent.

As we mentioned in the introduction if G is a free group, combinatorial methods developped in [1] permit to give an estimate for $P_{e,e}^{(2n)}$ as $n \longrightarrow +\infty$. We remember briefly this result below (see [4] for more details).

If there exist relations between the generators, we can apply arguments developped in [10] : G can be considered as the fundamental group of some compact Riemannian manifold M . Using the fact that the random walk on G has the same behavior as the Brownian motion on M , a result can be given using γ_n the number of different words was lentgth is less than or equal to n .

Consider $D(G) = (G,G) = \{ab\ ab, \forall(a,b) \in G \times G\}$; D(G) is a normal subgroup of G

<u>Definition</u> :

We say that G is solvable if there exist $r \in \mathbb{N}$ such that :

$G \supset D(G) \supset D_2(G) \supset \ldots \supset D_k(G) \supset \ldots \supset D_r(G) \supset \{c\}$ where

$D_k(G) = (D_{k-1}(G)\ ,\ D_{k-1}(G))$.

III.1.- Random walks on free groups :

If $A_i(z)$ (resp. $\overline{A_i}(z)$) is the generating function of the words was first letter is a_i (resp $\overline{a_i}$) and the last letter $\overline{a_i}$ (resp. a_i) we obtain : (cf [4])

$$A_i(z) = \sum_{n=o}^{\infty} A_i(2n)z^{2n} = \frac{P(a_i)P(a_i)z^2}{1+\overline{A_i}(z) - \sum_{j=1}^{k} [A_j(z)+\overline{A_j}(z)]}$$

and : $$G(z) = \sum_{n=o}^{\infty} P_{e,e}^{(2n)} z^{2n} = \frac{1}{1 - \sum_{j=1}^{k} [A_i(z)+\overline{A_i}(z)]}$$

It's easy to prove that : $A_i(z) = \overline{A_i}(z)$, $i \in \{1,2,\ldots,k\}$

Let $\alpha = P(a_1)P(a_i)$, $\beta = P(a_2)P(a_2)$

We obtain : for $k = 2$

$$z^2 = \frac{G(z)+1}{4\alpha_1\left(1-\frac{\alpha_2}{\alpha_1}\right) G^2(z)} \left\{ \left(1+\frac{\alpha_2}{\alpha_1}\right) [G(z)+1] -2\left[\frac{\alpha_2}{\alpha_1} G^2(z)+2\frac{\alpha_2}{\alpha_1} G(z)+1+\left(\frac{\alpha_2}{\alpha_1}\right)^2 - \frac{\alpha_2}{\alpha_1} \right]^{1/2} \right\}$$

$P_{e,e}^{(2n)}$ is given by the Cauchy formula :

$$P_{e,e}^{(2,n)} = \frac{1}{2\pi i} \int_C \frac{dG}{f^n(G)} \qquad \text{where } f(G) = z^2$$

and C is a closed path surronding $z = 1$.

The result is given by the saddel point method :

Let $$u(z) = \frac{1}{4\alpha_1 z^2}$$

we obtain :

$$P_{e,e}^{(2n)} \underset{n \to +\infty}{\sim} \frac{K(4\pi_1 d)^n}{n^{3/2}} \qquad (2)$$

Where : $d = u(z_o)$, z_o is the unique positive solution of $u'(z) = 0$

$0 < 4\alpha_1 d < 1$

$$K = \left[\frac{2\pi\, u''(z_o)}{u(z_o)} \right]^{-1/2}$$

(2) implies that : $U = \sum_{n=o}^{\infty} P_{e,e}^{(2n)} < +\infty$

II.2.- Random walks on finitely generated groups.

The idea is to relate the random walk on G to the Brownian motion on some compact Riemannian manifold.

First we summarize below some basic definitions and properties concerning Riemannian geometry (for more details see any handbook in this field).

II.2.1.- Riemannian geometry

Definition 1 :

A Riemannian manifold M is a differentiable manifold on which we can define a metric given by a positive definite quadratic form :

$TM_x \longrightarrow \mathbb{R}$ $\qquad$ TM_x is the tangent space on $x \in M$

$\xi \longrightarrow < \xi , \xi >$ $\qquad$ and $< \xi , \xi >$ is fixed on each TM_x

The metric permits to measure lengths, angles, areas, volumes, etc..
Now let $x_o \in M$, consider the set $\pi_1(M, x_o)$ of all closed oriented loops going from x_o to x_o .
On $\pi_1(M, x_o)$ we can define a product o :
$(\pi_1(M,x_o),o)$ is a group called the fundamental group at x_o .
This group is independent of x_o (modulo some isomorphism). For this reason we can write $\pi_1(M)$.

Example 2

If $M = S^1$ (circle), then $\pi_1(S^1) = \mathbb{Z}$.

Definition 3 :

Consider two manifolds M,N such that $\dim M = \dim N$ and $f : M \rightarrow N$ a mapping, f is a covering if $\forall y \in N, \exists V(y)$ neighbourhood of y such that :
$f^{-1}(V(y)) = U_1 \cup U_2 \cup \ldots$ and $f : U_k \longrightarrow V_j$ is a diffeomorphism.

Example 4

1) $M = \mathbb{R}$ and $N = S$, f defined by $f(t) = \exp(2\pi i t)$ is a covering

2) $M = S$, $N = S$, f defined by $f(z) = z^n$ with $z = 1$ is a covering.

Definition 5 :

f is a universal covering if $\pi_1(M) = id$

(this mean that M is simply connected).

Example 6 :

1) $f : \mathbb{R} \longrightarrow S$, $f(t) = \exp(2\pi i t)$ is a universal covering

2) this example is more interesting in the context of trees :

consider the homogeneous tree

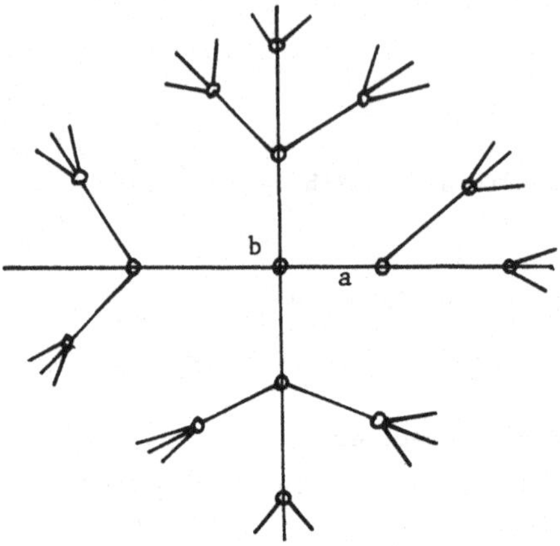

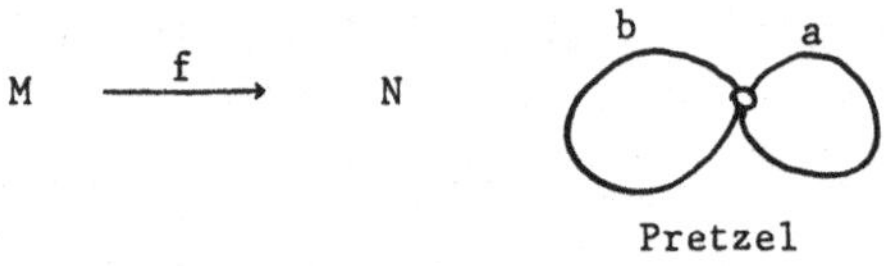

f is the universel covering

Let G be a discrete group generated by the finite set of generators $A = \{a_1, \dots, a_k\}$.

Any element $g \in G$ can be written as :

$$g = a_{i_1}^{\varepsilon_1} \, a_{i_2}^{\varepsilon_2} \dots a_{i_n}^{\varepsilon_n} \qquad \varepsilon_A = \pm 1 \qquad (1)$$

we shall denote by $|g| = \inf n$ the inf being taken under (1) and by $\gamma(n) = \mathrm{Card}\,\{g \in G \; ; \; |g| \leqslant n\}$ the growth function of G .

Definition 7 : Let M,N Riemannian manifolds.
$f : M \longrightarrow N$ is a normal Riemannian covering if there exist $H \triangleleft \pi_1(M)$ normal subgroup of $\pi_1(M)$ so that $G \simeq \pi_1(M)/H$ and that G is transitive on the fiber. (G is the group of deck transformations).

III.2.2.- Random walks on G .

If G is any finitely generated discrete group, it's possible then to find $M \longrightarrow N$ a normal Riemannian covering for which G is exactly the group of deck transformations. (we suppose that N is compact).

On the simply connected Riemannian manifold M we can define a Brownian mition $\alpha(t)$ (see [10]) and we apply here the following results :

i) $\alpha(t)$ is transient $\Longleftrightarrow \int_1^{+} \frac{xdx}{V_m(x)} < +\infty$ (2) where $V_m(x) = \mathrm{Vol}(B_x(m))$

is the Riemannian volume of the ball $B_x(m) = \{y \in M \; ; \; d(y,m) \leqslant x\}$ d is the metric.

ii) the Brownian notion on M is transient if and only if G is transient (i.e. : $\sum_{n=o}^{\infty} p_{e,e}(n) < +\infty$ for all probability measure p supported by $A \cup \bar{A}$).

Milnor [9] proved that :

Lemma 8 :

Let $f : M \longrightarrow N$ be the universal covering of the compact Riemannian manifold N , $T = \pi_1(N)$ the fundamental group of N

Then there is a constant $C > o$ such that for each $y \in M$ and all $r \geqslant 1$ one has :

$$\mathrm{Vol}(B_y(Cr+C)) \geqslant \gamma(r) \geqslant \mathrm{Vol}(B_y(C^{-1}r))$$

with $\gamma(r) = \mathrm{Card}\;\{g \in \pi_1(N) \; ; \; |g| \leqslant r\}$

From this lemma and the relation (2) we obtain that :

Theorem 9 :

A necessary condition for which $\sum_{n=o}^{\infty} p_n(e,e) < +\infty$ is that $\sum_{n=1}^{\infty} \frac{n}{\gamma(n)} < +\infty$

Under some additional condition concerning the curvature of N which is fulfilled if G is, for example, solvable the above condition is sufficient. Therefore :

Theorem 10 :

If G is finitely generated and solvable then :

$$\sum_{n=o}^{\infty} p_{e,e}(n) < +\infty \Longleftrightarrow \sum_{n=o}^{\infty} \frac{n}{\gamma(n)} < +\infty$$

Remark 11 :

. For a free abelian discrete group G generated by $\{a,b\}$ one has :

$\gamma(n) = 2n^2+2n+1$

$$\sum_{n=1}^{\infty} \frac{n}{\gamma n} \sim \sum_{n=1}^{\infty} \frac{1}{2n} = +\infty \qquad G \text{ is not transient}$$

. For a free group generated by $\{a,b\}$ one has :

$\gamma(n) = 2.3^n-1$

$$\sum_{n=1}^{\infty} \frac{n}{\gamma(n)} \sim \sum_{n=1}^{\infty} \frac{n}{2.3^n} < +\infty$$

. In this case we can prove by another method that

$$\sum_{n=o}^{\infty} p_{e,e}(n) < +\infty$$

IV.- Concluding remarks.

The asymptotic behavior of random walks on trees can be obtained in many cases with the help of combinatorial methods. For random walks on partially commutative monoïds and semi-groups the problem is still open but it's highly probable that a characterization using the growth function $\gamma(n)$ can also be given.

References :

[1] Flajolet P. (1981) : Analyse d'algorithmes de manipulation d'arbres et de fichiers - Cahier du B.U.R.O. 34-35.

[2] Flajolet P. (1984) : The evolution of two stacks in bounded space and random walks in a triangle. Publication I.N.R.I.A. Roquencourt.

[3] Françon J. (1985 : Une approche quantitative de l'exclusion mutuelle. Proceedings of the conference S.T.A.C.S.

[4] Françon J. and Schott R. (1984) : Multicontinued fractions and combinatorial of words on finitely generated groups. N.A.T.O. advanced workshop "combinatorial algorithms on words", Moratea, Italy, May 18-22.

[5] Guivarc'H Y., Keane M., Roynette B. : Marches aléatoires sur les groupes de Lie - Lecture Notes in mathematics n° 624, Springer Verlag

[6] Jonassen A. and Knuth D.E. (1978) : A trivial algorithm whose analysis isn't. Journal of Comp. and System Sc. 16, 323-332.

[7] Louchard G. (1983) : The Brownian notion a neglected tool for the complexity analysis of sorted tables manipulations. R.A.I.R.O., vol. 17, 4, 365-385.

[8] Louchard G. (1985) : Random walks, Gaussian processes and list structures Technical report 152. Université libre de Bruxelles. Laboratoire d'informatique théorique.

[9] Milnor J. (1968) : A note on curvature and fundamental group J. Diff. Geometry 2, 1-7.

[10] Vanopoulos N.Th. (1981) : Potential theory and diffusion on Riemannian manifolds. Zygmund 80th Birsthday Volume.

[11] Wall H.S. (1948) : Analytic theory of continued fractions Von Nostrand, Toronto.

[12] Wolf J.A. (1968) : Growth of finitely generated solvable groups and curvature of Riemannian manifolds. J. Diff. Geom. 2, 421-446.

INFINITE TREES, MARKINGS AND WELL FOUNDEDNESS

By

Ran Rinat, Nissim Francez and Orna Grumberg

Department of Computer Science
Technion- Israel Institute of Technology
Haifa, Israel.

December 1985

Abstract

A necessary and sufficient condition for a given marked tree to have no infinite paths satisfying a given formula is presented. The formulas are taken from a language introduced by Harel, covering a wide scale of properties of infinite paths, including most of the known notions of fairness. This condition underlies a proof rule for proving that a nondeterministic program has no infinite computations satisfying a given formula, interpreted over state sequences. We also show two different forms of seemingly more natural necessary and sufficient conditions to be inadequate.

1. Introduction

The problem of finding a sound and complete proof rule for proving that a given nondeterministic program terminates under a certain fairness assumption has been solved for various notions of fairness (e.g [AO 83, APS 84, FK 84, GFK 83, GFMR 81, LPS 81, PN 83]). In order for a program to be fairly terminating under any given notion of fairness, it has to admit no infinite fair computations, where the definition of a *fair computation* varies from one version of fairness to another and from one model of computation to another.

A most convenient way of defining the semantics of a nondeterministic program is by using a tree , the vertices of which correspond to the intermediate states of computations. Thus, all these notions of fairness can be viewed as conditions on paths in the computation tree. Therefore, a language in which general conditions on paths in trees can be expressed can generalize all the already discussed types of fairness. One such a language, called L, is introduced in [HA 84]. It is also shown there how three notions of fairness can be expressed within L. Many of other versions of fairness, along with (what is claimed there as) any sensible condition one can think of, can also be expressed in L.

An important result of [HA 84] is a recursive transformation that, given a tree T and a formula $\varphi \in L$, yields a tree T' , the infinite paths of which correspond to the infinite paths of T satisfying φ. Thus, using this transformation, proving that T has no infinite paths satisfying φ reduces to proving that T' has no infinite paths at all.

Returning to the issue of fair termination, two major approaches for proving it have been suggested in the literature (for a comprehensive discussion of these issues see [F 85]):

1. The method of *helpful directions* [GFMR 81, LPS 81] :

   According to this approach one defines a ranking of states by means of elements of a well-founded set. This ranking has to decrease according to rules derived directly from the fairness notion at hand.

2. The method of *explicit scheduler* [AO 83, APS 84, OA 84] :

   This approach is based on program transformation. By augmenting the given program using *random assignments*, an explicit fair scheduler is incorporated into the program. Thus, it remains to prove that the resulting program ordinarily terminates, for which a standard proof method exists.

Thus, a natural goal is to provide generalizations of both methods to the context of anguages like L: For any $\varphi \in L$, prove the absence (in a given program) of infinite comoutations satisfying φ.

The application of the *explicit scheduler* method to L is pursued in [DH 85], hough they use a different kind of explicit scheduler then used in [AO 83]. In this paper we pursue the alternative approach of *helpful directions*, directly connecting he computation trees and their specifications in L with decreasing well founded rankngs. We deal here with a subset of L called L^-, containing all formulas in L having no nfinite conjunctions or disjunctions. Weak and strong fairness can be expressed in L^- [HA 84]), but extreme fairness requires an infinite formula, and thus cannot be expressed in L^-. We remain in the level of trees, and the actual result concerns necesary and sufficient conditions, phrased in terms of decreasing well founded rankings, or a tree T to have no infinite paths satisfying a formula $\varphi \in L^-$. This condition is ntended to underly a syntactic proof rule for a programming language, having such rees as meaning for its programs.

Section 2 presents the language L^-. In section 3 the necessary and sufficient condition is presented. In section 4 we prove its correctness, and in section 5 we show the mpossibility of two, seemingly simpler and more natural, forms of necessary and ufficient conditions.

2. Basic definitions

We first define the trees, to which we refer. A *node* is a finite sequence of natural umbers (i.e an element of N^*) and a *tree* is a set of nodes (i.e a subset of N^*) closed nder the prefix operation. The *root* of the tree is ε, and a *path* is a maximal increasing equence of successive nodes (by the prefix ordering) starting at ε. Parts of a path are ermed *path−fragments*, or just *fragments*. A node is a *leaf* if it is the last element f a (finite) path. An example of a tree is shown in figure 1. A tree is *well founded* if all s paths are finite.

Let Σ be some fixed (possibly infinite) alphabet. A *Σ−marked tree* is one in which odes are marked with (possibly infinitely many) letters from Σ, i.e a tree T comes omplete with a marking predicate $M_T \subseteq T \times \Sigma$.

Throughout this paper, we refer to *recursive* marked trees, i.e marked trees for hich two algorithms exist: one that given an element of N^* decides whether it is a ode in the tree and of which kind (leaf or internal). The other decides, given a node v n the tree and a mark a, whether v is marked with a. Keeping in mind that the trees re to be regarded as denotations of programs, we discuss *only* recursive trees, and hus we do not specify this explicitly unless needed.

We now define the language L^- for stating properties of infinite paths in a marked ree (we repeat the definition of L from [HA 84] omitting infinite disjunctions and conunctions). An *atomic formula* is an expression of one of the forms $\exists_a$, $\forall_a$, $\exists^{\infty}_a$ or $\forall^{\infty}_a$, where $a \in \Sigma$ is a mark. L^- is the closure of the atomic formulas under finite conunctions and disjunctions. Note the absence of negation from L^- (and L).

We interpret the formulas of L^- over infinite paths in a marked tree as follows: Let be an atomic formula and let π be an infinite path. π *satisfies* φ ($\pi \models \varphi$) if either

a) $\varphi = \exists_a$ and there is a node on π marked with a.

b) $\varphi = \forall_a$ and all the nodes on π are marked with a.

c) $\varphi = \exists^{\infty}_a$ and there are infinitely many nodes on π marked with a.

d) $\varphi = \forall^{\infty}_a$ and there is a vertex in π from which all the nodes are marked with a.

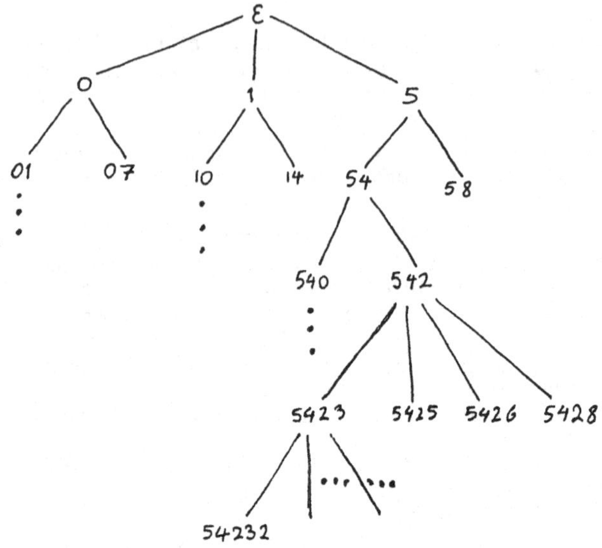

Figure 1. An example of a tree.

The interpretation of a general formula is obtained from the atomic formulas using the usual meaning of the boolean connectives. Note that formulas are interpreted over infinite paths only, and so the notion of a finite path satisfying a formula is undefined.

For example, consider playing chess on an infinite board (but with the standard set of 32 pieces) where moving rules are generalized in some reasonable way. An infinitely long game is a *draw* if both players call "check" infinitely often, otherwise it is a win for the player with the most calls. The game tree can be regarded as a marked tree with, say a and b marking nodes where player 1 or 2 checks respectively. The draw criterion is then given simply by the formula of L^-: $\exists^\infty{}_a \wedge \exists^\infty{}_b$.

We say that a tree T is *φ–avoiding*, or that *avoid*(T,φ) *holds*, if T has no infinite paths satisfying φ.

3. The proof rule

Our goal is to present a necessary and sufficient condition for a given marked tree to have no paths satisfying a given formula, based on a decreasing well founded ranking. Building on such condition, a higher level goal is to present a (sound and semantically complete) *proof rule* for proving that a nondeterministic program (having a tree as its operational semantics) has no infinite executions satisfying a given formula (properly interpreted). Keeping this higher goal in mind, we metaphorically refer to the condition itself as a proof rule. We would like the rule to follow the following scheme: given a marked tree T and a formula $\varphi \in L^-$, for which we want to prove *avoid*(T,φ) ,choose a well founded set $(W,<)$, and a variant (or rank) function ρ, mapping nodes in T to W. The variant ρ should satisfy certain restrictions, assuring that the tree T is *φ–avoiding*, i.e such W and ρ should exist iff the tree is *φ–avoiding*. These restrictions are, generally speaking, as follows: First, we require that the rank be nonincreasing, i.e if a node is mapped to some value then all of its descendents are mapped to lower or equal values. Second, we require that whenever a node's marking "leads" to the satisfaction of φ ,in a sense to be made precise, all of its descendents are mapped to strictly lower values. The determination whether a node leads or does not lead to the satisfaction of φ is formalized by means of an appropriate predicate on the nodes, providing this information. This predicate, which depends on the given formula, has to be such, that an infinite path contains infinitely many nodes satisfying the predicate iff this path satisfies the formula. Also, keeping in mind that this rule has to be applicable to programs, we would like the predicate to depend only on the markings of the nodes on the fragment from the root to the node itself.

However, the rule we present is of a more complicated form. The reason for this complication is that a rule of the above form is impossible, since no appropriate predicate can be found. This fact is made precise and proved in a later section. Another, more complicated, form of a rule, still more natural than the one we are about to present, is also shown to be impossible.

The difference between the simple scheme discussed above and the rule to be actually presented is, that instead of finding *one* well founded ranking for the whole tree, we have to find one for *each node* in T. Then, we use the same technique , but instead of referring to the whole tree, we refer to each subtree of T. Given a vertex v in the tree and a node u in the subtree of T rooted at v, we have the corresponding node-predicate, telling whether u leads to the satisfaction of φ *in the subtree rooted at* v. The exact meaning of this is explained after the definition of the predicate. Due to the nature of the rule , this predicate has to depend not only on the fragment from v to u, but also on the fragment from the root (of T) to v. There is also a need to distinguish between these two parts of the fragment from the root to u, and therefore the predicate has two arguments. Since its values depend only on *markings* found on a path, its arguments are elements of $(2^{\Sigma})^*$, where each symbol $x \in 2^{\Sigma}$ represents a set of marks from Σ.

Definitions: Let Σ be an alphabet of marks and let $\pi \in (2^{\Sigma})^*$.

1. A mark a *appears in* π if there exist $x \in 2^{\Sigma}$ and $u,v \in (2^{\Sigma})^*$, such that $\pi = uxv$ and $a \in x$. If $v=\varepsilon$ then a *appears in the last element of* π.
2. A mark a *appears in all* π if $\pi = x_1 \cdots x_n$, $x_i \in 2^{\Sigma}$ and $a \in x_i$ for all $1 \le i \le n$.

We now proceed to the definition of the predicate Q, which, intuitively, given the markings on the fragment from the root to v ,and the markings on the fragment from v to u, tells whether u leads to the satisfaction of φ in the subtree rooted at v.

Let $\varphi \in L^-$ be a formula over Σ. Define the predicate $Q_{\Sigma,\varphi}$: $(2^{\Sigma})^* \times (2^{\Sigma})^* \rightarrow \{T, F\}$ inductively as follows (we omit Σ and write Q_φ instead of $Q_{\Sigma,\varphi}$, where Σ is understood from the context):

Basis:

a) $Q_{\exists_a}(\pi_1, \pi_2)$ iff a appears in $\pi_1\pi_2$.

b) $Q_{\forall_a}(\pi_1, \pi_2)$ iff a appears in all $\pi_1\pi_2$.

c) $Q_{\exists^{\infty}_a}(\pi_1, \pi_2)$ iff a appears in the last element of $\pi_1\pi_2$.

d) $Q_{\forall^{\infty}_a}(\pi_1, \pi_2)$ iff a appears in all π_2.

Inductive step: Assuming that for $\varphi, \psi \in L^-$, $Q_\varphi(\pi_1, \pi_2)$ and $Q_\psi(\pi_1, \pi_2)$ have been defined for every π_1 and π_2, we have:

a) $Q_{\varphi \vee \psi}(\pi_1, \pi_2)$ iff $Q_\varphi(\pi_1, \pi_2)$ or $Q_\psi(\pi_1, \pi_2)$.

b) The definition of $Q_{\varphi \wedge \psi}(\pi_1, \pi_2)$ involves induction on the length of π_2:

- $Q_{\varphi \wedge \psi}(\pi_1, \varepsilon)$ holds for all π_1.

Let $\pi_2=\pi_2'x$ where $x \in 2^{\Sigma}$. Assume $Q_{\varphi\wedge\psi}(\pi_1, \pi_2'')$ has been defined for every π_2'' which is a prefix (not necessarily a proper one) of π_2'. Let $\bar{\pi}_2$ be the longest prefix of π_2' such that $Q_{\pi\wedge\psi}(\pi_1, \bar{\pi}_2)$ holds (such a prefix always exists because of the definition for $\pi_2=\varepsilon$). Then we have:

- $Q_{\varphi\wedge\psi}(\pi_1, \pi_2)$ iff there exist π, $\pi' \neq \varepsilon$ such that $\bar{\pi}_2\pi$ and $\bar{\pi}_2\pi'$ are prefixes of π_2 and both $Q_{\varphi}(\pi_1, \bar{\pi}_2\pi)$ and $Q_{\psi}(\pi_1, \bar{\pi}_2\pi')$ hold.

We now explain the intuition behind this definition. The argument π_1 of Q is to be identified with the word along the fragment from the root of T to v, which is the root of the subtree in mind. similarly, π_2 is to be identified with the fragment from v to the specific node u, for which Q has to determine whether it leads to the satisfaction of φ. Now, for all atomic formulas but $\forall^{\infty}_a$, Q refers to the concatenation of its two arguments. Hence for a given u, the determination whether it leads to satisfaction does not depend on v, i.e it is the same for all subtrees containing u.

For all the first three atomic formulas, Q determines leading for satisfaction in a natural way. For $\forall^{\infty}_a$, Q ignores the first argument, and relates only to the fragment from v to u, thus relating only to the subtree rooted at v. Restricted to this subtree, Q behaves as if the formula at hand is $\forall_a$. This fact can be justified by noticing, that a tree satisfies $\forall^{\infty}_a$ iff all its subtrees satisfy $\forall_a$. Thus, "leading to satisfaction within a given subtree" gets its meaning only through this formula, and the meaning is ,actually, leading to the satisfaction of $\forall_a$ in this subtree. Of course, all this is extended when connectives are introduced.

The definition of Q for disjunction is straightforward. In the conjunction case, Q always holds for the root of the subtree. For all other nodes u, Q holds in u iff there are nodes on the fragment from u's most recent ancestor for which Q holds, for which Q has determined leading to satisfaction of the two conjuncts of φ. Thus, for a given path, Q will induce infinitely many decrements iff there are infinitely many pairs of nodes for which Q has induced decrement with respect to the conjuncts of the conjunction. This happens iff Q has induced infinitely many decrements for each conjunct.

It is important to note that Q is recursive, i.e there is an algorithm to calculate Q from its two arguments.

Notations: Let T be a Σ-marked tree.

1. If v, u are vertices in the tree connected by $v=v_1-v_2-\cdots-v_{n-1}-v_n=u$, denote by $\pi_{v,u}$ the following element of $(2^{\Sigma})^*$: $\pi_{v,u}=x_1\cdots x_n$, where x_i is the set of marks of v_i. If v is the root of the tree, we write π_u instead of $\pi_{v,u}$.
2. For a vertex $v \in T$, denote by T_v the subtree of T rooted in v.
3. For a vertex v, denote by *children*(v) the set of all (direct) children of v.
4. For a vertex v, the predicate *leaf*(v) holds iff v is a leaf.

We can now present the rule:

Let T be a Σ-marked tree, and let $\varphi \in L^-$ be a formula over Σ. To prove $avoid(T, \varphi)$, choose for each $v \in T$ a well founded set $(W_v, <_v)$ and a predicate P_v: $W_v \times T_v \rightarrow \{T, F\}$ such that the following conditions hold:

(1) There exists $w \in W_v$ such that $P_v(w,v)$ holds.

(2) For all $u \in T_v$ and for all $w \in W_v$,

(a) $P_v(w,u) \wedge Q_\varphi(\pi_v, \pi_{v,u}) \rightarrow \forall s \in children(u) \exists w' < w: P_v(w', s)$.

(b) $P_v(w,u) \rightarrow \forall s \in children(u) \exists w' \leq w: P_v(w', s)$.

(3) For all $u \in T_v$, $leaf(u) \leftrightarrow P_v(0_v, u)$.

A proof rule for proving that T is φ*–avoiding*.

Due to technical reasons, we have chosen to use the parametrized predicate P instead of a variant function assigning values to nodes. Clause (1) of the rule assures that each root of a subtree has some value assigned to it. Clause (2)(a) assures that whenever Q induces decrement in a node u of a subtree rooted at v, then all of u's children are assigned strictly lower values. Clause (2)(b) assures that the values are nonincreasing. Clause (3) is included due to technical reasons, which will be effective when the rule is applied to programs. Its meaning is that a node is assigned 0_v iff it is a leaf. Here 0_v is a generic name for the minimal elements of W_v.

Note that since T and Q are recursive, then there is an algorithm which decides, given a node in the tree, whether its children should decrease in rank.

Example: Consider the tree T shown in figure 2(a), where all nodes but the ones on the "main" path (the infinite one) are marked with a, and the formula $\varphi = \exists^{\infty}_{a}$. Clearly, T is φ-avoiding since its only infinite path is not marked at all. To prove it, choose W_v to be the ordinal $\omega+1$ with the usual $\in$-ordering for all $v \in T$. For a node u in T_v, let $P_v(\omega,u)=T$ iff u is on the main path, and $P_v(n_u,u)=T$ iff u is not on the main path, and n_u is the distance between u to a leaf. The ranking for the root of T is shown in figure 2(b). It is easy to see that conditions (1)-(3) hold with the chosen P_v. Note that this example also demonstrates the fact, that the natural numbers (ω) do not always suffice, even when the tree is of a finite (and even bounded) degree.

4. Soundness and completeness of the rule

Throughout this section we let T be some fixed Σ-marked tree.

Definition: Let π be an infinite path in T, v a vertex on π, and $\varphi \in L^-$ a formula. π *is* φ*–decreasing starting at* v if there is an infinite sequence of nodes on π, $u_1, u_2, u_3 \cdots$ s.t u_1 is a descendent of v , each u_i is a descendent of u_{i-1} , and for all i, $Q_\varphi(\pi_v, \pi_{v,u_i})$ holds. We say that $\Delta(\pi,\varphi,v)$ *holds* to express the above notion. The sequence $u_1, u_2, u_3 \cdots$ is called a *decreasing sequence*.

Lemma 1: Let π be an infinite path in T, φ a formula , and v, v' vertices on π s.t v' is a descendent of v. If $\Delta(\pi,\varphi,v)$ holds then so does $\Delta(\pi,\varphi,v')$.

Proof: By induction on the structure of φ.

Basis:

(a) $\varphi = \exists^{\infty}_{a}$. Let $u_1, u_2 \cdots$ be a decreasing sequence causing $\Delta(\pi, \exists^{\infty}_{a}, v)$ to hold. By the definition of Q , this means that each u_i is marked a. Let j be the minimal

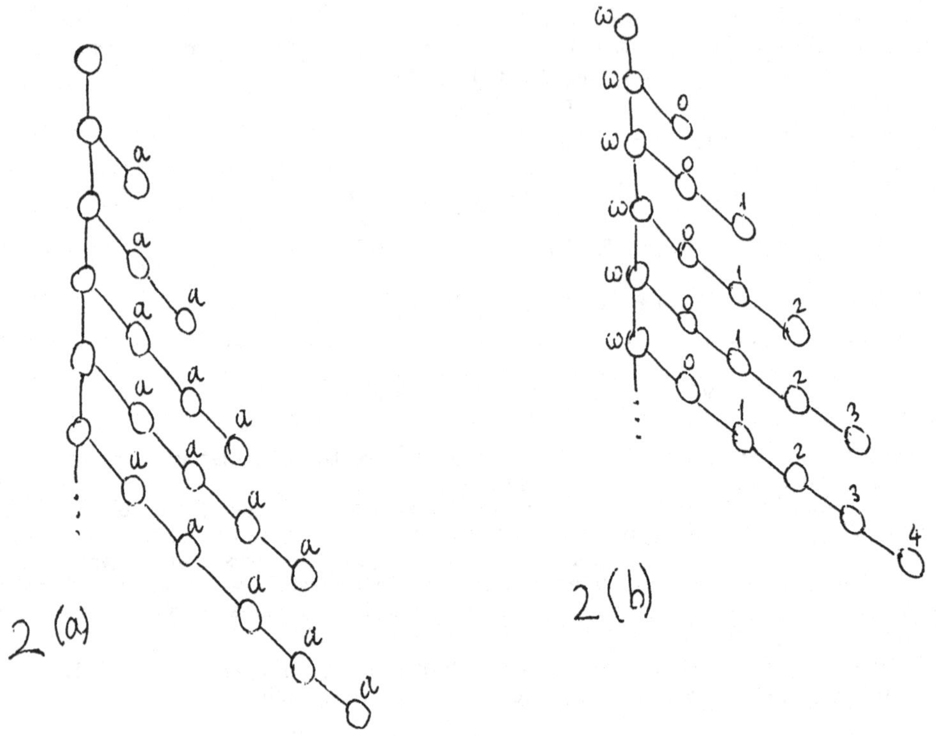

Figure 2. An example of a ranking function.

number s.t u_j is a descendent of v'. Then $Q_{\exists^{\infty}_a}(\pi_{v'}, \pi_{v',u_k})$ holds for all $k \geq j$, and so $\Delta(\pi, \varphi, v')$ holds.

(b) $\varphi = \exists_a$. Let $u_1, u_2, u_3 \cdots$ be as in (a) . Since $Q_{\exists_a}(\pi_v, \pi_{v,u_i})$ holds for all i, then a appears in $\pi_v \pi_{v,u_i}$ for all i. Let j be as in (a). Clearly, a appears in $\pi_{v'} \pi_{v',u_k}$ for all $k \geq j$, and thus $Q_{\exists_a}(\pi_{v'} \pi_{v',u_k})$ holds , and consequently so does $\Delta(\pi, \varphi, v')$.

(c) $\varphi = \forall_a$. The proof is similar to (a) and (b).

(d) $\varphi = \forall^{\infty}_a$. The proof is again similar to (a) and (b) (the fact that $Q_{\forall^{\infty}_a}(\pi_1, \pi_2)$ depends on π_2 only is immaterial).

Induction step:

(a) $\varphi = A \vee B$. Let $u_1, u_2, u_3 \cdots$ be as in the basis proofs. By the definition of Q and w.l.o.g we can assume that there is a sequence $i_1, i_2 \cdots$ s.t $Q_A(\pi_v, \pi_{v,u_{i_j}})$ holds for all j, and so $\Delta(\pi, A, v)$ holds. By the induction hypothesis , so does $\Delta(\pi, A, v')$. By the definition of Q , $\Delta(\pi, \varphi, v')$ holds.

(b) $\varphi = A \wedge B$. Let $u_1, u_2, u_3 \cdots$ be as in (a) .By the definition of Q, each u_i is associated with a pair of nodes s, s' s.t $Q_A(\pi_v, \pi_{v,s})$ and $Q_B(\pi_v, \pi_{v,s'})$ hold (actually, either s or s' must be u_i). Therefore, both $\Delta(\pi, A, v)$ and $\Delta(\pi, B, v)$ hold. By the hypothesis, the same is true for v'. Thus, there are infinitely many pairs of nodes s, s' s.t $Q_A(\pi_{v'}, \pi_{v',s})$ and $Q_B(\pi_{v'}, \pi_{v',s'})$ hold, and consequently so does $\Delta(\pi, A \wedge B, v')$.

□

Lemma 2: Let π be an infinite path in T and let φ be a formula. π satisfies φ *iff* there is a vertex v on π s.t $\Delta(\pi,\varphi,v)$ holds.

Proof: Assume that π satisfies φ. The proof is by induction on the structure of φ.

Basis:

(a) $\varphi=\exists^{\infty}{}_a$. Choose v to be the root (the first vertex of π).

(b) $\varphi=\exists_a$. Choose v to be the root.

(c) $\varphi=\forall_a$. Choose v to be the root.

(d) $\varphi=\forall^{\infty}{}_a$. Choose v to be the first vertex on π from which all vertices are marked a.

Induction step:

(a) $\varphi=A \vee B$. Assume w.l.o.g that π satisfies A. By the hypothesis, let v be a vertex s.t $\Delta(\pi,A,v)$ holds. By the definition of Q, so does $\Delta(\pi,A \vee B,v)$.

(b) $\varphi=A \wedge B$. Let v_A,v_B be vertices in π s.t $\Delta(\pi,A,v_A)$ and $\Delta(\pi,B,v_B)$ hold (by the hypothesis). Assume w.l.o.g that v_B is a descendent of v_A. By lemma 1, $\Delta(\pi,A,v_B)$ holds, and so, by Q's definition , so does $\Delta(\pi,A \wedge B,v_B)$. Thus, choose v to be v_B.

For the other direction, assume that $\Delta(\pi,\varphi,v)$ holds for some v in π. The proof is again by structural induction on φ.

Basis:

(a) $\varphi=\exists^{\infty}{}_a$. By the assumption, there is a sequence $u_1,u_2,u_3 \cdots$ s.t $Q_{\exists^{\infty}_a}(\pi_v,\pi_{v,u_i})$ holds for all i. This implies that each u_i is marked a, and thus π satisfies φ.

(b) $\varphi=\exists_a$. Let $u_1,u_2,u_3 \cdots$ be as in (a) .By definition, a appears in $\pi_v\pi_{v,u_1}$, and so π satisfies φ.

(c) $\varphi=\forall_a$. Let $u_1,u_2,u_3 \cdots$ be as in (a) . We have to show that each node in π is marked a. Let u be such a node. Let u_j be some vertex from the decreasing sequence appearing after u (i.e u_j is a descendent of u). Since $Q_{\forall_a}(\pi_v,\pi_{v,u_j})$ holds, we have that a appears in *all* $\pi_v\pi_{v,u_j}=\pi_{u_j}$. Since u is a node on π_{u_j}, we have that u is marked a.

(d) $\varphi=\forall^{\infty}{}_a$. By an argument similar to the previous one, we have that starting at v, π is everywhere marked a, and therefore π satisfies φ.

Induction step:

(a) $\varphi=A \vee B$. By the assumption, by Q's definition and w.l.o.g we can assume that $\Delta(\pi,A,v)$ holds. By the induction hypothesis, π satisfies A and thus also satisfies $A \vee B$.

(b) $\varphi=A \wedge B$. By Q's definition both $\Delta(\pi,A,v)$ and $\Delta(\pi,B,v)$ hold. By the hypothesis π satisfies A and satisfies B, thus also $A \wedge B$.

□

Theorem 1 (soundness of the proof rule): Let $\varphi \in L^-$ be a formula. If for all $v \in T$ there is a well founded set $(W_v,<_v)$ and a predicate $P_v: W_v \times T_v \rightarrow \{T, F\}$ s.t (1),(2) and (3) are satisfied then $avoid(T,\varphi)$ holds.

Proof: Assume that such W_v and P_v exist for each $v \in T$, and Assume , by way of contradiction, that there is an infinite path π in T that satisfies φ. By lemma 2, there is a vertex v on π s.t $\Delta(\pi,\varphi,v)$ holds. Let $u_1,u_2,u_3 \cdots$ be a decreasing sequence for which the above is true. By (1), there exists $w \in W_v$ s.t $P_v(w,v)$ holds. By (2) and (3) we have that if $P_v(w,u_i)$ holds for some w and i, then there exists $w' < w$ s.t $P_v(w',u_{i+1})$ holds, and this fact is still valid taking v as u_0. Thus, for each u_i,u_j s.t $i>j$ there are w_i,w_j s.t $w_i<w_j$ and $P_v(u_i,w_i)$, $P_v(u_j,w_j)$ hold. The sequence $w_1,w_2 \cdots$ is ,therefore, an infinite decreasing sequence of elements from W_v - a contradiction to the well foundness of W_v.

□

Theorem 2 (completeness of the proof rule): Let $\varphi \in L^-$ be a formula. If $avoid(T,\varphi)$ holds then For each $v \in T$ there exist $(W_v,<_v)$ and $P_v: W_v \times T_v \rightarrow \{T, F\}$ satisfying (1),(2) and (3) .

Proof: Assume that $avoid(T,\varphi)$ holds. We have to find appropriate W_v and P_v for each $v \in T$.

Let $v \in T$ be a vertex. Call a vertex u in T_v *decreasing* if $Q_\varphi(\pi_v,\pi_{v,u})$ holds and *steady* otherwise. From lemma 2 it follows that in T_v there is no infinite path with infinitely many decreasing nodes (since this would imply that this path satisfies φ , and thus $avoid(T,\varphi)$ would not hold).

Definition:

1. Let u be a vertex in T_v. *cone*(u) is the subtree of T_v rooted in u, truncated after each decreasing node, i.e *cone*(u) includes all path-fragments starting at u up to, and *including*, a decreasing node. If a fragment, finite or infinite, does not contain any decreasing nodes, then it is wholly contained in the cone. Note that u itself is always contained in *cone*(u) , thus a cone is never empty.
2. An *exit node* from a cone is a child of a decreasing node, after which the cone is truncated.

The above definition is illustrated in figure 3. Decreasing nodes are darkened. u_1,u_2,u_3,u_4 and u_5 are exit nodes from *cone*(v).

Now, inductively construct from T_v the tree T_v*, having cones as nodes as follows :

* The root of T_v* is *cone*(v).

* Assume that *cone*(u) is a node in T_v* for some $u \in T_v$.

- If *cone*(u) has no exit nodes then it is a leaf in T_v*.
- If *cone*(u) has exit nodes $u_1,u_2,u_3 \cdots$ (possibly countably many) then each *cone*(u_i) is a child of *cone*(u) in T_v*.

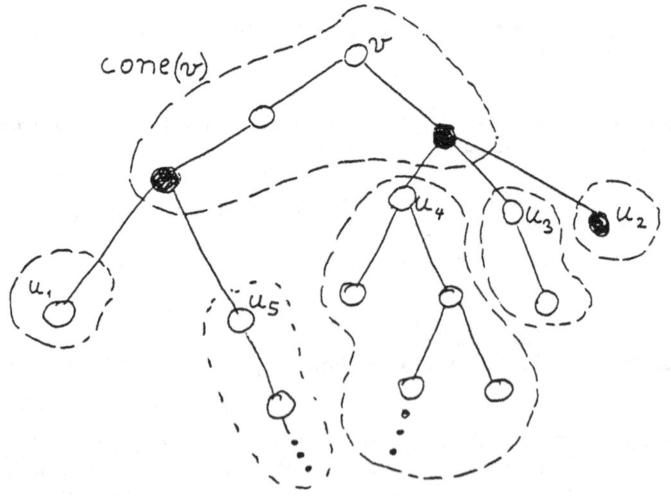

Figure 3. Cones and exit nodes.

Observations:

1. If $cone(u')$ is a child of $cone(u)$ then there is a path-fragment in T_v from a decreasing node in $cone(u)$ to u'.
2. From 1 it follows that $T_v{}^*$ is a tree (i.e it is acyclic) since a cycle in $T_v{}^*$ would imply one in T_v.
3. From 1 it also follows, that there is no infinite path in $T_v{}^*$, since this would imply an infinite sequence of decreasing nodes in T_v.
4. Each node in T_v is included in exactly one cone in $T_v{}^*$.

Now, since $T_v{}^*$ is well founded, there is a well founded set W_v that can rank $T_v{}^*$ in such a way that a parent has always a greater rank then each of its children. Also, w.l.o.g , we can assume that no vertex in $T_v{}^*$ is ranked 0_v (the minimal element of W_v). Denote the ranking of a vertex u in $T_v{}^*$ by $rank(u)$. We can now find $P_v: W_v \times T_v \to \{T, F\}$ as required. For all $u \in T_v$ and $w \in W_v$ we define

$P_v(w,u)=T \leftrightarrow (leaf(u) \text{ and } w=0_v) \text{ or } (rank(c[u])=w)$
where $c[u]$ is the cone in $T_v{}^*$ to which u belongs.

We now have to verify that (1), (2) and (3) hold with W_v and P_v :

(1) $P_v(rank(cone(v)),v)=T$.

(2) Let $u \in T_v$ and $w \in W_v$.

(a) $P_v(w,u) \wedge Q_\varphi(\pi_v,\pi_{v,u})$ means that u is decreasing and either $rank(c[u])=w$, or u is a leaf. If u is a leaf then the condition holds vacuously. Otherwise, each child u' of u is an exit node from $c[u]$, and therefore belongs to another cone, $cone(u')$, which is a child of $c[u]$. By the main characteristic of the ranking function , it follows that $rank(cone(u')) < rank(c[u])=w$, and thus $w'=rank(cone(u'))$ satisfies the requirements in case u' is not a leaf. If u' is a leaf, then $w'=0_v$ will do.

(b) $P_v(w,u)$ means that $rank(c[u])=w$, or u is a leaf. The leaf case leads again to vacuous satisfaction. Otherwise, if u is decreasing then the case is as in (a). If not, then each child of u is still included in $c[u]$, and therefore is either a leaf and is ranked 0_v, or has the same rank as u. At any case, the requirements are satisfied.

3. If u is not a leaf , then by the assumption on W_v(that no node in $T_v{}^*$ is ranked 0_v) we have that $P_v(0_v,u)$ does not hold.
If u is a leaf then by definition $P_v(0_v,u)$ holds.

□

5. Justification of the proof rule

As mentioned before, the suggested rule might seem more complicated then necessary, since it requires finding a well founded set for each $v \in T$. Moreover, we are ed to define Q having two arguments instead of one. It would be more natural if we could define a predicate $R_{\Sigma,\varphi}:(2^\Sigma)^* \to \{T, F\}$ in some way, adhering to the following proof rule:

Alternative proof rule 1: To prove $avoid(T,\varphi)$, find a well founded set $(W,<)$ and a predicate $P: W \times T \to \{T, F\}$ s.t the following conditions hold:

(1) There exists $w \in W$ such that $P(w,root)$ holds.

(2) For all $v \in T$ and for all $w \in W$,

(a) $P(w,v) \wedge R_\varphi(\pi_v) \to \forall s \in children(v) \exists w'<w: P(w',s)$.

(b) $P(w,v) \to \forall s \in children(v) \exists w' \leq w: P(w',s)$.

(condition (3) (the leaf condition) is omitted here since it was introduced for technical reasons only).

Unfortunately, however natural this rule is, it is impossible, since an appropriate predicate R cannot be found. We now prove that.

Theorem 3: There exists no recursive predicate R, with which the alternative proof rule is sound and complete.

Proof: By way of contradiction, let R be such a predicate. In order to proceed we need the following lemma:

Lemma 3: Let T be a tree composed of one path only, $v_1, v_2, v_3 \cdots$ where v_i is the descendent of v_{i-1} ,and let φ be a formula. T satisfies φ *iff* there are infinitely many vertices v on T for which $R_\varphi(\pi_v)$ holds (we refer to such nodes as *decreasing with respect to* φ).

Proof (of lemma 3): Assume that T satisfies φ, and assume, by way of contradiction, that there are only finitely many decreasing nodes on T. Let k be the maximal index s.t v_k is decreasing. Choose the well founded set $W=(\{0,1 \ldots k\}, \leq)$ and the predicate $P: W \times T \rightarrow \{T,F\}$ defined as follows:

$$P(w,v_i)=T \leftrightarrow (\, w=k+1-i \text{ and } 1 \leq i \leq k) \text{ or } (w=0 \text{ and } i>k).$$

It is easy to see that conditions (1) and (2) hold with W and P, which is a contradiction to the soundness of the rule.

Assume now, that there are infinitely many decreasing nodes in T, $v_{i_1}, v_{i_2} \cdots$. We show that there exist no W and P which satisfy the conditions of the proof rule. By contradiction, assume their existence. From conditions (1) and (2) it follows, that for all $j \geq 1$ there exists w_j s.t $P(w_j, v_{i_j})$ holds, and $w_j > w_{j+1}$. Therefore, the sequence $w_1, w_2 \cdots$ is an infinite decreasing sequence of elements from W - a contradiction. From the completeness assumption we have that $avoid(T,\varphi)$ does not hold, hence T satisfies φ.

□

We now proceed with the proof of theorem 3. Let T_0 be the following tree:

```
     a    a    a    a    a
T0:  O----O----O----O----O----....
```

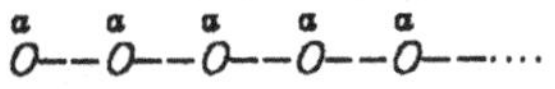

i.e a path everywhere marked by a. T_0 satisfies $\forall^{\infty}_a$, and so, by lemma 3, there are infinitely many decreasing nodes nodes on T with respect to $\forall^{\infty}_a$ (i.e there are infinitely many nodes v, for which $R_{\forall^{\infty}_a}(\pi_v)$ holds). Let v_1 be the first such node (the closest to the root), and let u_1 be its successor. The tree T_1 is the following:

```
     a    a    a         a    a
T1:  O----O----O----O----O----O----....
               v1   u1
```

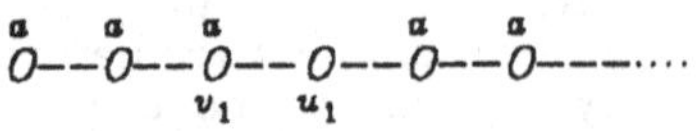

i.e, T_1 has *the same vertices as* T_0, only they are marked differently, and the difference is that u_1 has no mark. From the above, we have that v_1 is decreasing with respect to $\forall^{\infty}_a$. Also, T_1 satisfies $\forall^{\infty}_a$ and thus, by lemma 3, there are infinitely many decreasing nodes in it. Let v_2 be the first such node *after* u_1, and let u_2 be its successor. The tree T_2 is the following:

```
     a    a    a         a    a    a         a    a
T2:  O----O----O----O----O----O----O----O----O----O----....
               v1   u1             v2   u2
```

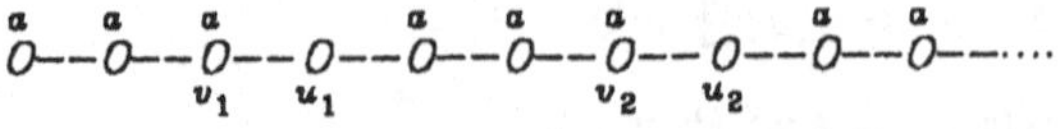

Again, we have that v_1 and v_2 are decreasing with respect to $\forall^\infty_a$, and also T_2 satisfies $\forall^\infty_a$. Now, continue to define T_i for all $i \geq 0$ in the same way, so that in each T_i there are exactly i *blanks* (nonmarked vertices), and $v_1, v_2 \cdots v_i$ are decreasing with respect to $\forall^\infty_a$.

Now, define T_∞ to be the tree having the same nodes as all the T_i, with all of them marked a, except the u_j, which are left blank. T_∞ has the following form:

$$T_\infty:\ \begin{array}{cccccccccccccc} a & a & a & & a & a & a & & a & a & & a & & a & a \\ O-- & O-- & O-- & O-- & O-- & O-- & O-- & O-- & O-- & O--\cdots & \cdots--O-- & O-- & O-- & O--\cdots \\ & & v_1 & u_1 & & & v_2 & u_2 & & & v_i & u_i & & \end{array}$$

By the definition of T_∞ we have that each v_i is decreasing with respect to $\forall^\infty_a$, and hence, by lemma 3, T_∞ satisfies $\forall^\infty_a$, which is not true, since it has infinitely many blanks (all the u_j) - a contradiction. This completes the proof of theorem 3.

Remark: T_∞ is a recursive tree since R is recursive. If we do not restrict the discussion to recursive trees only, then the recursiveness assumption on R is superfluous.

□

Although alternative proof rule 1 is impossible, one might think of another possibility, which still seems more natural then the suggested rule. Up to now, there has always been a predicate, which tells, given a vertex in a marked tree, whether or not its successors should decrease in rank. In case of a negative answer, its successors should have remained steady. However, it is also possible to allow increment of the rank, i.e, instead of a predicate, we could have a function $f_{\Sigma,\varphi}:(2^\Sigma)^* \to \{d,s,i\}$, where d,s and i mean *decrease*, *steady* and *increase* respectively, and then suggest the following proof rule:

Alternative proof rule 2: To prove $avoid(T,\varphi)$, find a well founded set $(W,<)$ and a predicate $P: W \times T \to \{T, F\}$ s.t the following conditions hold:

1) There exists $w \in W$ such that $P(w,root)$ holds.
2) For all $v \in T$ and for all $w \in W$,

   (a) $P(w,v) \wedge f_\varphi(\pi_v)=d \to \forall s \in children(v) \exists w'<w: P(w',s)$.

   (b) $P(w,v) \wedge f_\varphi(\pi_v)=s \to \forall s \in children(v) \exists w' \leq w: P(w',s)$.

   (c) $P(w,v) \wedge f_\varphi(\pi_v)=i \to \forall s \in children(v) \exists w': P(w',s)$.

In this rule, if f returns i for some node, then its successors can have any ranks, including higher ones. At first sight, this rule seems to work perfectly, since the "problematic" formula $\forall^\infty_a$ is no longer such if we define:

$$f_{\forall^\infty_a}(\pi) = \begin{cases} d & \text{if } a \text{ appears in the last element of } \pi \\ i & \text{else} \end{cases}$$

It can be easily verified, that alternative rule 2 is sound and complete for proving $avoid(T, \forall^\infty_a)$, if we adhere to the suggested f. We could also define f for all the other atomic formulas, in a way similar to the original proof rule (this is immediate, since in all the atomic formulas except $\forall^\infty_a$, the two arguments π_1 and π_2 are concatenated, and so we relate to the fragment from the root to the vertex only, and so we do not need two arguments). It therefore turns out, that one could come with a sound and complete proof rule of the last form for each atomic formula. It also turns out, that it is possible to define f for a conjunction of two formulas. The problem arises when we try to bring in disjunctions. We now prove that it is impossible to find such an f to solve the general case.

Theorem 4: There exists no recursive f with which alternative proof rule 2 is sound and complete.

Proof: By way of contradiction, let f be such a function. We refer to nodes v, for which $f_\varphi(\pi_v)=d$ as *decreasing with respect to φ*, and *steady* and *increasing* is used similarly. If φ is understood from the context we omit the "with respect to φ". We need two lemmas for the proof.

Lemma 4: Let T be a tree composed from one path only, $v_1,v_2,v_3\cdots$,and let φ be a formula. If T satisfies φ then there are infinitely many decreasing nodes in T.

Proof (of lemma 4): Similar to the proof of lemma 3. The fact that ranks can increase now is immaterial.

□

Lemma 5: Let T and φ be as in lemma 4. If there are infinitely many increasing nodes in T, then $avoid(T,\varphi)$ holds.

Proof (of lemma 5): Assume there are infinitely many increasing nodes in T. For each $k\geq 1$ let $n_k\geq 1$ be the minimal integer s.t v_{k+n_k} is a increasing vertex. i.e, for a vertex v_k, v_{k+n_k} is its first increasing descendent. Since there are infinitely many increasing nodes, n_k is defined for all k. Now, define $W=(N,\leq)$ and a predicate $P:W\times T\to\{T,F\}$ as follows:

$$P(w,v_k)=T \leftrightarrow (k=1 \text{ and } w=n_1+1) \text{ or } (k\geq 2 \text{ and } w=n_{k-1}).$$

It is easy to see that conditions (1) and (2) hold with W and P, and so, by the soundness of the rule, $avoid(T,\varphi)$ holds.

□

We now proceed with the proof of theorem 4. Let T_0 be as in the proof of theorem 1:

$$T_0:\ \overset{a}{0}\text{--}\overset{a}{0}\text{--}\overset{a}{0}\text{--}\overset{a}{0}\text{--}\overset{a}{0}\text{--}\cdots$$

i.e a path everywhere marked by a. T_0 satisfies $(\forall^\infty_a \vee \exists^\infty_b)$, and so, by lemma 4, there are infinitely many decreasing nodes in it (with respect to $(\forall^\infty_a \vee \exists^\infty_b)$). Let $v_{0,1}$ be the first such node, and let $u_{0,1}$ be its successor. We now define a few more trees, all of which have the same nodes as T_0, but are different in their markings. Let $T_{0,1}$ be the following:

$$T_{0,1}:\ \overset{a}{0}\text{--}\overset{a}{0}\text{--}\underset{v_{0,1}}{\overset{a}{0}}\text{ -- }\underset{u_{0,1}}{0}\text{ --}\overset{a}{0}\text{--}\overset{a}{0}\text{--}\cdots$$

i.e all nodes are marked a except $u_{0,1}$, which is blank. We know, that $v_{0,1}$ is decreasing with respect to $(\forall^\infty_a \vee \exists^\infty_b)$. Again, we have that $T_{0,1}$ satisfies $(\forall^\infty_a \vee \exists^\infty_b)$, and thus, by lemma 4, let $v_{0,2}$ be the first decreasing node after $u_{0,1}$. The tree $T_{0,2}$ is the following:

$$T_{0,2}:\ \overset{a}{0}\text{--}\overset{a}{0}\text{--}\underset{v_{0,1}}{\overset{a}{0}}\text{ -- }\underset{u_{0,1}}{0}\text{ --}\overset{a}{0}\text{--}\overset{a}{0}\text{--}\underset{v_{0,2}}{\overset{a}{0}}\text{ -- }\underset{u_{0,2}}{0}\text{ --}\overset{a}{0}\text{--}\overset{a}{0}\text{--}\cdots$$

i.e all nodes are marked a except $u_{0,1}, u_{0,2}$, which are left blank. Now continue and define $T_{0,i}$ for all i in the same way. In $T_{0,i}$ there are exactly i blanks (all the $u_{0,s}$ for all $s\leq i$), and $v_{0,1}\cdots v_{0,i}$ are decreasing.

Now, define $T_{0,\infty}$ to be the tree with nodes exactly as in all the $T_{0,i}$, all of them narked a, except all the $u_{0,i}$, which are left blank ($T_{0,\infty}$ relates to the $T_{0,i}$ exactly as T_∞ related to the T_i in the proof of theorem 1). In $T_{0,\infty}$ there are infinitely many blank nodes. Also there are infinitely many decreasing nodes with respect to $(\forall^\infty_a \vee \exists^\infty_b)$. Thus, there must be infinitely many increasing nodes in $T_{0,\infty}$, or else we would conclude, by a slight variation of lemma 3, that $T_{0,\infty}$ satisfies $(\forall^\infty_a \vee \exists^\infty_b)$, which is untrue. Let v_1 be the first increasing vertex, and let u_1 be its successor. T_1 is the following tree:

```
      ?   ?   ?   b   a   a
T1:   O---O---O---O---O---O----....
              v1  u1
```

.e all nodes up to, and including v_1 are marked *as in* $T_{0,\infty}$ (their marks are represented by an ?), u_1 is marked b, and all other nodes are marked a. v_1 is increasing with respect to $(\forall^\infty_a \vee \exists^\infty_b)$. Now, since T_1 satisfies $(\forall^\infty_a \vee \exists^\infty_b)$, there are, by emma 4, infinitely many decreasing nodes in it. Let $v_{1,1}$ be the first decreasing node after u_1, and let $u_{1,1}$ be its successor. $T_{1,1}$ is the following tree:

```
        ?   ?   ?   b   a   a   a            a   a
T1,1:   O---O---O---O---O---O---O ---  O ---O---O-----....
                v1  u1          v1,1   u1,1
```

.e $u_{1,1}$ is blanked, and all nodes following it are marked a. All nodes up to $v_{1,1}$ are marked as in T_1. Again, $v_{1,1}$ is decreasing, and $T_{1,1}$ satisfies $(\forall^\infty_a \vee \exists^\infty_b)$. Continue and define $T_{1,i}$ for all i analogously to the definition of $T_{0,i}$ before, so that each $T_{1,i}$ has exactly i blanks *after* u_1 (all the $u_{1,i}$), and $v_{1,1} \cdots v_{1,i}$ are decreasing.

Now, define $T_{1,\infty}$ from the $T_{1,i}$ analogously to the definition of $T_{0,\infty}$ from the $T_{0,i}$. In $T_{1,\infty}$ there are infinitely many decreasing nodes, and thus by an argument already used once, there are infinitely many increasing ones. Let v_2 be the first increasing vertex after u_1, and let u_2 be its successor. Define T_2 as follows:

```
      ?   ?   ?   b   ?   ?   ?   b   a   a
T2:   O---O---O---O---O---O---O---O---O---O----....
              v1  u1          v2  u2
```

.e all nodes up to, and including v_2 are marked as in $T_{1,\infty}$, u_2 is marked b, and all other nodes are marked a. Note that in T_2 there are exactly two nodes marked b (u_1, u_2).

Continue and define T_i for all i in the same way. Each T_i has exactly i nodes marked b ($u_1 \cdots u_i$), and has at least i increasing nodes ($v_1 \cdots v_i$). Define T_∞ as the tree having the same nodes as all the T_i, and marks which are obtained from them in the usual way (more formally, a vertex v is marked as in T_k , where k is the minimal integer s.t v_k is a descendent of v). In T_∞ there are infinitely many increasing nodes (the v_i), and so by lemma 5, $avoid(T_\infty, (\forall^\infty_a \vee \exists^\infty_b))$ holds, but then, there are infinitely many nodes marked b, which leads to the satisfaction of $(\forall^\infty_a \vee \exists^\infty_b)$ - a contradiction.

Remark: Again, T_∞ is recursive since f is such, and this assumption on f can be omitted if we do not restrict the discussion to recursive trees.

□

5. Conclusion

We have presented a necessary and sufficient condition for a marked tree to have no infinite paths satisfying a given formula, taken from a rather expressive language. Two other simpler forms of conditions were also discussed and were shown to be impossible. The completeness proof is a generalization of the "cone construction" [GFMR 81, ' 85].

Considering the case of fairness, we observe that the existing proof rules for termination under this assumption are close in spirit to the simpler conditions, those which were shown here to be inadequate. They are close in the sense that only *one* well founded set has to be found in order to complete the proof, rather then one for *each subtree*, as it is here. The ability to use a simpler rule in the fairness case stems from the fact, that it is a *very* special case of the general one, both in the markings allowed and in the formula used (see [Ha 84] for the translation of fairness to trees and formulas).

As mentioned before, we omitted the discussion of infinite conjunctions and disjunctions. It ,therefore, remains to close this gap. We conjecture that the kind of condition discussed here *cannot* be applied to the infinite formulas of L; We have, however, no proof of this. One natural attempt yeilds a nonrecursive tree as a counter example.

A most natural extension of the work is to apply the condition to programs, i.e to provide a syntactic proof rule for proving that a program has no infinite computations satisfying a given formula from L. A first stage of such a work would be to properly interpret markings and formulas when relating to programs.

Another intriguing direction of research could be characterizing the φ-*avoidance* of a marked tree in terms of convergent sequences of computation elements in appropriate metric spaces. This approach is discussed in [DM 84] for the case of fairness and three other liveness properties. Relating to fairness, it is shown there how to define a distance between computation elements of a program in such a way, that a given computation is fair iff it is a Cauchy sequence. This approach is further developed in [C 84], where Milner's *CCS* is used as the computational model. It would be of interest to define, for each $\varphi \in L$ a distance function d_φ in such a way, that a path in the tree satisfies φ iff it forms a Cauchy sequence (with respect to d_φ).

Acknowledgments

We acknowledge J.A.Makowsky's remarks, suggesting the need for this kind of investigation.

The part of the second author was partially supported by the fund for the promotion of research, the Technion.

REFERENCES

[AO 83] K.R. Apt, E.R. Olderog: "Proof rules and transformations dealing with fairness", SCP 3: 65-100, 1983.

[APS 84] K.R. Apt, A. Pnueli, J. Stavi: "Fair termination revisited with delay", TCS 33: 65-84, 1984. Also in: proc. 2nd Conference on foundations of software technology and theoretical computer science (FST-TCS), Bangalore, India, December 1982.

[C 84] G. Costa: "A metric characterization of fair CCS", TR CSR-169-84, dept. of comp. sci., Edinburgh university, October 1984.

[DH 85] I. Dayan, D. Harel: "Fair termination with cruel schedulers", manuscript, Weizman institute, June 1985.

[DM 84] P. Degano, U. Montanati: "Liveness properties as convergence in metric spaces. Proc. of 16th ACM STOC, 1984.

[F 85] N. Francez:"**Fairness**", in press 1985, Springer Verlag, Texts and monographs in computer science series (D. Gries- series editor).

[FK 84] N. Francez, D. Kozen: "Generalized fair termination", proc. 11th ACM-POPL , Salt Lake City, January 1984.

[GFK 83] O. Grumberg, N. Francez, S. Katz: "A complete proof rule for strong equifairness", proc. 2nd workshop on logics of programs, CMU, June 1983. In: LNCS 164 (E.Clarke, D.Kozen - eds.), Springer Verlag, 1983. To appear in JCSS.

[GFMR 81] O. Grumberg, N. Francez, J.A. Makowsky, W.P. de Roever: "A proof rule for fair termination of guarded commands", proc. of the Int. Symp. on Algorithmic languages, Amsterdam, October 1981, North-Holland, 1981. To appear in Information and Control.

[HA 84] D. Harel: "A general result on infinite trees and its applications" , proc. 16th ACM-STOC, May 1984.

[LPS 81] D. Lehmann, A. Pnueli, J. Stavi: "Impartiality, justice and fairness: the ethics of concurrent termination", proc. 8th ICALP, Acre, Israel, July 1981. In: LNCS 115 (O. Kariv, S. Even - eds.), Springer-Verlag, 1981.

[OA 84] E.R. Olderog, K.R. Apt: "Transformations realizing fairness assumptions for parallel programs", TR 84-8, LITP, University of Paris 7, February 1984.

[PN 83] A. Pnueli: "On the extremely fair treatment of probabilistic algorithms", proc. 15th ACM-STOC, Boston, April 1983.

Computable Directory Queries.

(Extended abstract)

E. Dahlhaus[1]
Department of Mathematics (FB 20)
Technical University Berlin (West Berlin)

and

J.A. Makowsky[2]
Department of Computer Science, Technion, Haifa, Israel

Abstract: We generalize relational data bases such as to include also directories of relations and directories of directories. In this framework we study computable directory transformations which generalize the computable queries introduced by A. Chandra and D. Harel. We introduce a transformation language **DL** and show its completeness. The language *DL* can serve as a basis for specification and correctness of directory transformations and also as a basis to study their complexity. We also introduce a sublanguage DL_0 of DL which is of equal expressive power but has an independent set of basic constructs. Our approach can be seen also in a broader context: It allows the general manipulation of "objects" (as in SMALLTALK or SETL) and adds to it a construct for parallelism (as in VAL). In this sense our results can be viewed as contributing to the mathematical foundation of object oriented programming.

---

[1] During the work on this paper visiting (Spring 1985) at the Department of Computer Science, Technion, Haifa, Israel, sponsored by Minerva Foundation.

[2] During work on this paper visiting at the Institut fuer Informatik (Winter 1984/85) and the Forschungsinstitut fuer Mathematik (Summer 1985), ETH-Zuerich, Switzerland.

1. Introduction

The relational model for data bases was introduced as a means to describe an appropriate user interface. It served to give semantics to concepts from data bases without taking into account the way the data basis was represented in a computer. The relational model was extremely successful (cf. [U82, M83]). When dealing with a file/directory system as well as a data basis the question arises if one can describe the resulting user interface in a similar way.

In the present paper we attempt to extend the relational model for data bases to allow directories. Directories are sets of (sets of sets of ...) relations, or, in the terminology of logic, higher order relations. The formal definition of this extension of the relational model is presented in section 3. In [CH80] queries are (partial) functions mapping finite sequences of relations (the data base state) into a new relation (the answer to the query). In their framework it is not possible to express what is a restructuring of a data basis or to deal with hierarchies of relations. In our model the analogue of a query is a directory transformation which maps directories into directories. Queries will be special cases of directory transformations. Directory transformations will be called *directory queries*. Other special cases are directory manipulation programs such as *tar* in *UNIX*, system programs reorganizing the division of a disk, or any other restructuring of entire data base systems.

Programming languages which manipulate higher order relations have been considered in various other contexts before. Mostly, the motivation behind such set oriented languages stems from the need to implement readily arbitrary, abstractly defined data structures. The purpose of very high level languages is to " provide high level abstract objects and operations between them, high level control structures and the ability to select data representation in an easy and flexible manner" [SSS79]. The most prominent example is **SETL** introduced by J. Schwartz [Sch75].

In some sense also "object oriented" programming can be viewed as set oriented. But, in contrast to set oriented programming, "objects" have two components: the static component (the object) and the dynamic component (the message passing mechanism). A prominent example of an object oriented programming language (or better environment) is **Smalltalk** [GR83]. A short introduction into **Smalltalk** can also be found in [Ho83]. The latter is also a good reference for concepts and implementations of programming languages in general.

Our paper can also be viewed as a contribution to the theoretical foundations of set oriented and (the static aspects of) object oriented programming.

The main problem we address in this paper is that of defining precisely the semantic notion of a *computable directory query* extending naturally the notion of computable queries. This is the content of section 3 and 4. In section 5 we give some examples of computable directory queries. With such a definition one can now define the semantics of various directory query languages.

In section 6 we define the syntax and in section 7 the semantics of a directory query language DL. DL is an extension of QL of [CH80] with various directory handling constructs. They correspond to the set theoretic operations union, complement, power set, singleton set and the replacement and induction principle. The induction principle also occurs in QL in the form of the **while**-construct. The replacement principle leads to a new programming construct **mkdir** y_i **from** y_j **in** $\sim y_k$ **by** P. This construct is very much in the spirit of parallel programming or of data flow languages. It is similar to the **for all** construct of *VAL* (cf. [Ho83]). It replaces the subdirectories of y_k simultaneously and puts them into the directory y_i. The construct also allows parallel query processing to be expressible in DL. As mentioned before, the programming language DL turns out to be an abstract and well defined sublanguage of **SETL** which is equivalent to **SETL** both in computing power and flexibility.

In section 8 we present the main results which include the *completeness* of DL and the *independence* of constructs of a complete sublanguage DL_0 of DL. Here we call a language L *complete* iff for every computable directory query there is a program in L corresponding to it, and we call the constructs of a language L to be *independent* iff the completeness fails if one of the constructs of L vanishes. In section 9 we give a brief comment on the basic idea of the proofs, and in section 10 we discuss possible extensions of our framework. The full proofs appear in [DM85].

The second author is indebted to M.Magidor and J.Stavi for valuable discussions.

2. Conclusions.

We see the main merits of this paper in the precise definition of the semantics of set oriented programming languages. Traditionally, in set theory, all mathematical objects are built from the empty set alone, though the use of urelements (elements which are not sets, i.e. which do not have elements themselves) was never completely rejected. In [Ba75] it was actually argued that avoiding urelements results in a conceptual loss. Our semantics is based on a set theory of hereditarily finite sets with urelements, which allow us to make the concept of user interface invariance (isomorphism invariance) precise. Our two main theorems (the completeness of DL and the independence of the constructs of DL_0) just illustrate that the chosen framework for our semantics is correct.

In this paper we were motivated by the analogy

file = relation

directory = collection of files and subdirectories.

In [DM85a] we present a similar framework, but properly based on sets with urelements. There we also characterize a programming language in which exactly the **NP**-recognizable objects can be created. It is straight forward to construct a similar sublanguage of DL, and also other sublanguages capturing complexity classes, as in [Im83].

3. The semantic model.

The purpose of this section is to define data bases of higher order. The traditional relational data bases are then first order data bases containing only relations. Higher order relational data bases also contain finite sets of finite relations which are called simple directories. More complicated directories can be formed by allowing directories to contain finite sets of both relations and directories of lower order. Relations are just structured files.

More formally, we start our definition as in [CH80]. Let U denote a fixed countable set, called the universal domain. Let $D \subset U$ be finite and nonempty, and let $R_1,\ldots,R_k$ for $k>0$, be relations such that for, for all i, $R_i \subset D^{a_i}$. $B=(D,R_1,\cdots,R_k)$ is called a ***relational first order data base of type*** a, where $a=(a_1,\cdots,a_k)$. R_i is said to be of rank a_i. We shall also call the relations ***directories of order 1***.
Let $V_1(D)$ be the set of all directories of order 1.

$$V_1(D)=\bigcup_{i\in\mathbf{N}} P_{fin}(D^i)$$

where $P_{fin}(X)$ denotes the set of all finite subsets of X.
$V_{j+1}(D)=V_j(D)\cup P_{fin}(V_j(D))$ and $V(D)=\bigcup_{j\in\mathbf{N}} V_j(D)$. $V(D)$ is the set of all directories and $V_j(D)$ is the set of directories of order at most j. The ***order of a directory*** $\delta\in V(D)$ is the smallest j such that $\delta\in V_j(D)$.
A data base of higher order (dbho) is an ordered tuple $B=(D,\Delta_1,\ldots,\Delta_k)$ where each Δ_i is a directory in $V(D)$.

Two directories $\Delta\in V_m(D)$ and $\Delta^\# \in V_m(D^\#)$ over domains D and $D^\#$ are ***similar*** if
(i) Δ and $\Delta^\#$ are of the same order;
(ii) If $\Delta\in V_1(D)$ then Δ and $\Delta^\#$ have the same rank.
(iii) Otherwise, there is a function $f:\Delta\to\Delta^\#$ which is 1−1, onto and such that for each $\delta\in\Delta$, δ and $f(\delta)$ are similar.
Each directory $\Delta\in V$ can be thought of as a labeled directed acyclic graph in the following way: The leaves are either relations (i.e. in $V_1(D)$) or the empty directory, which is in $V_2(D)$ and is denoted by ϕ_{dir}. In the first case their label is the rank of the relation. In the other case the label is −1. Here we have to remark that for each natural number k we have an ***empty relation*** ϕ_k ***of rank*** k. Two directories are similar if their labeled graphs are

isomorphic.

Let $B=(D,\Delta_1,\ldots,\Delta_k)$ and $B^\#=(D^\#,\Delta^\#_1,\ldots,\Delta^\#_k)$ be two dbho's and let $h:D \to D^\#$ a function between the two domains. We define an extension $\bar{h}:V(D) \to V(D^\#)$ in the following way:

(i) For $\delta \in V_1(D)$ a k-ary relation

$$\bar{h}(\delta)=\{(h(d_1),\ldots,h(d_k)):((d_1),\ldots,(d_k)) \in \delta\}$$

So $\bar{h}(\delta)$ is a k-ary relation in $V_1(D^\#)$.

(ii) For $\delta \in V_m(D)$ we put

$$\bar{h}(\delta)=\{\bar{h}(\alpha):\alpha \in \delta\}.$$

If h is one-one then $\bar{h}(\delta)$ is similar to δ. This is not true in general because we think of directories as sets, not as multisets.

h is an isomorphisms from B into $B^\#$ iff h is one-one and onto and for $0 \le i \le k$ $\bar{h}(\Delta_i)=\Delta^{\#_i}$.

Two dbho's $B=(D,\Delta_1,\ldots,\Delta_k)$ and $B^\#=(D,\Delta^\#_1,\ldots,\Delta^\#_k)$ are *similar* if each Δ_i is similar to $\Delta^\#_i$.

Two dbho's $B=(D,\Delta_1,\ldots,\Delta_k)$ and $B^\#=(D,\Delta^\#_1,\ldots,\Delta^\#_k)$ are *isomorphic* if they are similar and there is an isomorphism $h:B \to B^\#$.

In the case that each Δ_i is a relation this notion of isomorphisms coincides with the usual notion of isomorphism of relational data bases. In general it is a natural extension of this notion.

4. Computable directory queries and relations.

Let D be a finite set and $V(D)$ be the set of directories over D. An k-ary directory transformation is a function $T:V(D)^n \to V(D)$ such that for every bijection $h:D \to D$ and every $\delta_1,\ldots,\delta_k \in V(D)$ we have

$$T(\bar{h}(\delta_1),\ldots,\bar{h}(\delta_k))=\bar{h}(T(\delta_1,\ldots,\delta_k))$$

If we replace $V(D)$ by $Rel(D)$ this is just the isomorphism invariance of queries in [CH80].

Since all the elements of $V(D)$ are finite objects it makes sense to speak of a "standard" coding of $V(D)$ in the natural numbers **N**. This allows us to use freely the notion of computable functions over $V(D)$.

An k-ary directory transformation is *computable* if it is computable using the standard coding.

5. Examples:

(i) The computable queries are computable directory queries: If $B=(D,R_1,\ldots,R_k)$ is a relaional data base state and q is a computable query producing a relation Q we just regard each R_i as a directory of order 1 and put T_q to be the obvious k-ary directory transformation.

ii) Let δ be a directory and let $\{\delta\}$ be the directory containing δ as its only subdirectory. The ransformation $T_{singleton}$ which maps δ into $\{\delta\}$ is clearly a computable directory transformaion.

iii) Let δ_1, δ_2 be two directories and let $\delta_1 \cup \delta_2$ be the directory which contains exactly the subdirectories of δ_1 and δ_2 as its subdirectories. The transformation $T\cup$ which maps δ_1 and δ_2 into $\delta_1 \cup \delta_2$ is clearly a computable directory transformation.

iv) Let δ_1, δ_2 be two directories and let $\delta_1 - \delta_2$ be the directory which contains exactly the subdirectories of δ_1 which are not in δ_2 as its subdirectories. The transformation $T_{difference}$ which maps δ_1 and δ_2 into $\delta_1 - \delta_2$ is clearly a computable directory transformation.

v) Let δ be a directory and let $P(\delta)$ be the directory containing exactly each subset of subdirectories of δ as a subdirectory. The transformation T_{power} which maps δ into $P(\delta)$ is learly a computable directory transformation.

vi) Let δ be a directory and let $U(\delta)$ be the directory containing exactly each subdirectory of subdirectory of δ as a subdirectory. The transformation $T_{\bigcup}$ which maps δ into $U(\delta)$ is learly a computable directory transformation.

vii) Let R be an n-ary relation of power p. We associate with R a directory δ of order 2 containing p n-ary relations each ot which contains exactly one n-tuple of R and such that each n-tuple of R occurs in δ. Clearly, this defines a computable directory transformation.

viii) (Kuratowski pair) Set

$$Pair(\delta_1,\delta_2)=\{\{\delta_1\},\{\delta_1,\delta_2\}\}$$

learly *Pair* is a computable directory transformation.

ix) Let δ be a directory and let $HTC(\delta)$ be the the set of all directories and relations, which re in its *transitive closure under membership* (the hereditary transitive closure). Then learly HTC is a computable directory query.

The examples (i)-(vii) will be among the basic constructs of our directory transformation anguage *DL*, defined in the next section. The reader can easily find more examples. As an xercise for computable predicates we suggest comparison of relations via file length, arity of elations and testing whether a directory is in $V_k(D)$.

6. The directory query language *DL* (Syntax).

The directory query language *DL* we define is essentially a programming language computing finite higher order objects (directories) over some finite domain. As for *QL* from [CH80], its access to a directory, however, is only through a restricted set of operations: the operations from *QL* augmented by the operations from examples (i) - (vi) in the previous section. Let us now define *DL* formally. We include also a definition of *QL* to make the paper more self contained.

$y_1, y_2, \ldots$ are *variables* of *DL*. The set of *terms* of *DL* is inductively defined as follows:

(i) E is a term of *QL*; for $i \geq 1$ y_i are terms of *QL*; if dir_i is a directory name then dir_i is a term of *DL*; if rel_i is a relation name then it is a term of *QL*.

(ii) For any terms t_1, t_2 of *QL*

$$(t_1 \cap t_2), (-t_1), (t_1\downarrow), (t_1\uparrow) \text{ and } (t_1\sim)$$

are terms of *QL*.

(iii) All terms of *QL* are also terms of *DL*.

(iv) For any terms t_1, t_2 of *DL*

$$\{t_1\},\ U(t_1),\ P(t_1),\ Singl(t_1),\ (t_1 - t_2),\ (t_1 \cup t_2)$$

are terms of *DL*.

The set of *programs* of *DL* is inductively defined as follows:

(i) If t is a term of *DL* (*QL*) then $y_i := t$ is a program of *DL* (*QL*).

(ii) If P_1, P_2 is a program of *DL* (*QL*) then $(P_1; P_2)$ and **while** y_i **do** P_1 are programs of *DL* (*QL*).

(iii) All programs of *QL* are also programs of *DL*.

(iv) If P is a program of *DL* then

$$\textbf{mkdir } y_i \textbf{ from } y_j \textbf{ in } y_k \textbf{ by } P(y_1, \ldots, y_m)$$

is a program of *DL*. The variable y_j occurs here as a *bounded* variable similar to j in $\sum_j a_j$.

7. The directory query language *DL* (Semantics):

Let $B=(D,R_1,\ldots,R_k)$ be a dbho.

i) Let z be a function from the variables $y_1,y_2,\cdots$ into $V(D)$, the set of directories over D. Ve call such a function a ***directory assignment over*** B or ***assignment*** for short. We think of he set of all directory assignments over B as the set of *states* for our directory query. We lenote this set by $States(B)$.

ii) The *meaning* of a program P acting on B is a partial function $\mu(P){:}States(B) \rightarrow States(B)$.

'irst we define for every term t of DL inductively the meaning function $\mu_0(t){:}States(B) \rightarrow V(D)$ in the following way:

'or terms t in QL, $\mu_0(t)$ is defined as in [CH 80]. If t_1 and t_2 are terms in QL then:

$\mu_0(E)(z)=\{(x,x){:}x \in D\}$,

$\mu_0(y_i)(z)=z(y_i)$,

$\mu_0(r_i)(z)=R_i$,

$\mu_0(t_1 \cap t_2)(z)= \mu_0(t_1)(z) \cap \mu_0(t_2)(z)$, if $\mu_0(t_1)(z)$ and $\mu_0(t_2)(z)$ have the same arity, otherwise $\mu_0(t_1 \cap t_2)(z)=\phi_0$.

$\mu_0({*}(t_1))(z)= {*}'(\mu_0(t_1)(z))$, if $\mu_0(t_1)(z)$ is a relation, otherwise it is ϕ_0. $*$ stands here for $\neg$, $\downarrow$, $\uparrow$ r $\sim$. The meaning $*'$ of $*$ is complement, projection of all components except of the first, xtension of the relation by one last component, or cyclic permutation respectively.

'or the other terms in DL, μ_0 is defined inductively in the following way:

et t_1 and t_2 be terms in DL. Then for each $z \in States(B)$:

1) $\mu_0(\{t_1\})(z)=\{\mu_0(t_1)(z)\}$,

2) $\mu_0(P(t_1))(z)=$ *Powerset of* $\mu_0(t_1)(z)$

3) $\mu_0(U(t_1)(z)= \bigcup(\mu_0(t_1))(z)$, if all subdirectories of $\mu_0(t_1)(z)$ are relations of the same arity r all subdirectories of it are not relations, otherwise it is set to be ϕ_0.

4) $\mu_0(t_1 \cup t_2)(z)= \mu_0(t_1)(z) \cup \mu_0(t_2)(z)$, if $\mu_0(t_1)(z)$ and $\mu_0(t_2)(z)$ are both relations of the ame arity or both not relations, otherwise it is set to be ϕ_0.

5) $\mu_0(t_1 - t_2)(z)= \mu_0(t_1)(z)-\mu_0(t_2)(z)$

ere $X-Y$ is the set of all elements of X not being in Y. Note that $X-Y$ is a relation of arity k f X is a relation of arity k. Otherwise $X-Y$ is a nonrelational directory.

6) $\mu_0(Singl(t_1))(z)= \{\{x\} : x \in \mu_0(t_1)(z)\}$

Next we define for every program $P \in DL$ inductively the meaning function $\mu(P)$ in the ollowing way:

a) If P is of the form $y_i{:=}t$ then we put $\mu(P)(z)(y_j)=z(y_j)$ if $j \neq i$ and $\mu(P)(z)(y_i)=\mu_0(t)(z)$ therwise.

b) If P is $P_1;P_2$ then $\mu(P)(z)=\mu(P_2)(\mu(P_1)(Z))$. This is the usual composition of functions.

(c) If P is **while** y_j **do** P_1 then $\mu(P)(z)$ is defined in the usual way on a sequence of states $z_{i+1}=\mu(P_1)(z_i)$ with $z_0=z$. $\mu(P)(z)$ is the first z_i such that $z_i(y_j)$ is not an empty relation or directory.

(d) If P is **mkdir** y_i **from** y_j **in** y_k **by** $P_1(y_1, \ldots, y_m)$ then

$\mu(P)(z)(y_i)= \{\mu(P_1)(z_1)(y_j): z_1(y_l)=z(y_l)$ for $l \neq j$ and $z_1(y_j) \in z(y_k)\}$, if for all z_1, s.t $z_1(y_l)=z(y_l)$ for $l \neq j$ and $z_1(y_l) \in z(y_k)$ $\mu(P_1)(z_1)(y_j)$ is defined, otherwise $\mu(P)(z)(y_i)$ is undefined.

In words this says, for the case $m=j=1$, that the new directory y_i is obtained in the following way: one applies in *parallel* to all the subdirectories y_j of y_k the program P_1 and puts in to y_i all the results so obtained. If $j>m$ the new directory contains exactly one subdirectory of the form $t(y_1,\ldots,y_m)$. Otherwise, the directories $y_1,\ldots,y_{j-1},y_{j+1},\ldots$ are free parameters. Remember that y_j occurs here as a *bounded* variable. The reader acquainted with axiomatic set theory will easily recognize in this definition the *replacement axiom of Zermelo-Frankel set theory*.

8. Main results.

Let $B=(D,R_1,\ldots,R_k)$ be a dbho and $z_{initial}$ be the assignment with $z_{initial}(y_i)=R_i$ for all $i \leq k$ and $z_{initial}(y_i)=\phi_{-1}$ for all $i>k$. Given a program $P(y_1, \ldots, y_n) \in DL$ and a variable y_j we look at the function $T_{P,j}: V(D)^k \to V(D)$ $T_{P,j}(R_1, \ldots, R_k)=\mu(P)(z_{initial})(y_j)$.

Main Theorem: (i) For every program $P \in DL$ and each variable y_j the the function $T_{P,j}: V(D)^k \to V(D)$ is a computable directory query.

(ii) The directory query language DL is complete, i.e. for every computable directory query T there is a program $P_T \in DL$ computing it.

The proof of this theorem will be outlined in section 9. In the proof of the Main Theorem we shall use the main result of [CH80], which asserts that the query language QL is complete.

The natural question arises to whether the set of basic constructs is minimal, and if not, what are the exact interrelationships. It turns out that this is a rather delicate problem. In the following definition we introduce a sublanguage DL_0 of DL which has an independent set of constructs.

Definition: Let DL_0 be obtained from DL by restricting its definition to the constructs **while** and **mkdir** together with

$$(t\downarrow),\ (t\uparrow),\ (t\sim),\ E,\ U(t),\ Singl(t),\ (t_1 - t_2).$$

Theorem (Independence of the constructs):

For each construct c of DL_0 there is a computable directory query T which is not computable n $DL_0 - \{c\}$.

9. Outlines of the proofs.

The proof of the Main Theorem consists of three steps. In the first and third step we use a coding and decoding program *TAR* and TAR^{-1}. *TAR* is, inspired by the *UNIX* program of the same name, a program that takes directories of arbitrary order and makes one file from which the original directory can be uniquely reconstructed by TAR^{-1}.
The difficulty in writing *TAR* and, especially TAR^{-1} in *DL* comes from the fact that we may not use names and other information of the directory structures. In *UNIX* for example the corresponding programs use commands different from the shell language to achieve such a coding. In *DL* we can only write isomorphism preserving programs and, therefore have to construct auxiliary data structures within *DL*.
The programs *TAR* and TAR^{-1} allow us to reduce our completeness proof to the completeness proof for *QL* in [CH80]. This is the middle step in our proof.

The proof of the independence of the constructs of DL_0 is always based on variations of the same idea: Omitting a construct adds to the resulting sublanguage a closure property which does not hold for DL_0. For example, to show that the complement $\neg$ is needed we observe that complement-free programs commute with homomorphic images of the domain. The least non-trivial result here is the necessity of the **mkdir**-construct. Our proof uses the fact that we do not have the full power set operation among our basic constructs. The full power set, however, can be obtained using *Singl* and **mkdir**.

10. Extended directory queries.

When directory and data base systems are used in practice, several operations and predicates outside the formal relational and directory framework are useful, or even necessary, to turn the system into a practical and efficient model. Concerning the purely relational aspect of data bases, [CH80] addresses this issue and proposes the extended query language *EQL*. The main difference in [CL80] between computable and extended computable queries lies in the semantics. In the extended model they look at two sorted structures where an additional domain F is added, whose elements may be numbers, or any other set of terms, whose interpretations is fixed.

If we want to adapt this approach to our framework we should first examine what we really have in mind. The new objects to be introduced are really "names", i.e. interpretations of certain terms whose meaning is never changed and is part of the user interface. They can be words over some finite alphabet A (including natural numbers in some b-ary notation). They usually have some standard operations and relations on them, such as concatenation, arithmetical operations and/or a linear order. This makes the new universe with its functions into a Herbrand universe. It is easy to modify our framework for this purposes. We take the extended semantic model of [CH80] as our starting point, i.e. $V_1(D \cup F)$. Here D is a finite set of urelements, as before, and F is a possibly infinite set disjoint from D. There must be enough functions to make sure that every element of F is the interpretation of some term. Relations are always finite and their one-dimensional projections are always either in D or in F. The restrictions of isomorphisms on F are always the identity. The constructions of $V(D \cup F)$ is continued naturally. We leave it to the reader to formulate everything in detail.

In contrast to the case of [CH80], extending the directory model in this way does not give us increased expressive power. The universe of the natural numbers, e.g. does exist in $V(D)$, though it is not an element of any $V_k(D)$. Since we allow higher order relations, every ***finite set of natural numbers*** can be thought of being in some $V_k(D)$, and therefore, relations involving natural numbers can be coded in $V(D)$. The advantage of the extended approach lies in its inherent economy, both conceptually and computationally. Conceptually, we can now formulate various aspects of directory systems, which were only expressible before in a rather cumbersome way. Among these are time stamp labels, listing the names of the subdirectories of a directory (the ls-command in UNIX) with all its variations, and the introduction of arithmetical and statistical functions. The set of urelements D, however, is not assumed to be linearly ordered and cannot be linearly ordered within DL. In contrast to this, the directories and relations can be linearly ordered by the lexicographic order of the names.
The study of the relationship between complexity classes and various sublanguages of DL will be delayed to future research. It seems clear that various results of [CH82, Im82, Im83, HP84] have their analogues.

11.. References.

[Ba75] J.Barwise, Admissible sets and structures, Springer, Berlin 1975.

[CH80] A.K.Chandra and D.Harel, Computable queries for relational data bases, JCSS vol. 21.2 (1980) pp. 156-178.

[CH82] A.K.Chandra and D.Harel, Structure and complexity of relational queries, JCSS vol. 25, (1982) pp. 99-128.

[DM85] E. Dahlhaus and J.A.Makowsky, Computable Directory Queries, Technical report of the Institut fuer Informatik, ETH-Zuerich, No. 65 (September 1985).

DM85a] E. Dahlhaus and J.A.Makowsky, The choice of programming primitives for SETL-like programming languages, Technical report of the Institut fuer Informatik, ETH-Zuerich, No. 65 (September 1985), to appear in the Proceedings of ESOP '86, LNCS, Springer, Berlin 1986.

GR83] A. Goldberg and D. Robson, Smalltalk-80 : The language and its implementation, Addison-Wesley, Reading MA, 1983.

Ho83] E. Horowitz, Fundamentals of programming languages, Computer Science Press, Rockville 1983.

HP84] D. Harel and D. Peleg, On static logics, dynamic logics and complexity classes, Information and Control vol. 60, (1984) p. 86-102.

Im82] N. Immerman, Relational queries computable in polynomial time, 14th ACM Symposium on Theory of Computing, (1982) p. 147-152.

Im83] N. Immerman, Languages which capture complexity classes, 15th ACM Symposium on Theory of Computing, (1983) p. 347-354.

Le65] A, Levy, A hierarchy of formulas in set theory, Memoirs of the American Mathematical Society 57, Providence 1965.

M83] D.Maier, The theory of relational data bases, Computer Science Press, Rockville 1983.

SSS79] E. Schonberg, J.T. Schwartz and M. Sharir, Automatic data structure selection in SETL, 6. Annual ACM Symposium on Principles of Programming Languages (1979) pp.197-210.

Sch75] J.T. Schwartz, On programming: An interim report on the SETL project, 2nd ed., Courant Institute of Mathematical Sciences, New York 1975.

U82] J.D. Ullman, Principles of data base systems, Computer Science Press, Rockville 1982.

Zl82] M.M.Zloof, Office by Example: A business language that unifies data and word processing and electronic mail, IBM Systems Journal vol. 21.3 (1982) pp. 272-304.

RELATING TYPE-STRUCTURES
Partial variations on a theme of Friedman and Statman
(Preliminary version)

Andrea Asperti Giuseppe Longo
Dipartimento di Informatica, Corso Italia 40, I-56100 Pisa

Introduction.

In Friedman[1975] a result of semantic completeness for typed λ-calculus is given, by using the full type structure over ω. That is, two typed terms are shown to be provably equal iff they define the same functional of finite type over ω (i.e. the same morphism in the category of sets and all functions). The key to Friedman's result is a simple and elegant notion of homomorphism between type-structures. We wish to extend Friedman's notion and its consequences to more constructive settings.

As a matter of fact both typed and type-free λ-calculus have been primarely regarded as a formalization of the concept of effective process or computation. Indeed, λ-terms are extremely adequate to describe computable functions and functionals (see Barendregt[1984],Goedel[1958] and Troelstra[1973], say). Moreover, since Scott's work, the role of λ-calculus and its extensions in denotational semantics of programming languages is well-known, as it provides the core of functional programming languages.

For these reasons, we are interested in models which yield useful properties for the theory of programs.* For example, they should provide (effective) solutions to equations which recursively define programs and data types. These may be given by using the effectively given domains of

Reseach partially supported by Ministero P.I. (40%, Matematica).

Scott[1982]. Actually, one may take, as objects, the countable collections of the computable elements in the effectively given domains, that is one may consider the category **CD** of constructive domains (see Giannini/Longo[1984] for an introduction). **CD** is a model for typed λ-calculus, is closed under inverse limits and limits are also preserved by the functors which give higher type objects (product, exponentiation....). By this, the semantics of the recursive definitions of programs and data may be soundly and effectively given (see Kanda[1979], Smith[1977], Smith/Plotkin[1982]).

Note that the fully effective flavour of **CD** is due to the fact that morphisms and functors are represented by recursive functions over suitable indexing of objects. Indeed, **CD** is one of the several interesting subcategories of the category $\mathbf{PER}_{\omega}$, of countable (quotient) sets, we will consider. $\mathbf{PER}_{\omega}$ may be loosely viewed as the constructive counterpart of **Set**, the usual category of sets and functions (see Hyland[1982]). The choice of countable (and numbered) data types is a very natural one also for the purposes of Computer Science.

The aim of this paper is to compare type-structures, i.e. models of typed λ-calculus, when data types are taken to be (possibly structured) countable sets and morphisms to be effective transformations. This will be done by tools related to Friedman's homomorphisms and Plotkin-Statman's logical relations (see later or Friedman[1975], Plotkin[1980], Statman[1984]). More precisely, in S.1, we introduce and discuss the notion of "partial retraction system". The methods in S.1 are endebted to the work in Friedman[1975] and Plotkin[1982] (the latter was brought to our attention when this paper was in preparation). However, Friedman's technique relies on an highly non constructive use of the axiom of choice for sets. By an informal analogy, we may say that, in our approach, choice functions are replaced by retraction pairs, which are "effective" relatively to the intended category.

A key point in the present perspective, as much as in Friedman's; is the possibility of handling partial functions. In our case, though, we cannot rely on the intuitive notion of partiality for set-theoretic functions. Therefore, S.2 is devoted to an introduction to categories with partial morphisms and to "complete objects", which will turn out to be relevant notions for the proof of the main theorem. The content of S.2 is also motivated by the increasing interest in (typed) partial computations and in their denotational semantics (see Plotkin[1984]).

Finally, in S.3 we present the relation between PER_ω and the typed λ-calculus by retraction systems.

In view of the space limitations imposed by the publisher, several proofs (and comments) are omitted (see Asperti/Longo[1986]).

1. Partial retraction systems.

1.1 Definition. The collection of types **Tp** over a ground set At of atomic type simbols is inductively defined by:

i) $At \subseteq Tp$; ii) if $\sigma,\tau \in Tp$ then $\sigma\tau \in Tp$.

Given sets C_i' s and $C = \{C_i \mid i \in At\}$, $T_C = \{C_\sigma\}_{\sigma\in Tp}$ is a **pre-type-structure** over C if $\forall\sigma,\tau \in Tp \ C_{\sigma\tau} \subseteq C_\sigma \to C_\tau$ (= **Set**$[C_\sigma,C_\tau]$), the set of all functions from C_σ to C_τ . For $x \in C_{\sigma\tau}$ and $y \in C_\sigma$, we write xy for $x(y)$, if there is no ambiguity.

A **type-structure** $T_A = (\{A_\sigma\}_{\sigma\in Tp}, [\![\,]\!])$ is a pre-type-structure which is a model of typed λ-calculus. That is, given an **A-environment** $h : Var \to \cup_{\sigma\in Tp}\{A_\sigma\}$, one has:

Var) $[\![x^\sigma]\!]_h = h(x^\sigma) \in A_\sigma$

App) $[\![M^{\sigma\to\tau} N^\sigma]\!]_h = [\![M^{\sigma\to\tau}]\!]_h([\![N^\sigma]\!]_h) \in A_\tau$

β) $[\![\lambda x^\sigma.M^\tau]\!]_h(a) = [\![M^\tau]\!]_{h[a/x^\sigma]} \in A_\tau$ for $a \in A_\sigma$.

(Note : we may omit types, if there is no ambiguity; for $T_A = (\{A_\sigma\}_{\sigma\in Tp}, [\![\,]\!])$ $T_A \models M = N$ means that, for all A-environment h, $[\![M]\!]_h = [\![N]\!]_h$).

In particular one obtains type-structures from "concrete" categories, i.e.

from categories whose objects and morphisms may be viewed as sets and functions "in extenso" (see definition 2.1 below), provided that they are closed under formation of morphisms spaces. More formaly, let $\mathbf{K}$ be a Cartesian Closed Category (CCC) and $C = \{C_i \mid i \in At\} \subseteq Ob_K$. Then $\mathbf{K}_C = \{C_\sigma \mid \sigma \in Tp\}$ is the type-structure generated by $\mathbf{C}$ in $\mathbf{K}$, that is each $C_{\sigma\tau}$ coincides with $C_\tau{}^{C_\sigma}$, the representative of the collection $\mathbf{K}[C_\sigma,C_\tau]$ of morphisms from C_σ to C_τ (if needed and not ambiguous, we may identify $C_\tau{}^{C_\sigma}$ and $\mathbf{K}[C_\sigma,C_\tau]$). By well-known results relating CCC's and typed λ-calculus (see Lambek[1980], Scott[1980], Dibjer[1983], Poigne'[1985]), $\mathbf{K}_C$ is indeed a type-structure. The equivalence between the assumption on "concreteness" ("enough points", as in. 2.1) and rule (ξ) is tidely investigated in Berry[1979] and Poigne'[1985], for the typed case, and Koymans[1982], Barendregt[1984], for the type-free calculus.

1.2 **Examples.** The simplest type-structures are the full type-structures and the term model of typed $\lambda\beta\eta$. That is $\mathbf{Set}_{\mathbf{A}}$, where $\mathbf{A}$ is a collection of sets, and $\mathbf{Term} = ((Term_\sigma)_{\sigma\in Tp}, [\,])$, where $Term_\sigma$ is the set of terms of type σ modulo $\beta\eta$ convertibility. (If there is no ambiguity, we identify M^σ and $[M^\sigma] = \{N^\sigma \mid \lambda\beta\eta \vdash M^\sigma = N^\sigma\}$.)

1.3 **Definition.** Let $\mathbf{K}$ be a category. Then $B < A$ (B is a **retract** of A) if there exist $\alpha \in \mathbf{K}[A,B]$ and $\beta \in \mathbf{K}[B,A]$ such that $\alpha\cdot\beta = id$.

Given a CCC $\mathbf{K}$, there is a canonical way to inherit retractions at higher types; namely, if $B_\sigma < A_\sigma$ via $(\alpha_\sigma,\beta_\sigma)$ and $B_\tau < A_\tau$ via (α_τ,β_τ), then $B_{\sigma\tau} < A_{\sigma\tau}$ via

(1.3.1) $\quad \alpha_{\sigma\tau}(x) = \alpha_\tau\cdot x\cdot\beta_\sigma\,, \qquad \beta_{\sigma\tau}(y) = \beta_\tau\cdot y\cdot\alpha_\sigma\,.$

Clearly, $\alpha_{\sigma\tau} \in \mathbf{K}[A_{\sigma\tau},B_{\sigma\tau}]$, $\beta_{\sigma\tau} \in \mathbf{K}[B_{\sigma\tau},A_{\sigma\tau}]$ and $\alpha_{\sigma\tau}\cdot\beta_{\sigma\tau} = id$.

Retractions play a major role in the semantic investigation of type theory, as they provide a strong and precise notion of "subtype". A simple corollary of the main result in this section will be the following (see 1.7):

let $\mathbf{A} = (A_i)_{i \in At}$ and $\mathbf{B} = (B_i)_{i \in At}$ be collections of objects in a CCC $\mathbf{K}$; if $\forall i \in At \;\; B_i < A_i$, then

$$\mathbf{K_A} \models M = N \;\Rightarrow\; \mathbf{K_B} \models M = N .$$

This fact gives a strong consequence in any type, by some information on the ground types. However, it uses the assumption that both type-structures are built in the same category. That is, for all $\sigma,\tau \in Tp$, both $A_{\sigma\tau}$ and $B_{\sigma\tau}$ are exactly the morphism in $\mathbf{K}$ of the intended type. 1.12 and 1.14 will prove more by comparing type-structures which do not need to satisfy this strong assumption. That comparison will be made possible by an essential use of partial morphisms.

How does partiality come in? Given type-structures $\mathbf{T_A}$ and $\mathbf{T_B}$ assume that, for all $\sigma \in Tp$, A_σ and B_σ are objects of a CCC $\mathbf{K}$. Very roughly, the idea is to take the category $\mathbf{K}$ where the type structure with "more morphisms" , $\mathbf{T_A}$ say, is built in (or just the category **Set** of sets). Then, even if $A_{\sigma\tau} = A_\tau{}^{A_\sigma}$, $B_{\sigma\tau}$ may be smaller than $B_\tau{}^{B_\sigma}$, for some $\sigma,\tau \in Tp$. Thus, for $x \in A_{\sigma\tau}$, $\alpha_{\sigma\tau}(x) = \alpha_\tau \circ x \circ \beta_\sigma$ doesn't need to be in $B_{\sigma\tau}$, that is $\alpha_{\sigma\tau} : A_{\sigma\tau} \to B_{\sigma\tau}$ doesn't need to be defined on x .

For the purposes of this section we only need the classical notion of partiality for set-theoretic functions. In particular, we write $B <_p A$ if there exist partial functions $\alpha : A \to B$ and $\beta : B \to A$ s.t. $\alpha \circ \beta = id$. Clearly, then, α is a (possibly partial) surjection and β a total injection.

1.4 Definition. Let $(A_\sigma)_{\sigma \in Tp}$, $(B_\sigma)_{\sigma \in Tp}$ be pre-type-structures. Then $((\alpha_\sigma, \beta_\sigma))_{\sigma \in Tp}$ is a **partial retraction system** (p.r.s.) from $(A_\sigma)_{\sigma \in Tp}$ onto $(B_\sigma)_{\sigma \in Tp}$ if $\forall i \in At \;\; B_i <_p A_i$ via partial functions (α_i, β_i) and

<u>cond 1)</u> $\forall x \in A_{\sigma\tau} \;\; [\, \alpha_\tau \circ x \circ \beta_\sigma \in B_{\sigma\tau} \;\Rightarrow\; \alpha_{\sigma\tau}(x) = \alpha_\tau \circ x \circ \beta_\sigma \,]$

<u>cond 2)</u> $\forall z \in B_{\sigma\tau} \; \forall y \in dom\alpha_\sigma \;\; (\beta_{\sigma\tau}(z))y = \beta_\tau(z(\alpha_\sigma(y)))$.

1.5 Proposition. Let $(A_\sigma)_{\sigma \in Tp}$, $(B_\sigma)_{\sigma \in Tp}$ be pre-type-structures and let $((\alpha_\sigma, \beta_\sigma))_{\sigma \in Tp}$ be a p.r.s. from $(A_\sigma)_{\sigma \in Tp}$ onto $(B_\sigma)_{\sigma \in Tp}$. Then $\forall \sigma,\tau \in Tp$ one has:

i) $\alpha_\sigma \cdot \beta_\sigma = \text{id}$

ii) $(\beta_{\sigma\tau}(z))(\beta_\sigma(y)) = \beta_\tau(zy)$

Proof : If $\sigma \in At$, i) holds, by definition of p.r.s.. Suppose then that i) holds for σ, τ. One then has, for all $z \in B_{\sigma\tau}$ and $y \in B_\sigma$:

$$(\beta_{\sigma\tau}(z))(\beta_\sigma(y)) = \beta_\tau(z(\alpha_\sigma(\beta_\sigma(y)))) \quad \text{by } \underline{\text{cond2}} \text{ and induction on i)}$$
$$= \beta_\tau(zy) \quad \text{by induction on i) .}$$

This is ii) and, thus, for all $y \in B_\sigma$:

$$\alpha_\tau((\beta_{\sigma\tau}(z)(\beta_\sigma(y)) = \alpha_\tau(\beta_\tau(zy))$$
$$= zy \quad \text{by induction on i).}$$

Observe now that $(B_\sigma)_{\sigma \in Tp}$ is extensional and then

$$\alpha_\tau \cdot (\beta_{\sigma\tau}(z)) \cdot \beta_\sigma = z \ .$$

Thus cond1 applies and gives $\forall z \in B_{\sigma\tau}$ $\alpha_{\sigma\tau}(\beta_{\sigma\tau}(z)) = z$ or,equivalently, i) at the higher type. □

By i) in the proposition, for each $\sigma \in Tp$, $B_\sigma <_p A_\sigma$. Moreover, by ii) , the injection β_σ is a total "homomorphism", i.e. it preserves functional application. As this is a fundamental notion, we briefly survey the connections between p.r.s., homomorphisms and logical relations.

Let $\mathbf{T_A} = (A_\sigma)_{\sigma \in Tp}$ and $\mathbf{T_B} = (B_\sigma)_{\sigma \in Tp}$ be a pre-type-structure and $R_i \subseteq A_i x B_i$, $i \in At$. Define then a **logical relation** $(R_\sigma)_{\sigma \in Tp}$, with $R_{\sigma\tau} \subseteq A_{\sigma\tau} x B_{\sigma\tau}$, by

$$R_{\sigma\tau}(a,b) \iff \forall x,y\ (\ R_\sigma(x,y) \Rightarrow R_\tau(a(x),b(y))).$$

1.6 Proposition . Let $(R_\sigma)_{\sigma \in Tp}$ be a logical relation.

1) If $\forall i\ R_i$ is single valued and $\forall \sigma\ R_\sigma$ is surjective, then $\forall \sigma\ R_\sigma$ is single valued.

2) If $\forall \sigma\ R_\sigma$ is single valued, set $\alpha^R{}_\sigma(a) = b$ iff $R_\sigma(a,b)$. Then $(\alpha^R{}_\sigma)_{\sigma \in Tp}$ is an homomorphism, i.e. $\forall \sigma,\tau\ \alpha^R{}_{\sigma\tau}(a)(\alpha^R{}_\sigma(c)) = \alpha^R{}_\tau(a(c))$. Conversely, each surjective homomorphism $(\alpha'_\sigma)_{\sigma \in Tp}$ defines a logical relation $(R'_\sigma)_{\sigma \in Tp}$ such that $\forall \sigma\ R'_\sigma(a,b)$ iff $\alpha'_\sigma(a) = b$.

(The **proof** is easy).

Recall that also the β_σ's in 1.5 yield an homomorphism. The way β_σ's and the $\alpha^R{}_\sigma$'s relate is expressed by 1.7-8. Assume that a p.r.s. $(\alpha_\sigma,\beta_\sigma)_{\sigma\in Tp}$ is given from pre-type- structure $(A_\sigma)_{\sigma\in Tp}$ onto $(B_\sigma)_{\sigma\in Tp}$. Define a logical relation $(R_\sigma)_{\sigma\in Tp}$ as above over R_i, where $R_i(a,b)$ iff $\alpha_i(a) = b$. Then, by combined induction on types, one has:

1.7 Theorem . $\forall\sigma \in Tp$ 1) $\forall b \in B_\sigma \ R_\sigma(\beta_\sigma(b),b)$,

2) $\forall a \in A_\sigma \ \forall b \in B_\sigma \ R_\sigma(a,b) \Rightarrow \alpha_\sigma(a) = b$.

Thus any p.r.s. $(\alpha_\sigma,\beta_\sigma)_{\sigma\in Tp}$ gives a logical relation $(R_\sigma)_{\sigma\in Tp}$ which is surjective (1.7.1) and single valued (1.7.2) and, hence, a surjective homomorphism ($\alpha^R{}_\sigma$ in 1.6.2). In general, $\alpha^R{}_\sigma$ and α_σ are partial maps:

1.8 Corollary. $\forall\sigma \in Tp$ $\mathrm{range}\beta_\sigma \subseteq \mathrm{dom}R_\sigma = \mathrm{dom}\alpha^R{}_\sigma \subseteq \mathrm{dom}\alpha_\sigma$. Moreover, $\forall a \in \mathrm{dom}\alpha^R{}_\sigma \ \alpha_\sigma(a) = \alpha^R{}_\sigma(a)$ and, hence, $\alpha^R{}_\sigma \cdot \beta_\sigma = id$.

As it will be pointed out in 1.10 and 1.12, the existence of a surjective homomorphism between two type-structures has several consequences. The first, 1.10, has been recently communicated by Statman to the authors.

1.9 Definition. Let $\mathbf{T_A}$ be a type-structure. $f \in A_{\sigma\tau}$ is **n-piecewise-λ-definable** iff $\forall a_1,\dots,a_n \in A_\sigma \ \exists M \in Term_{\sigma\tau}$ (closed) $f(a_i) = [M]_h(a_i)$, for $1 \le i \le n$.

1.10 Proposition. Let $\mathbf{T_A}$ be a type-structure such that there exists a surjective homomorphism from $\mathbf{T_A}$ onto **Term** and let $f \in A_{\sigma\tau}$ be 2-piecewise-λ-definable. Then f is λ-definable.

1.11 Fundamental theorem on logical relations (Statman[1984]). Let $\mathbf{T_A}$ and $\mathbf{T_B}$ be type structures and $(R_\sigma)_{\sigma\in Tp}$ a logical relation. Then

$$\forall\sigma\ \ \forall h_A, h_B\ \ (\forall x^\sigma\ \ R_\sigma(h_A(x^\sigma), h_B(x^\sigma)) \Rightarrow \forall M\ \ R_\sigma([\![M]\!]_{h_A}, [\![M]\!]_{h_B}) .$$

1.12 Corollary (Friedman[1975]). If $(R_\sigma)_{\sigma\in Tp}$ is a surjective and single valued logical relation (a surjective homomorphism) from $\mathbf{T_A}$ onto $\mathbf{T_B}$, then $\mathbf{T_A} \models M = N \Rightarrow \mathbf{T_B} \models M = N$.

The main difficulty with logical relations is to construct singled valued and surjective ones, i.e. to find surjective homomorphisms. P.r.s.' may be viewed as a tool for defining them in a way which is constructive w.r.t. the intended category.

We give a proof of the relation between type-structures in 1.12 by a direct use of p.r.s.'. Recall that all typed terms posses a normal form; moreover, if M is in normal form and $M \equiv PQ$, then, for no P', $P \equiv \lambda x.P'$

1.13 Lemma. Let $\mathbf{T_A} = ((A_\sigma)_{\sigma\in Tp}, [\![\,]\!])$, $\mathbf{T_B} = ((B_\sigma)_{\sigma\in Tp}, [\![\,]\!]')$ be type-structures and $((\alpha_\sigma,\beta_\sigma))_{\sigma\in Tp}$ a p.r.s. from $\mathbf{T_A}$ onto $\mathbf{T_B}$. For each B-env. h define an A-env. h' by

$\forall\sigma \in Tp\ \ \forall x^\sigma\ \ \ h'(x^\sigma) = \beta_\sigma(h(x^\sigma))$

Then one has :

i) if M, of type σ, is in n.f.,

$M \equiv x$ or $M \equiv PQ \Rightarrow [\![M]\!]_{h'} = \beta_\sigma([\![M]\!]'_h)$,

$M \equiv \lambda x.P \Rightarrow \alpha_\sigma([\![M^\sigma]\!]_{h'}) = [\![M^\sigma]\!]'_h$;

ii) for all M, of type σ, $\alpha_\sigma([\![M]\!]_{h'}) = ([\![M]\!]'_h)$.

Proof : i) By induction on the structure of M.

- $M \equiv x^\sigma$: $[\![x]\!]_{h'} = h'(x) = \beta_\sigma(h(x)) = \beta_\sigma([\![x]\!]'_h)$

- $M \equiv PQ$: by the remark above, one has $P \equiv x$ or $P \equiv RS$.

By the inductive hypothesis, $[\![P]\!]_{h'} = \beta_{\sigma\tau}([\![P]\!]'_h)$ and $\alpha_\sigma([\![Q]\!]_{h'}) = [\![Q]\!]'_h$. Then

$[\![PQ]\!]_{h'}$ $= ([\![P]\!]_{h'})([\![Q]\!]_{h'})$ by definition of $[\![\,]\!]$

$= (\beta_{\sigma\tau}([\![P]\!]'_h))([\![Q]\!]_{h'})$ by induction

$= \beta_\tau([\![P]\!]_h(\alpha_\sigma([\![Q]\!]_{h'})))$ by <u>cond2</u>

$= \beta_T([\![P]\!]_h([\![Q]\!]'_h)$ by induction

$= \beta_T([\![PQ]\!]'_h)$ by definition of $[\![\,]\!]'$.

- $M \equiv \lambda x^\sigma.P^\tau$: for all $y \in B_\sigma$

$\alpha_T([\![\lambda x.P]\!]_{h'}(\beta_\sigma(y))) = \alpha_T([\![P]\!]_{h'}[\beta_\sigma(y)/x]))$ by def. of $[\![\,]\!]$

$= [\![P]\!]'_h[y/x]$ by induction

$= [\![\lambda x.P]\!]'_h(y)$ by definition of $[\![\,]\!]'$

Thus by extensionality:

$\alpha_T \cdot [\![\lambda x.P]\!]_{h'} \cdot \beta_\sigma = [\![\lambda x.P]\!]'_h$

and by cond1 :

$\alpha_{\sigma T}([\![\lambda x.P]\!]_{h'}) = [\![\lambda x.P]\!]'_h$

ii) Just observe that each λ-term M has a n.f. M' , and, as $\mathbf{T_A}$ and $\mathbf{T_B}$ are models,

$\alpha_\sigma([\![M]\!]_{h'}) = \alpha_\sigma([\![M']\!]_{h'})$

$= [\![M']\!]'_h$ by i) above and i) in 1.5

$= [\![M]\!]'_h$. □

It is now easy to derive the following facts.

1.14 Theorem. Let $\mathbf{T_A}$ and $\mathbf{T_B}$ be type-structures and suppose that there exists a p.r.s from $\mathbf{T_A}$ onto $\mathbf{T_B}$. Then $\mathbf{T_A} \models M = N \Rightarrow \mathbf{T_B} \models M = N$.

1.15 Corollary. Let $A = (A_i)_{i \in At}$ and $B = (B_i)_{i \in At}$ be collections of objects in a CCC $\mathbf{K}$; if $\forall i \in At\ \ B_i < A_i$, then $\mathbf{K_A} \models M = N \Rightarrow \mathbf{K_B} \models M = N$.

1.16 Corollary. (Completeness, Friedman[1975]). Let $A = (A_i \mid i \in At)$ be a collection of infinite sets. Then $Set_A \models M = N \Leftrightarrow \lambda\beta\eta \vdash M = N$.

Remark . The 1-section theorem of Statman[1980,1985] fully characterizes "complete" models, in the sense of 1.16, by a necessary and sufficient condition. However, the existence of a p.r.s. and, hence, of a surjective homomorphism from a type-structure onto **Term** is a stronger property which gives some important extra information (e.g. 1.10).

Note that, in 1.16, α_σ, β_σ are well defined as partial functions, for all $\sigma \in Tp$; indeed, this is all what we need, as **Set**, with the obvious partial morphisms, is used as "frame" category (by frame category we intend the category whose objects include the types of the considered type-structures). Also Plotkin[1982] uses a particular notion of partial retractions in our sense, in the special case of cpo's. The point then is to give a notion of partiality in arbitray categories, in order to apply theorem 1.14 to type-structures of more structured objects (numbered sets, domains, cpo's....). In these cases, the definition of p.r.s. cannot be given, in general, as simply as in **Set** and a p.r.s. must be constructed "effectively", in a sense which is relative to the intended frame category.

2. Partial morphisms and complete objects.

This section is devoted to a very short account on categories with partial morphisms. We survey and develop the notions introduced in Longo/Moggi[1984a], in particular the notion of "complete object". Proofs are omitted: for a more extensive presentation one should refer to Asperti/Longo[1986].

The natural setting for partial morphisms, in Category Theory, is the Theory of Toposes, as it is mostly motivated by the categorical treatment of wellknown set-theoretic notions (subset, inverse image...). We will not get into the details of Topos Theory and we sketch a more elementary approach to the interpretation of "divergence" in familiar categories. Moggi[1985] elegantly works out various aspects of a topos-theoretic approach to partial morphisms.

A stronger, but related, notion of Dominical Category may be found in DiPaola/Heller[1984]. In Dominical Categories the "domain" of a morphism (see below) is nicely given as an endomorphism; the results in this section are still valid in those categories. An updated overview on the topic may be found in Rosolini[1985].

We first need a notion of "concrete" category, which seems very sound for the purposes of denotational semantics (see Barendregt[1984]; Berry/Curien[1982] also discusses this notion and a different perspective).

2.1 Definition. A category **C** has **enough points** (short : e.p.) iff it contains a terminal object t and, for any $a \in Ob_C$, one has

$$\forall h \in C[t,a] \quad foh = goh \;\Rightarrow\; f=g .$$

The morphisms in a category with e.p. are functions or relations in extenso, for the morphisms from the terminal (or singleton) object to a given object may be considered as the points (or elements) of the later. One may look at them as subcategories of **Set** or $\mathbf{Set}_R$.

A sound requirement for a category, in order to allow partial morphisms, is the existence of everywhere divergent morphisms. These must correspond to everywhere divergent functions and, thus, behave like a zero w.r.t. right and left composition.

2.2 Definition. A category **C** has **partial morphisms** (is a pC) iff it has e.p. and

$$\forall b,c \in Ob_C \;\; \exists 0_{bc} \in C[b,c] \quad \forall a,d \in Ob_C \quad \forall f \in C[a,b] \quad \forall g \in C[c,d]$$
$$0_{bc} o f = 0_{ac} \quad \text{and} \quad g o 0_{bc} = 0_{cd} .$$

By 2.2, the "domain of definition" of a morphism $f \in C[a,b]$ is given by $\mathbf{dom}(f) = \{h \in C[t,a] \mid foh \neq 0_{tb}\}$.

2.3 Definition. Let **C** be a pC and t be a terminal object. Then the category of **total morphims**, C_T, has the same objects as **C** and morphisms defined as follows:

$C_T[a,b] = \{ f \in C[a,b] \mid \forall h \in C[t,a] (foh = 0_{tb} \Rightarrow h = 0_{ta})\}$.

2.4 Remark. Note that: 1) f, g total ⇒ fog total,
2) fog total ⇒ g total,
3) all monomorphisms are total.

Examples I. **pSet** is the category of sets with partial maps as morphisms. **pR** = **PR** is the monoid of the Partial Recursive functions (PR). **pEN** is the pC whose objects are pairs $\underline{a}$ = (a,e), where a is a countable set and $e : \omega \to a$ (onto) is an enumeration of a. Moreover, f is in **pEN**$[\underline{a},\underline{a}']$ iff there exists f' ∈ PR such that $f \cdot e = e' \cdot f'$. Clearly, $(\mathbf{pSet})_T$ = **Set**, $\mathbf{PR}_T$ = **R** and $(\mathbf{pEN})_T$ = **EN**, that is Malcev's category of numbered sets.

Since the beginning of denotational semantics of programming languages, the basic notions of approximation and continuity suggested the introduction of posets with a least element ⊥ . The bottom ⊥ provides the meaning to diverging computations over non trivial mathematical structures. This is mathematically very clear in several specific categories, such as continuous or algebraic lattices or cpo's, Scott's domains...... . It is not so obvious in interesting categories for computations such as **EN**, say see Asperti/Longo[1986]). (For typographical reasons, we write a^* instead of $a^{\perp}$; lifting and complete objects below were first defined, under different names, in Longo/Moggi[1984a]).

2.5 Definition. Let C be a pC. Then the **lifting** of $a \in Ob_C$ is the object a^* such that the functors $C[\_,a]$, $C_T[\_,a^*]$: C → **Set** are naturally isomorphic.

It is easy to chek that a^* defined as in 2.5 is unique, if it exists. The idea for a^* is very simple: in a pC any partial morphism may be uniquely extended to a total one, when the target object is "lifted" (and when the lifting exists).

2.6 Proposition. Let **C** be a pC and assume that for each $a \in Ob_C$ there exists the lifting a^*. Then there is a (unique) extension of the map $a \vdash a^*$ to a functor $\_^* : \mathbf{C} \to \mathbf{C_T}$ (the **lifting functor**).

Examples II. The lifting functor for **pSet** is obvious. It can be easely guessed also for the category **pPo** of posets and partial monotone functions with upward closed domains: just add a fresh least element and the rest is easy by applying proposition 2.6. The category **pCPO** is given by defining complete partial orders under the assumption that directed sets are not empty. Thus, the objects of **pCPO** do not need to have a least, bottom, element. As morphisms, take the partial continous functions with open domains following in Plotkin[1984]. Clearly, the lifting functor is defined as for **pPo.**

2.7 Definition. Let **C** be a pC. Then $a \in Ob_C$ is a **complete** object iff $a < a^*$ in $\mathbf{C_T}$.

The intuition should be clear. An object is complete when it "already contains", in a sense, the extra $\perp$. Think of an object d of **pCPO** and take its lifting d^* , i.e. add a least element $\perp$ to d. Then d is complete ($d < d^*$ via (i,j), say) iff d already contained a least element, $j(\perp)$ to be precise. Obviously, the objects of **CPO** are exactly the complete objects of **pCPO** . The following characterization gives the main motivation for the invention of complete objects: exactly on complete objects as targets all partial morphisms may be extended to total ones, *with the same target.* In a pC **C**, write

$\underline{f} \in \mathbf{C_T}[b,a]$ **extends** $f \in \mathbf{C}[b,a]$ iff $\forall h \in dom(f)\ \underline{f} \circ h = f \circ h$.

2.8 Theorem. Let **C** be a pC and a^* be the lifting of $a \in Ob_C$. Then

$$a < a^{\circ} \quad \Leftrightarrow \quad \forall b\ \forall f \in \mathbf{C}[b,a]\ \exists \underline{f} \in \mathbf{C_T}[b,a]\ \ \underline{f} \text{ extends } f\ .$$

2.9 Remark. (i) $b < a < a^{\circ} \Rightarrow b < b^{\circ}$. Note also that $a^{\circ} < a^{\circ\circ}$.
(ii) (informal) Let τ: $\mathbf{C}[b,a] \simeq \mathbf{C_T}[b,a^{\circ}]$; if $a < a^{\circ}$, then there exists a canonical retraction (in,out) where $in = \tau(id_a)$ is the "injection". Moreover, $\underline{f}$ in 2.8 is given by $\underline{f} = out \circ \tau(f)$.

The point with the categorical approach to complete objects is that their properties may be inherited at higher types.

2.10 Definition. A **partial Cartesian Closed Category** (pCCC) **C** is a pC which has all products and, for $a, b \in Ob_{\mathbf{C}}$, a (unique) object $b^a{}_p$, the **representation of partial morphisms**, such that there is a natural isomorphism $\Lambda : \mathbf{C}[\_ \times a, b] \simeq \mathbf{C_T}[\_, b^a{}_p]$.

In a pCCC we write b^a, if it exist, for the representation of total morphisms. That is, for the (unique, if it exits) object b^a such that there is a natural isomorphism $\Lambda_T : \mathbf{C_T}[\_ \times a, b] \simeq \mathbf{C_T}[\_, b^a]$.
Clearly, $\mathbf{C_T}$ is a CCC iff b^a exists always.

2.11 Proposition. Let **C** be a pCCC. Then
1) $\_^t{}_p : \mathbf{C} \to \mathbf{C}$ is the lifting functor in **C**,
2) for all $a, b \in Ob_{\mathbf{C}}$, $b^a{}_p$ is a complete object.

The next theorem "internalizes" the operation which extends each partial morphism to a total one, when the target is a complete object. It even gives a retraction $b^a < b^a{}_p$ in $\mathbf{C_T}$.

2.12 Theorem. Let **C** be a pCCC. Let $a,b \in Ob_{\mathbf{C}}$ be such that b^a exists. Then, if b is complete and τ is the isomorphism $\mathbf{C}[b^a{}_p \times a, b] \simeq \mathbf{C_T}[b^a{}_p \times a, b^{\circ}]$, one has

$$b^a < b^a{}_p \quad \text{via} \quad (\text{in}',\text{out}') ,$$

where $\text{out}' = \Lambda_T(\text{out}\circ\tau(\text{eval}))$ and $\text{in}' = \Lambda(\Lambda_T{}^{-1}(\text{id}_{b^a}))$. Moreover,

$$\forall c \in Ob_C \quad \forall f \in C[c\times a,b] \qquad \text{out}'\cdot\Lambda(f) = \Lambda_T(\text{out}\cdot\tau(f)) .$$

2.13 Remark. By 2.12, 2.11 and 2.9, completeness is inherited at higher types, also for total morphisms. That is, if b is complete and b^a exists, b^a also is complete.

Note that, as we deal with categories with e.p., for any $f \in C[t\times a,b]$ $\Lambda(f)$ is the "point" which represents f in $b^a{}_p$. Similarly for Λ_T w.r.t. b^a. Thus, if we "identify" $f \in C[a,b]$ with $\Lambda(f) \in b^a{}_p$ (and $g \in C_T[a,b]$ with $\Lambda_T(g) \in b^a$), theorem 2.12 gives

(2.13.1) $\quad \text{out}'(f) = \text{out}\circ\tau(f).$

This is exactly the definition of the extension $\underline{f}$ of f in 2.8.

3. Types as quotient sets.

The data types one is usually dealing in effective computations are countable sets, possibly structured by an order or similar relations. Indeed, since its origin, denotational semantics was based on the idea of interpreting (higher type) computations by countable approximations of (possibly infinite) processes. Thus even uncoutable sets for the interpretation of formal types have a countable and effective core. This is the leading idea for the various categories of Scott's domains (algebraic cpo's, effectively given domains....).

Countable or, more precisely, numbered sets may be viewed as quotients over the set ω of natural numbers. That is, each $\underline{A} = (A,e_a)$, where $e_A : \omega \to A$ is an onto map (numbering), defines an equivalence relation on ω by $n\underline{A}m$ iff $e_A(n) = e_A(m)$ (and conversely). Thus each element of A corresponds exactly to an equivalence class in ω and we may view at

Malcev-Ershov category of numbered sets as the category $\mathbf{ER}_\omega$ of equivalence relations on ω, whose morphisms are defined as follows. Let (P)R be the set of (partial) recursive functions. Then $f \in \mathbf{ER}_\omega[\underline{A},\underline{B}]$ iff there exists $f' \in R$ such that $f \cdot e_A = e_B \cdot f'$.

$\mathbf{ER}_\omega$ is a cartesian category (products are obvious), but not a CCC. It contains however several interesting full subCCC, such as Scott's constructive domains (see later) and a lot of higher type recursion theory may be carried on within it, see Ershov[1976] (or Longo/Moggi[1984]).

Clearly, given numbered sets $\underline{A}$ and $\underline{B}$, not all $f' \in R$ induces an $f \in \mathbf{ER}_\omega[\underline{A},\underline{B}]$, as f' must preserve $\underline{A}$-equivalences, that is $n\underline{A}m \Rightarrow f'(n)\underline{B}f'(m)$. This suggests a way to introduce higher type objects and thus to define a cartesian closed extension of $\mathbf{ER}_\omega$. Let $\{\varphi_i\}_{i \in \omega}$ be any acceptable goedel-numbering of PR. Define then

(Quot.) $\quad p\underline{B}^{\underline{A}}q$ iff $n\underline{A}m \Rightarrow \varphi_p(n)\underline{B}\varphi_q(m)$.

$\underline{A}^{\underline{B}}$ is a partial equivalence relation on ω, as it is defined on a subset of ω. Indeed, $\mathrm{dom}(\underline{B}^{\underline{A}}) = \{p \mid p\underline{B}^{\underline{A}}p\} \subseteq \omega$, and a partial numbering π_{AB}: $\mathrm{dom}(\underline{B}^{\underline{A}}) \to \underline{A}^{\underline{B}}$ is given by $\pi_{AB}(n) = \{m / n\underline{B}^{\underline{A}}m\}$.

In general, each partial surjective $\pi: \omega \to C$ uniquely defines a partial equivalence relation (and conversely).

3.1 Definition. The category $\mathbf{PER}_\omega$ of partial equivalence relations on ω has as objects the subsets of ω modulo an equivalence relation. Given objects $\underline{a} = (a,\pi_a)$ and $\underline{b} = (b,\pi_b)$, where π_a, π_b are partial numberings, $f \in \mathbf{PER}_\omega[\underline{a},\underline{b}]$ iff there exists $f' \in PR$ such that the f.d.c., where $\mathrm{dom}(\pi_x) = \omega_x$:

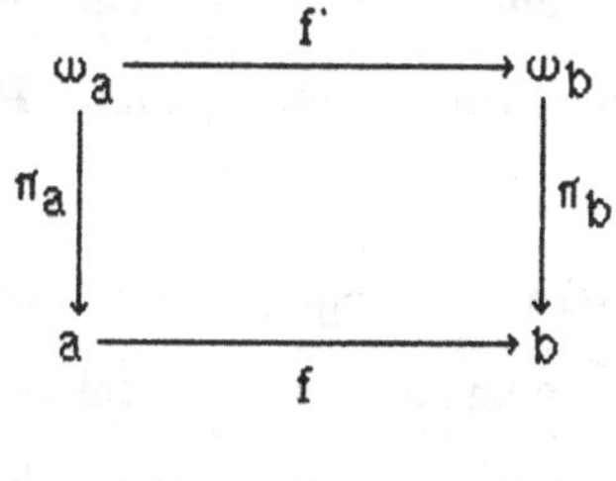

Clearly, $\mathbf{ER}_\omega$ is a full subcategory of $\mathbf{PER}_\omega$. Moreover, $\mathbf{PER}_\omega$ is a CCC.

Note that the representative $\underline{b}^{\underline{a}}$ of $\mathbf{PER}_\omega[\underline{a},\underline{b}]$ is partially enumerated by the quotient subset of ω determined by the partial relation $\underline{b}^{\underline{a}}$ (see (Quot.) above). That is, $\pi_{ab}(i) = f$ iff $f \cdot \pi_a = \pi_b \cdot \varphi_i$.

In Computer Science, $\mathbf{PER}_\omega$ is also known as the quotient set semantics of types over ω , following the ideas in Scott[1976] on λ-models (see Longo/Moggi[1984p] for details and further work on arbitrary (partial) combinatory algebras).

Classical computability suggests now a natural way to extend $\mathbf{PER}_\omega$ to a category with partial morphisms. Note that, by definition, if $p\underline{b}^{\underline{a}}p$ then φ_p is a (partial) recursive function which is total on dom($\underline{a}$), as we are defining an ordinary category with total morphisms. Just drop this condition and define $\mathbf{pPER}_\omega$ exactly as $\mathbf{PER}_\omega$ by allowing $f \in \mathbf{pPER}_\omega[\underline{a},\underline{b}]$ to be partial. More formally,

$$f \in \mathbf{pPER}_\omega[\underline{a},\underline{b}] \quad \text{iff} \quad \exists f' \in PR \; (\; n\underline{a}m \quad \underline{\text{and}} \quad f'(n)\downarrow \;\Rightarrow\; f'(m)\downarrow) \quad \underline{\text{and}} \;\; f\cdot\pi_a(n) = \pi_b \cdot f'(n) \; .$$

By checking the condition in 2.10 one may actually prove:

3.2 Theorem. $\mathbf{pPER}_\omega$ is a pCCC.

Clearly, $(\mathbf{pPER}_\omega)_T = \mathbf{PER}_\omega$.

(Notation : 1 - We will keep writing f', M', α' ... for the functions in PR defining the morphism f , M , α ... in $(\mathbf{p})\mathbf{PER}_\omega$.

2 - Set π_{abp} for the enumeration of $(\underline{b}^{\underline{a}})_p$).

$\mathbf{pPER}_\omega$ is the "effective" frame category which we are going to use in order to extend and apply theorem 1.14. It will serve as a tool for comparing type-structures of quotient sets within $\mathbf{PER}_\omega$ to the typed term model **Term** .

Observe first that **Term** = $((Term_\sigma)_{\sigma\in Tp}, [])$ is a collection of countable sets. Indeed each $Term_\sigma$ can be numbered by an injective $e_\sigma : \omega \to Term_\sigma$

Just code all terms in normal form and set $e_\sigma(i) = [N]$ if i is the code of the $\beta\eta$-normal form of N . Thus we may view $Term_\sigma$ as an object (Term $_\sigma$,e $_\sigma$) in $\mathbf{ER}_\omega$ and, hence, in $\mathbf{PER}_\omega$.

Moreover, $Term_{\sigma\tau} \subseteq \mathbf{PER}_\omega[Term_\sigma, Term_\tau]$, since for each $M \in Term_{\sigma\tau}$ there exixts $M' \in R$ such that $M \cdot e_\sigma = e_\tau \cdot M'$.

Clearly, M' depends uniformely effectively on M . That is,

(3.2.1) $\exists l \in R \;\; e_{\sigma\tau}(i) = M \;\Rightarrow\; \varphi_{l(i)} = M'$.

This will be used in theorem 3.4, jointly with the "inverse" property. The later property is formalized in 3.3 and gives some new information on typed terms and their relation to computable functions.

3.3 Main lemma. There exist a partial recursive function f such that for all $i \in \omega$, if there is a term $M^{\sigma\tau}$ for which one has $M^{\sigma\tau} \cdot e_\sigma = e_\tau \cdot \varphi_i$, then $M^{\sigma\tau} = e_{\sigma\tau}(f(i))$.

That is, for all $i \in \omega$ we can uniformely effectively find (an index for) M such that the following diagram commutes, if M exists:

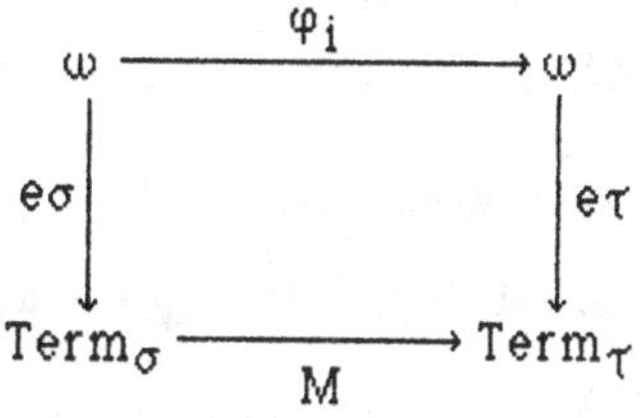

Proof : Clearly, (Diag.M) commutes iff

(1) $\exists M \in Term_{\sigma\tau} \;\; e_\tau(\varphi_i(e_\sigma^{-1}(x^\sigma))) = Mx^\sigma$, for x^σ not in FV(M)

Claim Let $M^{\sigma\tau}$ be in n.f., and $x^\sigma \neq y^\sigma$ two variables of type σ.

$Mx^\sigma[y^\sigma/x^\sigma] = My^\sigma \;\Leftrightarrow\; x^\sigma$ is not in FV(M).

(The proof is easy).

By definition, for all $N \in Term_\sigma$, $e_\tau(\varphi_i(e_\sigma^{-1}(N))) \in Term_\tau$. Then the algorithm which defines f goes as follows: given i , find x^σ and y^σ, if any, such that

(2) $$e_\tau(\varphi_i(e_\sigma^{-1}(x^\sigma)))[y^\sigma/x^\sigma] = e_\tau(\varphi_i(e_\sigma^{-1}(y^\sigma))) .$$

If and when the variables in (2) are found, set f(i) equal to the $\sigma\tau$-number of $\lambda x^\sigma.e_\tau(\varphi_i(e_\sigma^{-1}(x^\sigma)))$, that is

$$e_{\sigma\tau}(f(i)) = \lambda x^\sigma.e_\tau(\varphi_i(e_\sigma^{-1}(x^\sigma))) .$$

Observe now that (1) implies (2). (The converse doesn't need to hold). Therefore, if (1) applies, $e_{\sigma\tau}(f(i)) = \lambda x^{\sigma}.e_{\tau}(\varphi_i(e_{\sigma}^{-1}(x^{\sigma}))) = \lambda x^{\sigma}.Mx^{\sigma} = M$ by axiom (η). □

3.4 Theorem. Let $a = \{\underline{a}_i \mid i \in At\}$ be a collection of complete objects in $\mathbf{pPER}_{\omega}$ s.t. $\mathrm{Term}_i <_p \underline{a}_i$, for all $i \in At$. Then there exists a p.r.s. $(\alpha_{\sigma},\beta_{\sigma})_{\sigma\in Tp}$ in $\mathbf{pPER}_{\omega}$ from $(\mathrm{pPER}_{\omega})_a$ onto **Term**.

Proof: Let $\underline{a}_i >_p \mathrm{Term}_i$ via (α_i,β_i). Assume by induction that $(\alpha_{\sigma},\beta_{\sigma})$, $(\alpha_{\tau},\beta_{\tau})$, partial retractions in $\mathbf{pPER}_{\omega}$, have been defined. We will first construct $\beta_{\sigma\tau} \in \mathbf{PER}_{\omega}[\mathrm{Term}_{\sigma\tau}, a_{\sigma\tau}]$ satisfying <u>cond1</u> in 1.4.

Let $M \in \mathrm{Term}_{\sigma\tau}$. By definition, the f.d.c. :

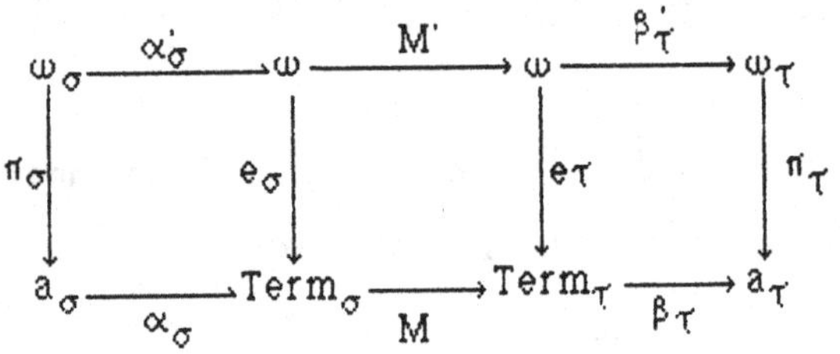

Define now $h_{\sigma\tau} : \mathrm{Term}_{\sigma\tau} \to (\underline{a}_{\tau}{}^{\underline{a}_{\sigma}})_p$ by

$$h_{\sigma\tau}(M) = \beta_{\tau}\cdot M\cdot\alpha_{\sigma} \in \mathbf{pPER}_{\omega}[a_{\sigma}, a_{\tau}].$$

Clearly $h_{\sigma\tau}$ is a well defined total map.

Claim : $h_{\sigma\tau} \in \mathbf{PER}_{\omega}[\mathrm{Term}_{\sigma\tau}, (\underline{a}_{\sigma}{}^{\underline{a}_{\tau}})_p]$.

Proof: see Asperti/Longo[1986]

As pointed out in 2.12, $a_{\sigma\tau} < (a_{\tau}{}^{a_{\sigma}})_p$ via (in',out') in $\mathbf{PER}_{\omega}$. Set then

$$\beta_{\sigma\tau} = \mathrm{out}' \cdot h_{\sigma\tau} \in \mathbf{PER}_{\omega}[\mathrm{Term}_{\sigma\tau}, a_{\sigma\tau}].$$

By 2.13.1, $\mathrm{out}'\cdot h_{\sigma\tau}(M) = \mathrm{out}'(h_{\sigma\tau}(M)) = \mathrm{out}\cdot\tau(h_{\sigma\tau}(M))$ and this is exactly the definition of the extension of $h_{\sigma\tau}(M) \in \mathbf{pPER}_{\omega}[a_{\sigma},a_{\tau}]$, in the sense of 2.8, to a morphism in $\mathbf{PER}_{\omega}[a_{\sigma},a_{\tau}]$. Indeed, by assumption on a and 2.13, $\forall\tau \in Tp$, a_{τ} is a complete object. Thus, if $h_{\sigma\tau}(M)(y)$ converges (i.e. $y \in \mathrm{dom}\alpha_{\sigma}$), $\beta_{\sigma\tau}(M)(y) = h_{\sigma\tau}(M)(y) = \beta_{\tau}\cdot M\cdot\alpha_{\sigma}(y)$ and this is exactly <u>cond2</u> for $\beta_{\sigma\tau}$.

We now need to define $\alpha_{\sigma\tau} \in \mathbf{pPER}_{\omega}[a_{\sigma\tau}, \mathrm{Term}_{\sigma\tau}]$. By definition the f.d.c.:

$$\begin{array}{ccccccc}
\omega & \xrightarrow{\beta'_\sigma} & \omega_\sigma & \xrightarrow{\varphi_i} & \omega_\tau & \xrightarrow{\alpha'_\tau} & \omega \\
\downarrow e_\sigma & & \downarrow \pi_\sigma & & \downarrow \pi_\tau & & \downarrow e_\tau \\
\mathrm{Term}_\sigma & \xrightarrow[\beta_\sigma]{} & a_\sigma & \xrightarrow[\pi_{\sigma\tau}(i)]{} & a_\tau & \xrightarrow[\alpha_\tau]{} & \mathrm{Term}_\tau
\end{array}$$

Define then $g_{\sigma\tau} : (a_{\sigma\tau}) \to (\mathrm{Term}_\tau{}^{\mathrm{Term}\,\sigma})_p$ by

$g_{\sigma\tau}(\pi_{\sigma\tau}(i)) = \alpha_\tau \cdot \pi_{\sigma\tau}(i) \cdot \beta_\sigma \in \mathbf{pPER}_{\omega}[\mathrm{Term}_\sigma, \mathrm{Term}_\tau]$.

Clearly $g_{\sigma\tau}$ is a well defined total map.

Claim : $g_{\sigma\tau} \in \mathbf{PER}_{\omega}[a_{\sigma\tau}, (\mathrm{Term}_\tau{}^{\mathrm{Term}\,\sigma})_p]$.

Proof: see Asperti/Longo[1986].

Let f be the function given by lemma 3.3. Define then $\alpha_{\sigma\tau}$ by

$$\alpha_{\sigma\tau}(\pi_{\sigma\tau}(i)) = e_{\sigma\tau}(f(r(i))) \quad \text{for all} \quad i \in \omega .$$

Note now that, by definition, f above preserves indexes, that is $\varphi_i = \varphi_j \Rightarrow e_{\sigma\tau}(f(i)) = e_{\sigma\tau}(f(j))$. Thus $\alpha_{\sigma\tau} \in \mathbf{pPER}_{\omega}[a_{\sigma\tau}, \mathrm{Term}_{\sigma\tau}]$ with $\alpha'_{\sigma\tau} = f \cdot r$. We only have to show that $\alpha_{\sigma\tau}$ satisfies <u>cond1</u>. That is, suppose that

(**) $\exists M \in \mathrm{Term}_{\sigma\tau} \quad \alpha_\tau \cdot \pi_{\sigma\tau}(i) \cdot \beta_\sigma = M$

(this is the assumption in <u>cond1</u>). Recall that, by (*) above,

$$\alpha_\tau \cdot \pi_{\sigma\tau}(i) \cdot \beta_\sigma \cdot e_\sigma = g_{\sigma\tau}(\pi_{\sigma\tau}(i)) \cdot e_\sigma = e_\tau \cdot \varphi_{r(i)}$$

and, hence, condition (**) can be rewritten as:

(***) $\exists M \in \mathrm{Term}_{\sigma\tau} \quad e_\tau \cdot \varphi_{r(i)} = M \cdot e_\sigma$.

Finally, by the construction of f , if (***) holds, one has:

$M = e_{\sigma\tau}(f(r(i)) = \alpha_{\sigma\tau}(\pi_{\sigma\tau}(i))$ and this proves <u>cond1</u> . □

We conclude this section by a very loose summary of applications of the previous results and their consequences (see S.1) to some interesting type-structures in denotational semantics.

The category **CD** of constructive domains may be naturally defined from Scott's effectively given domains (see Giannini/Longo[1984] or Ershov[1976] for a detailed definition). In short, each constructive domain is the collection of countable elements of some effectively given domain. By generalized Myhill-Shepherdson theorem **CD** is a full subCCC of $\mathbf{ER}_\omega$. Indeed, it may be embedded into $\mathbf{PER}_\omega$ by a full and faithfull functor which preserves products and function spaces. Clearly all objects in **CD** are complete as they possess a least, bottom, element.

As already pointed out , the usual ground types of data are trivially in **CD**, e.g. all flat cpo's (such as $\omega^\perp$). Moreover, given $i \in At$, all what is required in order to have a partial retraction, from a constructive domain a onto $Term_i$, is that a contains as many incompatible elements as the cardinality of $Term_i$. Thus, one may also take as ground types the following objects in **CD** and 3.4 applies: the partial recursive functions, the effectively generated trees on some countable alphabet...... Interesting instances of the later example of data type are the (possibly infinite) parenthesized expressions in a language (e.g. LISP S-expressions).

Aknowledgements We are greatly endebted to Eugenio Moggi for a few stimulating discussions which provided the main hints for the solution of some problems we had. Moreover, Eugenio's recent work on partial morphisms set the basis and the guidelines for the results stated here on that matter.

References

Asperti A., Longo G. [1986] "Categories of partial morphisms and the relation between type-structures", Dip. di Inf., Pisa.

Barendregt H. [1984], **The lambda calculus; its syntax and semantics**, Revised edition, North Holland.

Berry G. [1979] "Modèles complètement adequats et stables des lambda-calculus types", Thèse de Doctorat, Universite' Paris VII.

Berry G., Curien P.L. "Sequential Algorithms on Concrete Data Structures" **Theor. Comp. Sci. 20**, (265-321).

Dibjer P. [1983], "Category-theoretic logics and algebras of programs," Ph.D. Thesis, Chalmers University, Goeteborg.

Di Paola A., Heller A. [1984] "Dominical Categories," City Univ. New York.

Ershov Yu. L. [1976] "Model C of the partial continuous functionals," **Logic Colloquium 76** (Gandy, Hyland eds.) North Holland, 1977.

Friedman H. [1975], "Equality between functionals," **Logic Colloquium** (Parikh ed.), LNM 453, Springer-Verlag.

Giannini P., Longo G. [1984] "Effectively given domains and lambda calculus semantics," **Info.&Contr.** 62, 1 (36-63).

Gödel K. [1958] "Ueber eine bicher noch nicht benuetze Erweiterung des finiten Standpuntes," **Dialectica**, vol.12, pp.280-287.

Hyland M. [1982] "The effective Topos," *in* **The Brouwer Symposium**, (Troelstra, Van Dalen eds.) North-Holland.

Lambek J. [1980] "From lamba-calculus to cartesian closed categories," *in* Hindley/Seldin[1980].

Longo G., Moggi E. [1984] "The Hereditary Partial Recursive Functionals and Recursion Theory in higher types," **J. Symb. Logic**, vol. 49, (pp. 1319-1332).

Longo G., Moggi E. [1984p] "Goedel-numberings, principal morphisms, Combinatory Algebras: a category-theoretic characterization of functional completeness," *(prelim. version)* **Math. Found. Comp. Sci.**, Prague 1984 (Chytil, Koubek eds.) LNCS 176 , Springer-Verlag, 1984 (pp. 397-406).

Longo G., Moggi E. [1984a] "Cartesian Closed Categories of Enumerations and effective Type Structures," **Symposium on Semantics of Data Types** (Khan, MacQueen, Plotkin eds.), LNCS 173, Springer-Verlag, (pp. 235-247).

Moggi E. [1985] "Partial Morphisms in Categories of effective objects," Comp. Sci. Dept., Univ. of Edinburgh.

Plotkin G. [1980/84] "Domains", lecture notes, C.S. Dept. Edinburgh

Plotkin G. [1982] "Notes on completeness of the full continous type hierarchy", Manuscript, M.I.T.

Poigne' A. [1985] "On specification Theories and Models with higher types", to appear in **Info.&Contr.**

Rosolini G. [1985] "Domains and dominical categories" , C.S. Dept., C.M.U.

Scott D. [1976] "Data types as lattices," **SIAM Journal of Computing**, 5 (pp. 522-587).

Scott D. [1980] "Relating theories of the lambda-calculus," *in* To H.B. Curry: essays in lambda-calculus...Hindley, Seldin (eds.), Academic Press.

Scott D. [1982] "Domains for denotational semantics," (preliminary version), Proceedings ICALP 82, **LNCS** 140, Springer-Verlag).

Smyth M. [1977] "Effectively Given Domains", **Theoret. Comput. Sci. 5,** pp 255-272.

Smyth M., Plotkin G. [1982] "The category-theoretic solution of recursive domain equations" **SIAM Journal of Computing** 11, (pp.761-783).

Statman R. [1980] "Completeness, invariance and lambda-definability," **J. Symb. Logic** 47, 1 , (pp. 17-26).

Statman R. [1984] "Logical relations and typed lambda-calculus" **Info&Contr.** (to appear).

Statman R. [1985] "Equality between functionals revisited" *in* **H. Friedman's research on the found. of Math.** (Harrington et al. eds), North Holland.

Troelstra A. [1973] "Metamathematical investigation of Intuitionistic Arithmetic and Analysis," **LNM 344** , Springer-Verlag, Berlin.

ON APPLICATIONS OF ALGORITHMIC LOGIC

G.Mirkowska
A.Salwicki
University of Warsaw
Dept. of Mathematics and Informatics
PkiN room 850
00-950 Warsaw POLAND

ABSTRACT

In this survey we contained a short presentation of algorithmic logic AL and two sections devoted to selected applications of algorithmic logic. First, we argue that the goal of formal definition of a programming language can be fully achieved with the tools offered by AL. Next, we show possible applications of AL in production of modular software. The interaction between theoretical tools (AL) and software ones (LOGLAN programming language) seems to provide a reasonable way toward creation of reliable software from prefabricated modules and their specifications. It seems that in this way one can speed up the task of production of software.

key words:
program, computation, semantical properties, expressivity, algorithmic logic, axioms, inference rules, completeness, specification, verification, implementation, modules, extension, LOGLAN

AL algorithmic logic denotes in fact a family of logics. Every algorithmic logic consists of its language and of its axiomatic system. The language is an extension of a language of algorithms or a programming language. The system contains axioms and inference rules. Every algorithmic language contains three subsets: programs, terms and formulas. Formulas enable to express semantical properties of programs in this way that for a semantical property of program(s) there exists a formula β such that the property holds iff the formula β is valid. Next, the task of validation of formula can be replaced by a proof of it. In order to do so one need a sufficiently complete set of axioms and inference rules.

It turns out that this approach has many applications:
- one can study behaviour of programs a priori, before computation, by proving corresponding formulas that express semantical properties of programs,
- certain data structures can be specified as those algebraic systems which satisfy corresponding set of non-logical, specific axioms. This may seem too simple remark. However, in our case we allow that axioms can be algorithmic formulas beyond the set of first order formulas. This proposal has many consequences. On one hand it enables to axiomatize data structures which can not be axiomatized in first-order logic. Semantical property of a program can determine a data structure up to isomorphisms. On the other hand it simplifies the proofs of semantical properties, for it allows to hide the induction on the structure of elements of the data type and replaces it with inference rules which transform semantical property of a program into another semantical property.

1. AL - a short presentation

In this section we shall present logic of deterministic iterative programs. In spite of its simplicity it has many applications in more advanced fields like e.g. abstract data types.

LANGUAGE

In fact we shall deal with a class of languages. All languages have the same grammar. They can differ only due to the different sets of functional and relational signs. A language is a pair $L = (A, WFE)$ where

A - the alphabet is the set of admissible signs,

WFE - is the set of well formed experssions, it consists of terms, formulas and programs. We shall omit boring details (c.f. AL) for strict definitions.

The structure of the set of programs will be seen below.

SEMANTICS

We assume that the semantics of terms and classical formulas is known (c.f. RS). We shall recall only denotations. Let L be a fixed algorithmic language. Let R be an algebraic system, we shall call it also a data structure. such that the signatures of L and R agree (are similar). By a valuation we mean any mapping which for every variable assigns a value. Value of apropositional variable must be logical, value of individual variable is an element of the universe of R.
For a term τ by $\tau_R(v)$ we denote the value of the term in the data structure R at the valuation v. Similarly, for a formula α by $\alpha_R(v)$ we denote the value of the formu α in R at v. The meaning of programs can be defined on many various ways. Here, it will be defined with help of the notion of computation i.e. a sequence of configuratio

Configuration is an ordered pair

$$\langle v, I_1; \dots, I_n \rangle$$

where v is a valuation of variables i.e. state of memory

$I_1 \dots I_n$ are instructions, finite sequence of them represents state of control.

The pair (v,K) is called the initial configuration of the program K at the valuation v.

We assume the following transformation rulesi.e. transition relation in the set of configurations. Let REM denote any finite sequence of instructions.

(v, x:=ω;REM) $\mapsto$ (v',REM) where $v'(z) = v(z)$ for $z \neq x$
$v'(x)$ = value of ω at v, $\omega_R(v)$

(v, begin K;M end; REM) $\mapsto$ (v, K; R; REM)

(v, if γ then K else M fi;REM) $\mapsto$ (v,K; REM) if $\gamma_R(v)$ = true

(v, if γ then K else M fi; REM) $\mapsto$ (v,M; REM) if $\gamma_R(v)$ = false

(v, while γ do K od; REM) $\mapsto$ (v, REM) if $\gamma_R(v)$ = false

(v, while γ do K od; REM) $\mapsto$ (v,K; while do K od;REM) if $\gamma_R(v)$ = true

By a computation we understand any sequence of configurations such that for every two consecutive configurations $c_i \mapsto c_{i+1}$ holds.
A computation is finite if it reaches a configuration of the form (v', $\emptyset$) where the list of instructions is empty.
For every program K, we define a partial mapping K_R putting

$K_R(v) = v'$ iff the unique computation which starts at the configuratio (v,K) is finite and ends at (v', $\emptyset$) .

We shall say that the valuation v ' is the result of computation of K at data v.

SEMANTICAL PROPERTIES OF PROGRAMS - a sample

Halting property.

Stop K at v $\equiv$ the computation starting at the configuration (v,K) is finite.

Stop K $\equiv$ all computations of K (for every data v) are finite.

Correctness with respect to a given pre- and post-conditions

Corr K w.r.t. α, β $\equiv$ for every valuation v, if the formula α (precondition) is satisfied by v then the computation starting at (v,K) is finite and its result v' satisfies formula β.

Partial correctness

PCorr K w.r.t. α, β $\equiv$ for every valuation v, if the formula α is satisfied by v and if the computation starting at (v,K) is finite then its result v' satisfies β.

Weakest precondition of a formula α w.r.t. a program M
is any formula β such that for every data structure R the following two conditions hold

(i) if an initial valuation v (data) satisfies formula β then the computation of M is finite and the result satisfies α ,

(ii) let β' be another formula with the property (i), then the implication ($\beta' \Rightarrow \beta$) is valid in R.

Strongest postcondition of a formula α w.r.t. a program M
is any formula δ such that for every data structure R the following two conditions hold

i) if data satisfy the formula α and program M halts then the result v' satisfies the formula δ,

(ii) let δ' be another formula with the property (i) then the implication ($\delta \Rightarrow \delta'$) is valid in R.

Equivalence with respect to a postcondition β

We say that two programs M and K are equivalent with respect to a formula β iff for every valuation v either both computations starting at (v,M) and (v,K) respectively are infinite , or both are finite and both results of computations satisfy β.

We say that two programs are equivalent if they are equivalent with respect to every formula β.

EXPRESSIVITY

Our nearest goal is to find a language in which we shall be able to express semantical properties of programs. We are doing so by definition of a language which is a common extension of the language of first-order formulas and the language of algorithms(programs). The extension contains every expression of the form $K\alpha$ where K is a program and α is a formula of the extended language. We demand that the set of formulas is closed with respect to the usual formation rules.

The meaning of a formula $K\alpha$ is determined as follows: for every data structure R and a given valuation v we put the value of K at v as equal to the value of the formula at the resulting valuation v' of the computation that starts at (v,K) if the computation is finite, otherwise we define the value of K as false. In other words we demand that the following diagram commute

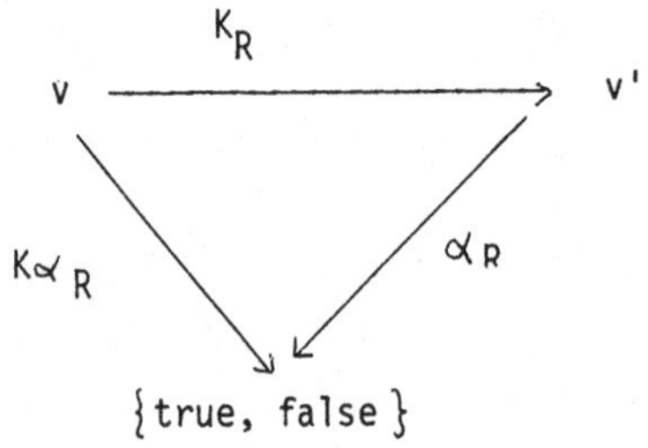

It is evident that Stop K property is expressed by the formula Ktrue. Similarly, correctness is expressed by

$$\alpha \Rightarrow K\beta$$

and partial correctness by

$$(\alpha \wedge K\,\text{true} \Rightarrow K\beta) \qquad \text{or} \qquad K\,\text{true} \Rightarrow (\alpha \Rightarrow K\beta)$$

Two programs K and M are equivalent with respect to a postcondition β iff the formula $K\beta \Leftrightarrow M\beta$ holds.

After a moment of reflection we realize that $K\alpha$ is the weakest precondition. Other semantical properties e.g. the strongest postcondition can be expressed too.

Our next goal is to find a deductivesystem in order to be able to prove valid algorithmic formulas.

LOGICAL SYSTEM

consists of the following set of axioms and inference rules.

axioms

(i) all formulas of the form of tautologies of classical propositional calculus cf. t_1 - t_{12} in (RS)

(ii) decomposition of programs

$(x := \omega)\gamma \equiv \gamma(x/\omega)$ where γ is a quantifier and program free formula, $\gamma(x/\omega)$ is again a formula obtained as a result of simultaneous replacement of all accurrences of x by ω

$\underline{\text{begin}}\ K;\ M\ \underline{\text{end}}\ \alpha \equiv K\,M\,\alpha$

$\underline{\text{if}}\ \gamma\ \underline{\text{then}}\ K\ \underline{\text{else}}\ M\ \underline{\text{fi}}\ \alpha \equiv ((\gamma \wedge K\alpha) \vee (\neg\gamma \wedge M\alpha))$

$\underline{\text{while}}\ \gamma\ \underline{\text{do}}\ M\ \underline{\text{od}}\ \alpha \equiv ((\neg\gamma \wedge \alpha) \vee (\gamma \wedge M\ \underline{\text{while}}\ \gamma\ \underline{\text{do}}\ M\ \underline{\text{od}}\ \alpha))$

(iii) distributivity axioms

$M(\alpha \wedge \beta) \equiv M\alpha \wedge M\beta$

$M(\alpha \vee \beta) \equiv M\alpha \vee M\beta$

$M\neg\alpha \Rightarrow \neg M\alpha$

$M\ \text{true} \Rightarrow (\neg M\alpha \Rightarrow M\neg\alpha)$

$M(\alpha \Rightarrow \beta) \Rightarrow (M\alpha \Rightarrow M\beta)$

$M\ \text{true} \Rightarrow ((M\alpha \Rightarrow M\beta) \Rightarrow M(\alpha \Rightarrow \beta))$

inference rules

$$\frac{\alpha,\ \alpha \Rightarrow \beta}{\beta} \qquad \frac{\alpha \Rightarrow \beta}{M\alpha \Rightarrow M\beta}$$

$$\frac{\{M(\underline{\text{if}}\ \gamma\ \underline{\text{then}}\ K\ \underline{\text{fi}})^i(\neg\gamma \wedge \alpha) \Rightarrow \beta\}_{i \in N}}{M\ \underline{\text{while}}\ \gamma\ \underline{\text{do}}\ K\ \underline{\text{od}}\ \alpha \Rightarrow \beta}$$

Logical axioms and inference rules allow to define the notions of formal proof and of consequence operation C. A formula α is a consequence of a set Z of formulas iff it has a proof from Z, in signs, $Z \models \alpha$.

Every triple $\langle L, C, A \rangle$ consisting of an algorithmic language L, a consequence operation C and of additional set A of non-logical axioms will be called a formalized algorithmic theory.

COMPLETENESS

The following theorem shows that the set of axioms and inferencerules is chosen properly

Theorem (on completeness Mir71)
Let $T = \langle L, C, A \rangle$ be a consistent algorithmic theory. Let α be a formula.
The following conditions are equivalent:
(i) the formula α has a formal proof from A, i.e. $\alpha \in C(A)$,
(ii) α is valid in every model of the theory T.

The next theorem is important in studies of abstract data types.

Theorem (on existence of models)
Every consistent algorithmic theory has a model.

Theorem (an analogue of downward Skolem-Löwenheim theorem)
If a theory has an infinite model than it has a denumerable model.

The results on completeness and on the existence of models enable to transform questions concerning semantical properties of programs and of data structures into syntactical questions about existence of proofs.

The questions of decidability were studied by various authors, A.Kreczmar (Kre74 estimated complexity of most important properties: Stop, Pcorr, Corr, W.Dańko (Dan80) found a criterion of undecidability of an algorithmic theory.

II Axioms of AL define semantics.

In this section we shall examine the following question:
Is it possible to develop a formal definition of a real programming language making use of AL as a base?

We shall arrive to an affirmative answer in a few steps, we show that the following elemnts of a programmming language can be axiomatized:

- meaning of program constructions(assignment, composition begin ... end, branching if ... then... else...fi, iteration while... do ... od),
- meaning of primitive data types e.g. integers,
- meaning of composed data types e.g. arrays,
- meaning of the notion of reference.

Hence a PASCAL-like language can be supplied together with the collection of axioms defining entire language, all its facets. We claim that a similar procedure can be applied to a more advanced programming language e.g. LOGLAN.

MEANING OF PROGRAM CONNECTIVES IS DEFINED BY AXIOMS OF AL

Def. 1 By an execution method I we shall mean a function which to every program of algorithmic language L assigns a binary relation in the set of all valuations in a given data structure.

Def.2 By standard execution method we understand a mapping I_s which to every program K assigns the relation

$I_s(K) = \{(v,v')$: there exists a finite computation leading from initial configuration (v,K) to the final one $(v', \emptyset)\}$

Def.3 We say that an execution method is proper for algorithmic logic iff the satisfiability relation which is based on it alllows to prove the soundness of AL axiomatization.

Theorem II.1
The standard execution method is proper for AL.

This in fact, is the adequacy theorem which was proved on the way toward the completeness result.

Def.4 By a semantic structure for L we shall mean the triple $\langle A, I \models \rangle$ where A is a data structure for L (i.e. an algebraic system of signature equal to that of L), I is an execution method and $\models$ is a satisfiability relation.

In the sequel we shall restrict our considerations to the class of semantic structures $\langle A, I \models \rangle$ such that

(1) the data structure A is normalized, i.e. for arbitrary valuations v,v' in A

$$\text{if } (\forall \alpha)\ \alpha_A(v) = \alpha_A(v') \quad \text{then } v = v'$$

(different valuations can be distinguished by means of a formula),

(2) the satisfiability relation $\models$ satisfies the following requirements

$$(\alpha \vee \beta)_A(v) = \text{true} \quad \text{iff} \quad \alpha_A(v) = \text{true} \text{ or } \beta_A(v) = \text{true}$$

$$(\alpha \wedge \beta)_A(v) = \text{true} \quad \text{iff} \quad \alpha_A(v) = \text{true} \text{ and } \beta_A(v) = \text{true}$$

$$(\neg\alpha)_A(v) = \text{true} \quad \text{iff} \quad \alpha_A(v) = \text{false}$$

$$(M\alpha)_A(v) = \text{true iff } (\exists v')\ (v,v') \in I(M) \text{ and } \alpha_A(v') = \text{true}$$

for arbitrary formulas α, β, program M and valuation v.

Without loss of generality we can assume that the algorithmic language L contains binary predicate of identity =. By a Herbrand semantical structure we shall mean any semantical structure which satisfies the following condition:

for every element a of the data structure A there exists a term τ_a such that

$$a = \tau_{aA}(v) \quad \text{for every valuation v in A.}$$

Theorem II.2

If a semantic structure is a Herbrand structure and if it is proper for AL then its execution method is identical with the standard execution method I_s.

PRIMITIVE DATA TYPES CAN BE AXIOMATIZED

It is well known that the set of natural numbers is the only up to isomorphism model of the following three formulas:

$$(\forall x) \quad \neg\, s(x) = 0$$

$$(\forall x,y) \quad s(x) = s(y) \Rightarrow x = y$$

$$(\forall x)\ \underline{\text{begin}}\ x := 0;\ \underline{\text{while}}\ x \neq y\ \underline{\text{do}}\ x := s(x)\ \underline{\text{od}}\ \text{true}$$

In real programming languages we encounter data types integer and real. The other primitive data types are easy to define. In order to define the primitive data type integer we have to axiomatize the class of finite subsets of the set Z of integers such that an element of the class consists of all integers between -M and M for certain integer M. For the lack of space we present a shorter axiomatization of unsymmetric set of positive integers modulo M+1. The formulas

$s(M) = 0$ here M is a constant added to the language

$s(x) = s(y) \Rightarrow x = y$

<u>while</u> $x \neq y$ <u>do</u> $x := s(x)$ <u>od</u> true

axiomatizes the class of all finite initial segments of the set of natural numbers. More precisely we are able to prove that every model of the above axioms is of the form

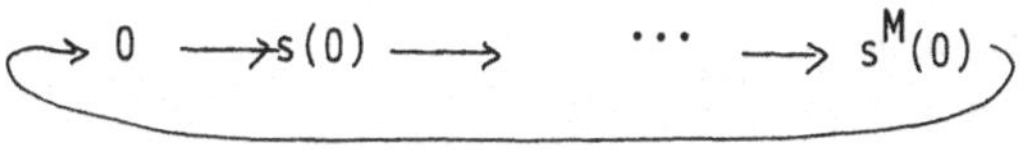

COMPOSED DATA TYPES. COROLLARY.

Composed data types are defined as arrays, records or structures with pointers. In her Ph.D. thesis H.Oktaba (Okt81) constructed algorithmic theories of arrays and of memory management systems. She proved representation theorems for both theories. In this way we obtained a characterization of the notion of reference.

Putting all these axioms and theories together we obtain a consistent algorithmic theory which has only standard models. It is possible to use these axioms in two directions:

a) as a criterion of correctness of an implementation of a programming language and/ or

b) as a set of axioms from which we shall attempt to prove semantical properties of programs written in the axiomatized language.

III AL in the process of software production.

The need for the tools that enable fast production of reliable software is unquestionable. We share the opinions which foresee industrial methods of software production. Here we shall discuss certain questions concerning with development of modules of software in a way that enables exchange and multiple use of modules. Our remarks take as a departure point the principle of factorization first time formulated in (Hoa72). In the majority of cases we are to develop a piece of software which performs certain operations not available in a moment. In other words, our future program is to be executed in a data structure other than supplied by hardware and system software. In 1972 C.A.R.Hoare remarked that in such case one should factori the goal onto two subgoals:

i) to specify and implement a data structure,

ii) to design, analyze and use an "abstract" program.

According to this advice we should develop two modules

Abstract program		Implementing module

The only link between these two pieces of software should consist of

Specification of data structure

The team developping abstract program should base on the specification only. The implementing team uses the specification as a criterion of correctness of an implementation.

The virtues of this method are manyfold. The principle of factorization makes possible to execute the abstract program in the presence of different implementing mo- dules. But a correct program need not to be adjusted! We can gain (or loose) on efficiency of computations depending on our choice of implementation for data structure. Another advantage of the method consist in the possibility of multiple applications of once created implementing module. Such a module realizes a problem-oriented language.

The work should have at least three visible stages:

a) formulation of a specification i.e. an axiomatic description of data structure

b) design of abstract program and its verification basing upon specification,

c) realization of the data structure and verification of its correctness (also basing on specification, we verify the validity of axioms).

EXAMPLE _ SIMULATION OF BANK DEPARTMENT

In this case it is natural to develop two modules

BANK	SIMULATION of BANK DEPARTMENT
defines notions & operations	uses notions &operations
type bankCustomer	A,B: bankCustomer
teller	C,D,E: teller
stand	F,G : stand
proc CustomerComes	call tellerServes
decides	call decides(..)
tellerServes	new bankCustomer

However, more detailed analysis leads to the conclusion that we should factorize several times in order to abstract and determine following modules - data structures.

BANK	OFFICE	SIMULATION KIT	PRIORITY_QUEUES	QUEUES

Remark.The last two notions are completely different ones (c,f, AHU74).

A kit for simulation should contain at least two general notions, of simulated process and of event_notice since we shall conduct our simulation putting these notices on "time axis". In fact we need that processes can stay in a queue, hence the notion of simprocess should be an extension of the notion of element of a queue. Event notices and time axis form another structure of priority queues.

What are queues and priority queues?

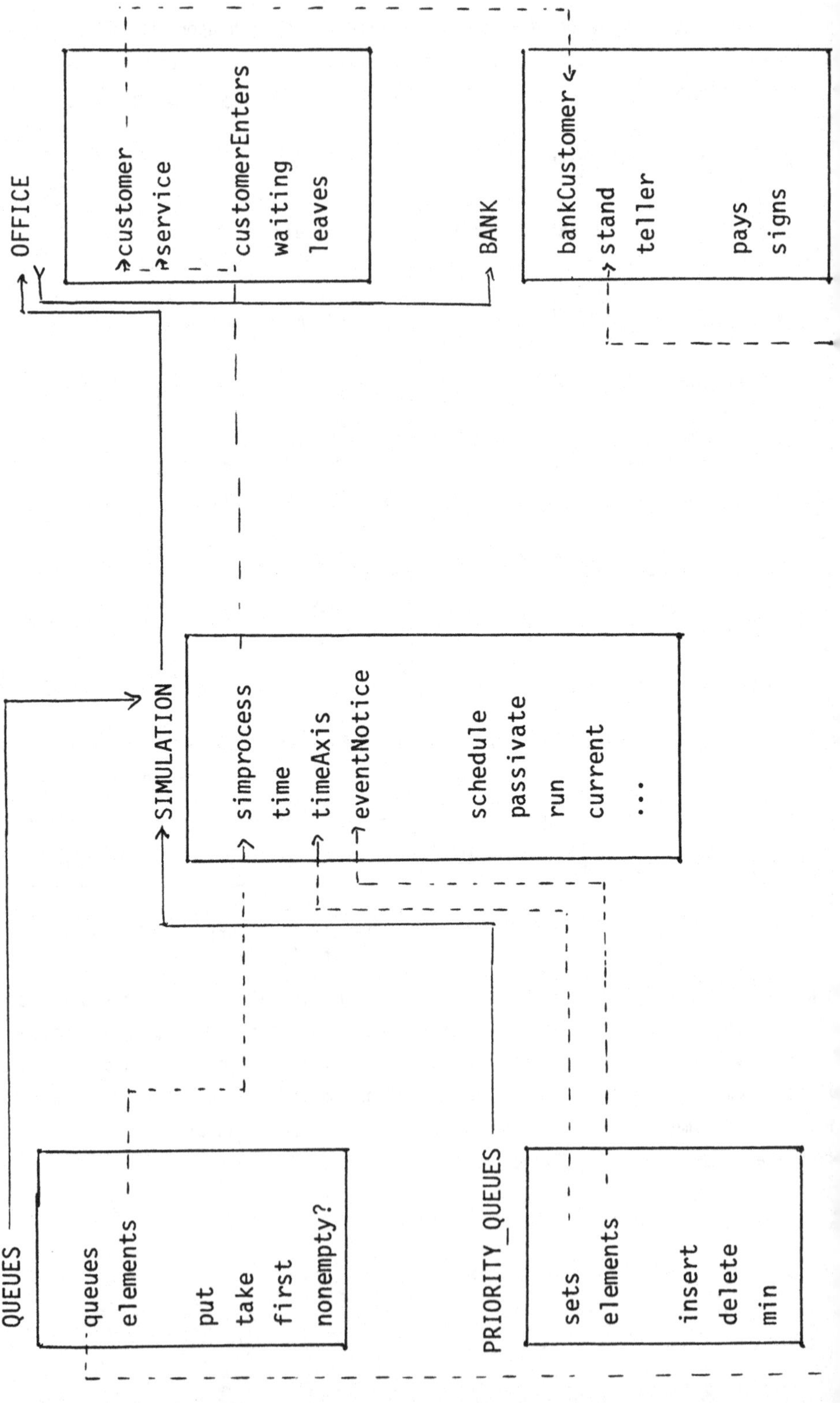
OFFICE
customer
service
customerEnters
waiting
leaves
BANK
bankCustomer
stand
teller
pays
signs
SIMULATION
simprocess
time
timeAxis
eventNotice
schedule
passivate
run
current
...
QUEUES
queues
elements
put
take
first
nonempty?
PRIORITY_QUEUES
sets
elements
insert
delete
min

QUEUES

An abstract formal specification of the notion of queue should be free of implementation details since they can look differently in various implementations.

We propose to consider structure with two sorts of elements $E \cup Q$

with operations $w: E \times Q \to Q$ and relations

$u: Q \to Q$ $n? : Q \to \{true, false\}$

$p: Q \to E$ $=$ (identity relation)

A structure which has this signature

$$\langle E \cup Q;\ w,\ u,\ p,\ n?,\ = \rangle$$

is called a queue structure iff it satisfies the following axioms

$(\forall e)\ (\forall q)$ $n?(w(e, q))$

$(\forall e)\ (\forall q)$ $-n?(q) \Rightarrow p(w(e, q)) = e$

$(\forall e)\ (\forall q)$ $n?(q) \Rightarrow p(w(e, q)) = p(q)$

$(\forall e)\ (\forall q)$ $-n?(q) \Rightarrow u(w(e, q)) = q$

$(\forall e)\ (\forall q)$ $n?(q) \Rightarrow u(w(e, q)) = w(e, u(q))$

$(\forall q)$ while $n?(q)$ do $q := u(q)$ od true

The last axiom is an algorithmic formula stating that every queue q can be emptied in a finite time, hence it is finite.

Theorem

Every data structure of similar signature e.g.

$$\langle A_1 \cup A_2;\ f1,\ f2,\ f3,\ r1,\ = \rangle$$

which satisfies the above axioms is isomorphic to so called standard structure of finite sequences of elements from A_1

$$\langle A_1 \cup A_1^* ;\ d,\ o,\ p,\ z,\ = \rangle$$

where operation d(a,b) denotes adding element a at the end of sequence b,
operation o(b) denotes a sequence which results from b by rejecting first el.
operation p(b) denotes first element in the sequence b ,
relation z consits of empty sequence only.

By this representation theorem we convince ourselves that the specification is consistent and contains all necessary properties of queues. It is in some sense complete set of properties of queues, their elements and operations on them. From implementors one should demand that their implementation meets the criteria listed in the axioms. For a user this set is sufficiently complete to prove any true property of program.

PRIORITY QUEUES

A system of priority queues enables operating on finite subsets of a certain linearly ordered set E of elements. We shall consider the following operations: insert an elemnet e to a set s, check whether e is an element of set s, delete e from s, show the least element in s.

Formally, by a priority queue we understand any algebraic system

$$\langle E \cup S \; ; \; i, d, min, mb, em, \leq \;, = \rangle$$

where

$i : E \times S \to S$

$d : E \times S \to S$

$min : S \to E$

$em : S \to \{true, false\}$

$: S \times S \to \{true, false\}$

$mb : E \times S \to \{true, false\}$

which satisfies the following axioms

A1) the set E is linearly ordered by the relation

A2) $- em(s) \Rightarrow [(\forall e)\; mb(e,s) \Rightarrow min(s) \leq e]$

A3) $[s' := i(e,s)]\{ mb(e,s) \wedge [e' \neq e \Rightarrow (mb(e',s) = mb(e',s'))]\}$

A4) $[s' := d(e,s)]\{ -mb(e,s) \wedge [e' \neq e \Rightarrow (mb(e',s) = mb(e',s'))]\}$

A5) <u>while</u> -em(s) <u>do</u> s := d(min(s),s) <u>od</u> true

A6)
```
mb(e,s) = begin s1 := s; bool := false;
                while -em(s1)∧ -bool do
                   e1 := min(s1);
                   bool := (e1 =e);
                   s1 := d(e1,s1)
                od
          end bool
```

This set of axioms is consistent and has a model.

Theorem (on representation)
Every model of the axioms A1) - A6) is isomorphic to the system of all finite subsets of the set E with the corresponding set-theoretical operations on it.

SIMULATION

In this section we present a collection of properties of data structure SIMULATION. While we know they are consistent, we can not present at the moment arguments for their completeness; we are unable in the day we are writing this to prove representation theorem.

By a system of simulated processes we shall understand the following algebraic system

$\langle SP \cup SQS \cup EVN \cup T$; idle?, terminated?, current, time, schedule, run, hold, passivate $\rangle$

where SP stands for simulated processes, SQS for "timeAxis" c.f. SIMULA67, EVN for eventNotices, T for data structure of time (here real).

current : $SQS \to SP$	idle? : $SP \to B_0$
time : $SQS \to T$	terminated? : $SP \to B_0$
schedule : $SP \times T \to SQS$	run : $SP \to SQS$
hold : $T \to SQS$	passivate : $SP \to SP$

which satisfies the axioms listed below

AS1) SQS is a priority queue structure $\langle EVN \cup SQS$; ins, del, min, mb, $\leq, = \rangle$

AS2) $EVN = SP \times T$

AS3) time (sqs) = min(sqs).projection on T , current(sqs) = min(sqs).proj on SP

AS4) $(\forall$ sqs) -idle(p) $\Leftrightarrow$ $(\exists$ t) mb((p,t),sqs)

AS5) mb((p,t),sqs) $\wedge$ mb((p,t'9,sqs) $\Rightarrow$ t =t'

AS6) sqs = o $\wedge$ idle(p) $\wedge$ -terminated(p) $\Rightarrow$ [SQS.schedule(p,t)]{ -idle(p) $\wedge$ sqs = ins(p,t),o }

AS7) sqs =o $\wedge$ -idle(p) $\wedge$ -terminated(p) $\Rightarrow$ [sqs.schedule(p,t)]{ -idle(p) $\wedge$ sqs = ins((p,tP,del((p,time),o))

AS8) terminated (p) $\Rightarrow$ [schedule (p,t)] { ERROR }

AS9) hold(t) α $\Leftrightarrow$ schedule (current,time + t) α , for every formula α

AS10) current = p' $\wedge$ -terminated (p) $\Rightarrow$ { [run(p)] α = schedule(p,time) α } for every formula α

AS11) sqs = o $\wedge$ mb((p,t'),sqs) = [run(p)]{current = p $\wedge$ sqs=ins((p,time),del((p,t'),o))

AS12) p=current $\wedge$ sqs=o $\Rightarrow$ [passivate]{idle(p) $\wedge$ sqs = del((p,time),o) }

s13) sqs = o $\Rightarrow$ [cancel(p)] { idle(p) $\wedge$ sqs=del((p,t),o) $\wedge$ terminated(p) }

IMPLEMENTATION TOOLS LOGLAN'82

In order to implement the whole system we have to construct five modules and put them together in a way that reflects our postulates that simulated processes are to be extensions of the notion of queue element and that time axis with event_notice notion should be a priority queue. In other words, it is not enough to put these modules together. We wish to put their submodules together, too. There are two levels of extending modules in our picture. SIMULATION is an extension of both structures QUEUE and PRIORITY QUEUES. But inside these modules we find smaller ones which are also exten sions , or refinements, of more general notions e.g. simprocess notion is an extension of the notion of element of a queue, event-notice notion is an extension of the notion of element of a priority queue etc.

In order to implemnet such a structure one needs appropriate software tools. They exists in SIMULA67 and LOGLAN'82 programming languages. LOGLAN offers a mode enab ling to inherit properties of amodule by certain rule ofcomposition of modules. The operation is also called prefixing. Prefixing is a two-argument operation on modules of programs. The prefix should be a class, the prefixed module can be of any kind: class, procedure, function or block. Roughly speaking the result of prefixing is the module obtained by concatenation of the declarative parts of two modules and their statements respectively. This form of program construction has an unexpectedly broad spectrum of applications.

Prefixing is also a challenge for the research on semantics specification and on efficient implementation (c.f. KKLS).

FINAL REMARKS

It is not possible to convey all the ideas and results which come as the result of putting together: theoretical methods of algorithmic logic(AL) and practical softwa tools supplied by LOGLAN82.

We have not place here to show how to use the algorithmic axioms in the verification of properties of programs, how to verify correctness of program modules.

We claim that the methods presented here and in (AL) allow to develop the modules of software together with the arguments on their correctness.

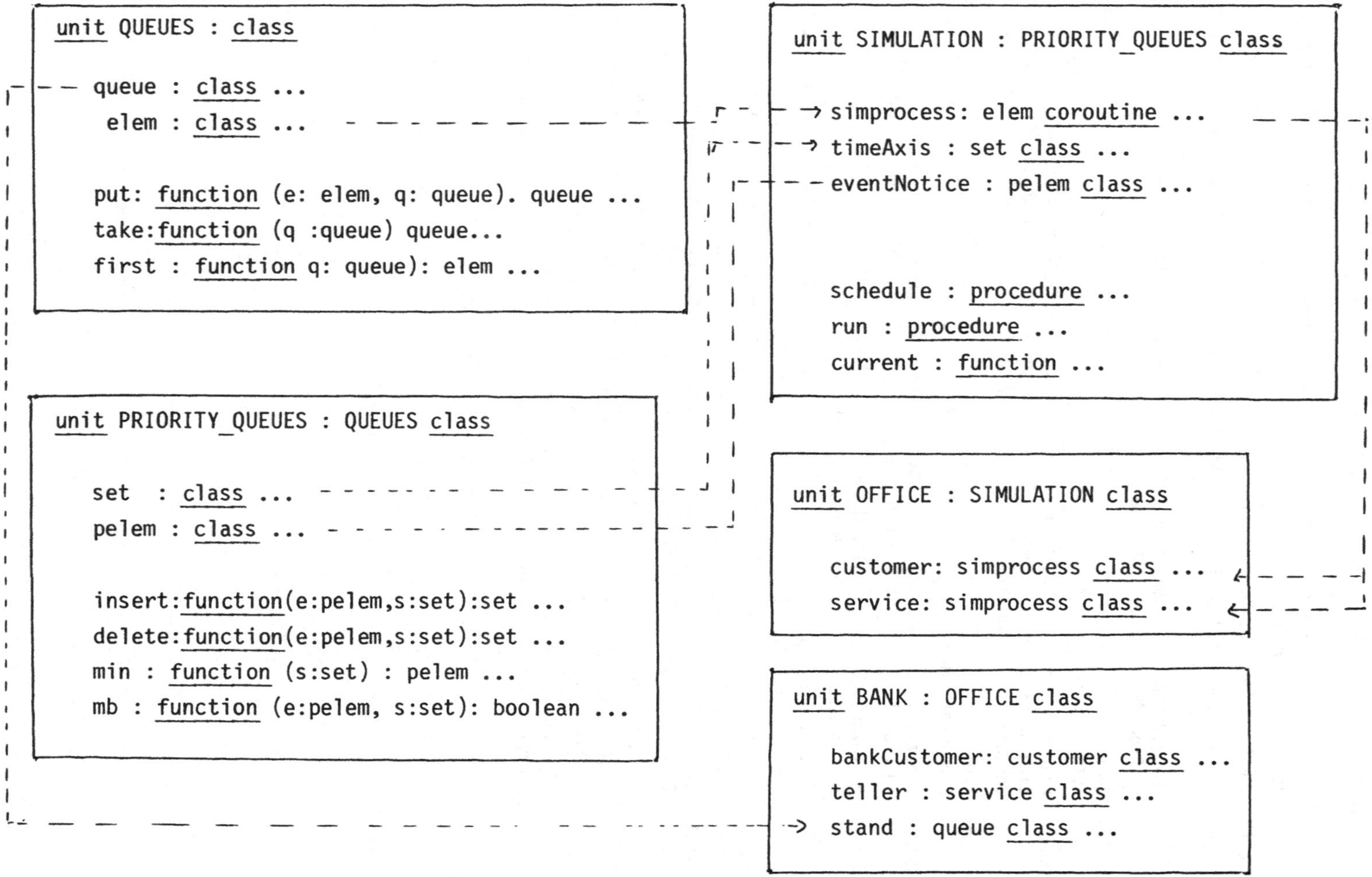

unit QUEUES : class
queue : class ...
elem : class ...
put: function (e: elem, q: queue). queue ...
take:function (q :queue) queue...
first : function q: queue): elem ...
unit PRIORITY_QUEUES : QUEUES class
set : class ...
pelem : class ...
insert:function(e:pelem,s:set):set ...
delete:function(e:pelem,s:set):set ...
min : function (s:set) : pelem ...
mb : function (e:pelem, s:set): boolean ...
unit SIMULATION : PRIORITY_QUEUES class
simprocess: elem coroutine ...
timeAxis : set class ...
eventNotice : pelem class ...
schedule : procedure ...
run : procedure ...
current : function ...
unit OFFICE : SIMULATION class
customer: simprocess class ...
service: simprocess class ...
unit BANK : OFFICE class
bankCustomer: customer class ...
teller : service class ...
stand : queue class ...

REFERENCES

(AL) Mirkowska,G.,Salwicki,A., Algorithmic logic, PWN Warszawa & D.Reidel Utrecht, to appear

(LOGLAN) LOGLAN'82 report, PWN Warszawa 1983

(AHU74) Aho,A.,Hopcroft,J.,Ullman,J., The design and analysis of computer algorithms, Addison-Wesley 1974

(Dan80) Dańko,W., A criterion of undecidability of algorithmic theories in Proc MFCS'80, LNCS 88, Springer, 1980

(Hoa72) Hoare,C.A.R., Proof of correctness of data representation Acta Informatica, 1972, 271-281

(KKLS) Krause,M.,Kreczmar,A.,Langmaack,H.,Salwicki,A., Specification and implementa tion problems of programming languages proper for hierarchical data types, rep.8410 Christian Albrechts University Kiel, 1984

(Kre74) Kreczmar,A., Effectivity problems of algorithmic logic Fundamenta Informaticae 1(1977) 19-32 also ICALP'74 LNCS14 Springer abbr.

(KS83) Kreczmar,A.,Salwicki,A., Concatenable type declarations ... in Proc IFIP TC Conf. Dresden 1983, North Holland, Amsterdam 1983,29-37

(Mir71) Mirkowska,G., On formalized systems of algorithmic logic Bull.Acad.Pol.Scie. Ser.Math.Astr. 19(1971) , 421-428

(Mir77) Mirkowska,G., Algorithmic logic and its applications in the theory of programs Fundamenta Informaticae 1(1977) 1-17,147-167

(MS86) Mirkowska,G., Axiomatic definability of programming language semantics manuscript 1986

(Okt81) Oktaba,H., Algorithmic theories of memory management systems, in polish, Ph.D. thesis, Uni Warsaw 1981

(RS) Rasiowa,H.Sikorski,R., Mathematics of metamathematics, PWN Warszawa 1963

(Sal78) Salwicki,A., On algorithmic theory of stacks, Fundamenta Informaticae3(1980) , 311-332 also abbr. in Proc MFCS'78 LNCS64 Springer

(Sal82) Salwicki,A., Algorithmic theories of data structures in Proc. ICALP'82, LNCS143 Springer

Vol. 167: International Symposium on Programming. Proceedings, 984. Edited by C. Girault and M. Paul. VI, 262 pages. 1984.

'ol. 168: Methods and Tools for Computer Integrated Manufacturing. Edited by R. Dillmann and U. Rembold. XVI, 528 pages. 1984.

/ol. 169: Ch. Ronse, Feedback Shift Registers. II, 1–2, 145 pages. 984.

/ol. 171: Logic and Machines: Decision Problems and Complexity. Proceedings, 1983. Edited by E. Börger, G. Hasenjaeger and). Rödding. VI, 456 pages. 1984.

/ol. 172: Automata, Languages and Programming. Proceedings, 984. Edited by J. Paredaens. VIII, 527 pages. 1984.

/ol. 173: Semantics of Data Types. Proceedings, 1984. Edited by 3. Kahn, D.B. MacQueen and G. Plotkin. VI, 391 pages. 1984.

/ol. 174: EUROSAM 84. Proceedings, 1984. Edited by J. Fitch. XI, 396 pages. 1984.

'ol. 175: A. Thayse, P-Functions and Boolean Matrix Factorization, VII, 248 pages. 1984.

'ol. 176: Mathematical Foundations of Computer Science 1984. Proceedings, 1984. Edited by M.P. Chytil and V. Koubek. XI, 581 ages. 1984.

ol. 177: Programming Languages and Their Definition. Edited by C.B. Jones. XXXII, 254 pages. 1984.

'ol. 178: Readings on Cognitive Ergonomics – Mind and Computers. Proceedings, 1984. Edited by G.C. van der Veer, M.J. Tauber, T.R.G. Green and P. Gorny. VI, 269 pages. 1984.

'ol. 179: V. Pan, How to Multiply Matrices Faster. XI, 212 pages. 1984.

'ol. 180: Ada Software Tools Interfaces. Proceedings, 1983. Edited by P.J.L. Wallis. III, 164 pages. 1984.

'ol. 181: Foundations of Software Technology and Theoretical Computer Science. Proceedings, 1984. Edited by M. Joseph and R. Shyamasundar. VIII, 468 pages. 1984.

'ol. 182: STACS 85. 2nd Annual Symposium on Theoretical Aspects f Computer Science. Proceedings, 1985. Edited by K. Mehlhorn. II, 374 pages. 1985.

ol. 183: The Munich Project CIP. Volume I: The Wide Spectrum anguage CIP-L. By the CIP Language Group. XI, 275 pages. 1985.

'ol. 184: Local Area Networks: An Advanced Course. Proceedings, 983. Edited by D. Hutchison, J. Mariani and D. Shepherd. VIII, 497 ages. 1985.

'ol. 185: Mathematical Foundations of Software Development. Proeedings, 1985. Volume 1: Colloquium on Trees in Algebra and Programming (CAAP'85). Edited by H. Ehrig, C. Floyd, M. Nivat and . Thatcher. XIV, 418 pages. 1985.

'ol. 186: Formal Methods and Software Development. Proceedings, 985. Volume 2: Colloquium on Software Engineering (CSE). Edited by H. Ehrig, C. Floyd, M. Nivat and J. Thatcher. XIV, 455 pages. 1985.

'ol. 187: F.S. Chaghaghi, Time Series Package (TSPACK). III, 05 pages. 1985.

'ol. 188: Advances in Petri Nets 1984. Edited by G. Rozenberg with the cooperation of H. Genrich and G. Roucairol. VII, 467 pages. 985.

'ol. 189: M.S. Sherman, Paragon. XI, 376 pages. 1985.

'ol. 190: M.W. Alford, J.P. Ansart, G. Hommel, L. Lamport, B. Liskov, G.P. Mullery and F.B. Schneider, Distributed Systems. Edited by M. Paul and H.J. Siegert. VI, 573 pages. 1985.

'ol. 191: H. Barringer, A Survey of Verification Techniques for Parallel Programs. VI, 115 pages. 1985.

ol. 192: Automata on Infinite Words. Proceedings, 1984. Edited by M. Nivat and D. Perrin. V, 216 pages.1985.

ol. 193: Logics of Programs. Proceedings, 1985. Edited by R. Parikh. , 424 pages. 1985.

Vol. 194: Automata, Languages and Programming. Proceedings, 1985. Edited by W. Brauer. IX, 520 pages. 1985.

Vol. 195: H.J. Stüttgen, A Hierarchical Associative Processing System. XII, 273 pages. 1985.

Vol. 196: Advances in Cryptology: Proceedings 1984. Edited by G. R. Blakley and D. Chaum. IX, 491 pages. 1985.

Vol. 197: Seminar on Concurrency. Proceedings, 1984. Edited by S.D. Brookes, A. W. Roscoe and G. Winskel. X, 523 pages. 1985.

Vol. 198: A. Businger, PORTAL Language Description. VIII, 186 pages. 1985.

Vol. 199: Fundamentals of Computation Theory. Proceedings, 1985. Edited by L. Budach. XII, 533 pages. 1985.

Vol. 200: J.L.A. van de Snepscheut, Trace Theory and VLSI Design. VI, 0–140 pages. 1985.

Vol. 201: Functional Programming Languages and Computer Architecture. Proceedings, 1985. Edited by J.-P. Jouannaud. VI, 413 pages. 1985.

Vol. 202: Rewriting Techniques and Applications. Edited by J.-P. Jouannaud. VI, 441 pages. 1985.

Vol. 203: EUROCAL '85. Proceedings Vol. 1, 1985. Edited by B. Buchberger. V, 233 pages. 1985.

Vol. 204: EUROCAL '85. Proceedings Vol. 2, 1985. Edited by B. F. Caviness. XVI, 650 pages. 1985.

Vol. 205: P. Klint, A Study in String Processing Languages. VIII, 165 pages. 1985.

Vol. 206: Foundations of Software Technology and Theoretical Computer Science. Proceedings, 1985. Edited by S.N. Maheshwari. IX, 522 pages. 1985.

Vol. 207: The Analysis of Concurrent Systems. Proceedings, 1983. Edited by B.T. Denvir, W.T. Harwood, M.I. Jackson and M.J. Wray. VII, 398 pages. 1985.

Vol. 208: Computation Theory. Proceedings, 1984. Edited by A. Skowron. VII, 397 pages. 1985.

Vol. 209: Advances in Cryptology. Proceedings, 1984. Edited by T. Beth, N. Cot and I. Ingemarsson. VII, 491 pages. 1985.

Vol. 210: STACS 86. Proceedings, 1986. Edited by B. Monien and G. Vidal-Naquet. IX, 368 pages. 1986.

Vol. 211: U. Schöning, Complexity and Structure. V, 99 pages. 1986.

Vol. 212: Interval Mathematics 1985. Proceedings, 1985. Edited by K. Nickel. VI, 227 pages. 1986.

Vol. 213: ESOP 86. Proceedings, 1986. Edited by B. Robinet and R. Wilhelm. VI, 374 pages. 1986.

Vol. 214: CAAP '86. 11th Colloquium on Trees in Algebra and Programming. Proceedings, 1986. Edited by P. Franchi-Zannettacci. VI, 306 pages. 1986.

This series reports new developments in computer science research and teaching – quickly, informally and at a high level. The type of material considered for publication includes preliminary drafts of original papers and monographs, technical reports of high quality and broad interest, advanced level lectures, reports of meetings, provided they are of exceptional interest and focused on a single topic. The timeliness of a manuscript is more important than its form which may be unfinished or tentative. If possible, a subject index should be included. Publication of Lecture Notes is intended as a service to the international computer science community, in that a commercial publisher, Springer-Verlag, can offer a wide distribution of documents which would otherwise have a restricted readership. Once published and copyrighted, they can be documented in the scientific literature.

Manuscripts

Manuscripts should be no less than 100 and preferably no more than 500 pages in length.
They are reproduced by a photographic process and therefore must be typed with extreme care. Symbols not on the typewriter should be inserted by hand in indelible black ink. Corrections to the typescript should be made by pasting in the new text or painting out errors with white correction fluid. Authors receive 75 free copies and are free to use the material in other publications. The typescript is reduced slightly in size during reproduction; best results will not be obtained unless the text on any one page is kept within the overall limit of 18 x 26.5 cm (7 x 10½ inches). On request, the publisher will supply special paper with the typing area outlined.
Manuscripts should be sent to Prof. G. Goos, GMD Forschungsstelle Karlsruhe, Haid- und Neu-Str. 10–14, Postfach 6380, 7500 Karlsruhe 1, Germany, Prof. J. Hartmanis, Cornell University, Dept. of Computer-Science, Ithaca, NY/USA 14850, or directly to Springer-Verlag Heidelberg.

Springer-Verlag, Heidelberger Platz 3, D-1000 Berlin 33
Springer-Verlag, Tiergartenstraße 17, D-6900 Heidelberg 1
Springer-Verlag, 175 Fifth Avenue, New York, NY 10010/USA
Springer-Verlag, 37-3, Hongo 3-chome, Bunkyo-ku, Tokyo 113, Japan

---

ISBN 3-540-16443-X
ISBN 0-387-16443-X